D1750082

MEMBRANE BIOGENESIS
AND
PROTEIN TARGETING

New Comprehensive Biochemistry

Volume 22

General Editors

A. NEUBERGER
London

L.L.M. van DEENEN
Utrecht

ELSEVIER
Amsterdam · London · New York · Tokyo

Membrane Biogenesis and Protein Targeting

Editors

WALTER NEUPERT and ROLAND LILL

*Institut für Physiologische Chemie, Physikalische Biochemie und Zellbiologie,
Ludwig-Maximilians-Universität München, Goethestraße 33,
8000 München 2, Germany*

1992
ELSEVIER
Amsterdam · London · New York · Tokyo

Elsevier Science Publishers B.V.
P.O. Box 211
1000 AE Amsterdam
The Netherlands

ISBN 0 444 89638 4 (volume 22)
ISBN 0 444 80303 3 (series)

Library of Congress Cataloging-in-Publication Data

```
Membrane biogenesis and protein targeting / editors, Walter Neupert
  and Roland Lill.
      p.    cm. -- (New comprehensive biochemistry ; v. 22)
  Includes bibliographical references and index.
  ISBN 0-444-89638-4
  1. Membrane proteins--Physiological transport.   I. Neupert,
Walter.  II. Lill, Roland.  III. Series.
QD415.N48  vol. 22
[QP552.M44]
574.87'5--dc20                                               92-24428
                                 .                               CIP
```

© 1992 Elsevier Science Publishers B.V All rights reserved.

No part of this publication may be reproduced, stored in a retrieval system or transmitted in any form or by any means, electronic, mechanical, photocopying, recording or otherwise, without the written permission of the Publisher, Elsevier Science Publishers B.V., Copyright & Permissions Department, P.O. Box 521, 1000 AM Amsterdam, The Netherlands.

No responsibility is assumed by the Publisher for any injury and/or damage to persons or property as a matter of products liability, negligence or otherwise, or from any use or operation of any methods, products, instructions or ideas contained in the material herein. Because of the rapid advances in the medical sciences, the Publisher recommends that independent verification of diagnoses and drug dosages should be made.

Special regulations for readers in the USA – This publication has been registered with the Copyright Clearance Center Inc. (CCC) Salem, Massachusetts. Information can be obtained from the CCC about conditions under which photocopies of parts of this publication may be made in the USA. All other copyright questions, including photocopying outside of the USA, should be referred to the copyright owner, Elsevier Science Publishers B.V., unless otherwise specified.

This book is printed on acid-free paper

Printed in The Netherlands

List of contributors

H. Alefsen, 299
Botanisches Institut, Universität München, Menzinger Straße 67, 8000 München 19, Germany.

J. Beckwith, 49
Department of Microbiology and Molecular Genetics, Harvard Medical School, 200 Longwood Avenue, Boston, MA 02115, U.S.A.

B. Böckler, 299
Botanisches Institut, Universität Kiel, Olshausenstraße 40, 2300 Kiel 1, Germany.

E. Breukink, 85
Department of Biomembranes, Centre for Biomembranes and Lipid Enzymology, University of Utrecht, Padualaan 8, 3584 CH Utrecht, The Netherlands.

S. Caplan, 329
Howard Hughes Medical Institute, and Department of Genetics, Yale University School of Medicine, 333 Cedar Street, New Haven, CT 06510, U.S.A.

H.-L. Chiang, 149
Department of Molecular and Cell Biology, Howard Hughes Medical Institute, University of California at Berkeley, 401 Barker Hall, Berkeley, CA 94720, U.S.A.

H.H.J. de Jongh, 85
Department of Biomembranes, Centre for Biomembranes and Lipid Enzymology, University of Utrecht, Padualaan 8, 3584 CH Utrecht, The Netherlands.

A.I.P.M. de Kroon, 85
Department of Biomembranes, Centre for Biomembranes and Lipid Enzymology, University of Utrecht, Padualaan 8, 3584 CH Utrecht, The Netherlands.

B. de Kruijff, 85
Department of Biomembranes, Centre for Biomembranes and Lipid Enzymology, University of Utrecht, Padualaan 8, 3584 CH Utrecht, The Netherlands.

R.A. Demel, 85
Department of Biomembranes, Centre for Biomembranes and Lipid Enzymology, University of Utrecht, Padualaan 8, 3584 CH Utrecht, The Netherlands.

B. Dobberstein, 105
Europäisches Laboratorium für Molekularbiologie, Meyerhofstraße 1, 6900 Heidelberg, Germany.

L.J. Garrard, 209
Department of Pharmacology, University of Texas Southwestern Medical Center, 5323 Harry Hines Boulevard, Dallas, TX 75235-9041, U.S.A.

J.M. Goodman, 209
Department of Pharmacology, University of Texas Southwestern Medical Center, 5323 Harry Hines Boulevard, Dallas, TX 75235-9041, U.S.A.

C. Harris, 9
Department of Biochemistry and Molecular Biology, Harvard University, Cambridge, MA 02139, U.S.A.

F.-U. Hartl, 329
Program in Cellular Biochemistry & Biophysics, Rockefeller Research Laboratory, Sloan-Kettering Institute, 1275 York Avenue, New York, NY 10021, U.S.A.

E. Hartmann, 119
Institut für Molekularbiologie, Robert-Rössle-Straße 10, 1115 Berlin-Buch, Germany.

C. Hergersberg, 265
Institut für Physiologische Chemie, Physikalische Biochemie und Zellbiologie der Universität München, Goethestraße 33, 8000 München 2, Germany.

S. High, 105
European Laboratory for Molecular Biology, 6900 Heidelberg, Germany.

C. Hikita, 66
Institute of Applied Microbiology, The University of Tokyo, Yayoi, Bunkyo-ku, Tokyo 113, Japan.

V. Hines, 241
Chiron Corporation, 4560 Horton Street, Emeryville, CA 94608, U.S.A.

J. Höhfeld, 185
Institut für Physiologische Chemie, Medizinische Fakultät der Ruhr-Universität Bochum, 4630 Bochum, Germany.

A.L. Horwich, 329
Department of Human Genetics, Yale University School of Medicine, 333 Cedar Street, New Haven, CT 06510, U.S.A.

W. Jordi, 85
Department of Biomembranes, Centre for Biomembranes and Lipid Enzymology, University of Utrecht, Padualaan 8, 3584 CH Utrecht, The Netherlands.

M. Kato, 66
Institute of Applied Microbiology, The University of Tokyo, Yoyoi, Bunkyo-ku, Tokyo 113, Japan.

K. Keegstra, 279
Department of Botany, University of Wisconsin, Madison, WI 53706, U.S.A.

R.C.A. Keller, 85
Department of Biomembranes, Centre for Biomembranes and Lipid Enzymology, University of Utrecht, Padualaan 8, 3584 CH Utrecht, The Netherlands.

B. Kerber, 279
Fachrichtung Botanik, Universität des Saarlandes, 6600 Saarbrücken, Germany.

J.A. Killian, 85
Department of Biomembranes, Centre for Biomembranes and Lipid Enzymology, University of Utrecht, Padualaan 8, 3584 CH Utrecht, The Netherlands.

P. Klappa, 137
Zentrum Biochemie/Abteilung Biochemie II der Universität Göttingen, Goßlerstraße 12d, W-3400 Göttingen, Germany.
A. Kuhn, 33
Abteilung für Angewandte Mikrobiologie, Universität Karlsruhe, Kaiserstraße 12, W-7500 Karlsruhe, Germany
W.-H. Kunau, 185
Institut für Physiologische Chemie, Medizinische Fakultät der Ruhr-Universität Bochum, 4630 Bochum 1, Germany
T. Kurihara, 309
Department of Molecular Biology, Princeton University, Princeton, NJ 08544, U.S.A.
R. Kusters, 85
Department of Biomembranes, Centre for Biomembranes and Lipid Enzymology, University of Utrecht, Padualaan 8, 3584 CH Utrecht, The Netherlands.
P.B. Lazarow, 231
Department of Cell Biology and Anatomy, The Mount Sinai Medical Center, One Gustave L. Levy Place, New York, NY 10029-6574, U.S.A.
J.-M. Li, 253
Department of Biochemistry, McIntyre Medical Sciences Building, McGill University, Montreal, Que. H3G 1Y6, Canada
H.-M. Li, 279
Department of Botany, University of Wisconsin, Madison, WI 53706, U.S.A.
R. Lill, 265
Institut für Physiologische Chemie, Physikalische Biochemie und Zellbiologie der Universität München, Goethestraße 33, 8000 München 2, Germany.
S.-I. Matsuyama, 21
Institute of Applied Microbiology, The University of Tokyo, Yoyoi, Bunkyo-ku, Tokyo 113, Japan.
M.T. McCammon, 209
Department of Pharmocology, University of Texas Soutwestern Medical Center, 5323 Harry Hines Boulevard, Dallas, TX 75235-9041, U.S.A.
D. Mertens, 185
Institut für Physiologische Chemie, Medizinische Fakultät der Ruhr-Universität Bochum, 4630 Bochum, Germany.
D.G. Millar, 253
Department of Biochemistry, McIntyre Medical Sciences Building, McGill University, Montreal, Que. H3G 1Y6, Canada.
J.D. Miller, 129
Department of Biochemistry and Biophysics, University of California, Medical School, San Francisco, CA 94143-9448, U.S.A.
S. Mizushima, 21, 63
Institute of Applied Microbiology, The University of Tokyo, 1-1-1, Yayoi, Bunkyo-ku, Tokyo 113, Japan.

H.W. Moser, 231
Kennedy Institute and Departments of Neurology and Pediatrics, The Johns Hopkins University, Baltimore, MD 21205, U.S.A.

G. Müller, 137
Zentrum Biochemie/Abteilung Biochemie II der Universität Göttingen, Goßlerstraße 12d, W-3400 Göttingen, Germany.

W. Neupert, 265
Institut für Physiologische Chemie, Physikalische Biochemie und Zellbiologie der Universität München, Goethestraße 33, 8000 München 2, Germany.

S. Nothwehr, 165
Institute of Molecular Biology, University of Oregon, Eugene, OR 97403, U.S.A.

J.M. Nunnari, 129
Department of Biochemistry and Biophysics, University of California, Medical School, San Francisco, CA 94143-0448, U.S.A.

S.C. Ogg, 129
Department of Biochemistry and Biophysics, University of California, Medical School, San Francisco, CA 94143-0448, U.S.A.

S.E. Perry, 279
Department of Botany, University of Wisconsin, Madison, WI 53706, U.S.A.

M. Pilon, 85
Department of Biomembranes, Centre for Biomembranes and Lipid Enzymology, University of Utrecht, Padualaan 8, 3584 CH Utrecht, The Netherlands.

T.A. Rapoport, 119
Institut für Molekularbiologie, Robert-Rössle-Straße 10, 1115 Berlin-Buch, Germany.

C.K. Raymond, 165
Institute of Molecular Biology, University of Oregon, Eugene OR 97403, U.S.A.

C.J. Roberts, 165
Institute of Molecular Biology, University of Oregon, Eugene OR 97403, U.S.A.

C. Robinson, 279
Department of Biological Sciences, University of Warwick, Conventry, CV4 7AL, United Kingdom.

M. Salomon, 299
Botanisches Institut, Universität München, Menzinger Straße 67, 8000 München 19, Germany.

M.J. Santos, 231
Department of Cell Biology, Universidad Catolica de Chile, Santiago, Chile.

R. Schekman, 149
Department of Molecular and Cell Biology, Howard Hughes Medical Institute, University of California at Berkeley, 401 Barker Hall, Berkeley, CA 94720, U.S.A.

G. Schlenstedt, 137
Zentrum Biochemie/Abteilung Biochemie II der Universität Göttingen, Goßlerstraße 12d, W-3400 Göttingen, Germany.

H. Schneider, 265
Institut für Physiologische Chemie, Physikalische Biochemie und Zellbiologie der Universität München, Goethestraße 33, 8000 München 2, Germany.

G.C. Shore, 253
Department of Biochemistry, McIntyre Medical Sci. Bldg, McGill University, 3655 Drummond Street, Montreal Que, H3G 1Y6, Canada.

P.A. Silver, 309
Department of Molecular Biology, Princeton University, Princeton, NJ 08544, U.S.A.

J. Soll, 299
Botanisches Institut und Botanischer Garten, Christian-Albrechts-Universität Kiel, Olshausenstraße 40, 2300 Kiel 1, Germany.

T. Söllner, 265
Institut für Physiologische Chemie, Physikalische Biochemie und Zellbiologie der Universität München, Goethestraße 33, 8000 München 2, Germany.

T.H. Stevens, 165
Institute of Molecular Biology, University of Oregon, Eugene, OR 97403, U.S.A.

R. Stuart, 265
Institut für Physiologische Chemie, Physikalische Biochemie und Zellbiologie der Universität München, Goethestraße 33, 8000 München 2, Germany.

S. Subramani, 221
Department of Biology, University of California, San Diego, 0322 Bonner Hall, La Jolla, CA 92093, U.S.A.

P.C. Tai, 9
Department of Biology, Georgia State University, University Plaza, Atlanta, GA 30302-4010, U.S.A.

K. Tani, 66
Institute of Applied Microbiology, The University of Tokyo, Yayoi, Bunkyo-ku, Tokyo 113, Japan.

H. Tokuda, 21
Institute of Applied Microbiology, The University of Tokyo, Yayoi, Bunkyo-ku, Tokyo 113, Japan.

B. Traxler, 49
Department of Microbiology and Molecular Genetics, Harvard Medical School, 200 Longwood Avenue, Boston, MA 02115, U.S.A.

D. Troschel, 33
Department of Applied Microbiology, University of Karlsruhe, 7500 Karlsruhe, Germany.

R. van 't Hof, 85
Department of Biomembranes, Centre for Biomembranes and Lipid Enzymology, University of Utrecht, Padualaan 8, 3584 CH Utrecht, The Netherlands.

C.A. Vater, 165
Institute of Molecular Biology, University of Oregon, Eugene, OR 97403, U.S.A.

G. von Heijne, 75
Department of Molecular Biology, K 87, Huddinge University Hospital, S-141 86 Huddinge, Sweden.

K. Waegemann, 299
Botanisches Institut, Universität Kiel, Olshausenstraße 40, 2300 Kiel 1, Germany.

J.S. Wall, 329
Brookhaven National Laboratory, Biology Department, Upton, NY 11973, U.S.A.

P. Walter, 129
Department of Biochemistry and Biophysics, University of California, Medical School, San Francisco, CA 94143-0448, U.S.A.

B. Wickner, 3
Molecular Biology Institute and Department of Biological Chemistry, University of California, Los Angeles, CA 90024-1570, U.S.A.

F.F. Wiebel, 185
Institut für Physiologische Chemie, Medizinische Fakultät der Ruhr-Universität Bochum, 4630 Bochum, Germany.

H. Wiech, 137
Zentrum Biochemie/Abteilung Biochemie II der Universität Göttingen, Goßlerstraße 12d, W-3400 Göttingen, Germany.

R. Zimmermann, 137
Zentrum Biochemie/Abteilung Biochemie II der Universität Göttingen, Goßlerstraße 12d, W-3400 Göttingen, Germany.

Contents

List of contributors V

Part A Bacteria

Chapter 1. Where are we in the exploration of Escherichia coli translocation pathways?
Bill Wickner 3

1. Protein translocation pathways 4
2. Open questions 6
References 7

Chapter 2. Components involved in bacterial protein translocation
Chris Harris and Phang C. Tai 9

1. Introduction 9
2. The minimal length of a prokaryotic signal peptide 10
3. The N(m) element: an export requirement for low basicity as well as low polarity 12
4. Is SecY/PrlA essential for protein translocation? 14
5. In vitro suppression of defective signal peptides 15
6. Roles of SecD and SecF in protein translocation 16
7. An inhibitor of protein translocation 17
8. Perspective 18
References 18

Chapter 3. Molecular characterization of Sec proteins comprising the protein secretory machinery of Escherichia coli
Shoji Mizushima, Hajime Tokuda and Shin-ichi Matsuyama 21

1. Introduction 21
2. Overproduction of Sec proteins 22
3. Purification of Sec proteins 23
4. Estimation of the numbers of Sec proteins and of the secretory machinery in one *E. coli* cell 24
5. Functions of SecA in protein translocation 25
6. Functions of SecE and SecY 27
7. Functions of SecD and SecF 28
8. Discussion 28
References 30

Chapter 4. Distinct steps in the insertion pathway of bacteriophage coat proteins
Andreas Kuhn and Dorothee Troschel 33

1. Introduction 33
2. Results and discussion 34
   2.1. Pf3 coat protein requires no leader sequence for membrane insertion . . 34
   2.2. Hybrid coat proteins of M13 and Pf3 35
   2.3. M13 procoat protein first binds electrostatically to the membrane surface 36
   2.4. Both hydrophobic regions are required for the partitioning of the M13 procoat protein into the membrane 38
   2.5. Translocation of the negatively charged periplasmic region of the M13 procoat protein is not primarily an electrophoretic event 39
   2.6. M13 procoat protein as a model substrate for leader peptidase 41
   2.7. From membrane to phage 42
   2.8. The Sec-independent insertion pathway is limited to simple membrane translocation domains 42
3. Conclusions 45
References 46

Chapter 5. Steps in the assembly of a cytoplasmic membrane protein: the MalF component of the maltose transport complex
Beth Traxler and Jon Beckwith 49

1. Introduction 49
   1.1. Issues in the study of membrane protein assembly and structure 49
   1.2. The MalF protein as a model system 51
   1.3. The mechanism of insertion into the membrane of MalF 52
2. The nature of topogenic signals in MalF 53
3. Kinetics of assembly of MalF in the cytoplasmic membrane 56
4. Assembly of MalF into the quaternary MalF–MalG–MalK structure 57
5. Summary 59
References 60

Chapter 6. Structural characteristics of presecretory proteins: their implication as to translocation competency
Shoji Mizushima, Katsuko Tani, Chinami Hikita and Masashi Kato 63

1. Introduction 64
2. Amino terminal positive charge of the signal peptide 64
3. Central hydrophobic stretch of the signal peptide 65
4. Function of the positive charge can be compensated for by a longer hydrophobic stretch 67
5. Carboxyl terminal region of the signal peptide including the cleavage site . . 68
6. Roles of charged amino acid residues in the mature domain in protein translocation 70
7. Chemical structure of the mature domain tolerated by the secretory machinery 72
References 73

Chapter 7. Sequence determinants of membrane protein topology
Gunnar von Heijne 75

1. Introduction 75
2. Results 76
   2.1. Signals and topologies 76
   2.2. The positive inside-rule 78
   2.3. Positively charged residues control membrane protein topology 78
   2.4. A membrane protein with pH-dependent topology 80
   2.5. Position-specific charge-pairing can affect the topology 80
   2.6. Sec-dependent versus sec-independent assembly 81
3. Discussion 82
References 83

Chapter 8. Lipid involvement in protein translocation
B. de Kruijff, E. Breukink, R.A. Demel, R. van 't Hof, H.H.J. de Jongh,
W. Jordi, R.C.A. Keller, J.A. Killian, A.I.P.M. de Kroon, R. Kusters
and M. Pilon 85

1. Introduction 85
2. Results and discussion 86
   2.1. Prokaryotic protein secretion 86
   2.2. Mitochondrial protein import 91
   2.3. Chloroplast protein import 97
3. Concluding remarks 99
References 99

Part B   Endoplasmic reticulum

Chapter 9. Membrane protein insertion into the endoplasmic reticulum:
signals, machinery and mechanisms
Stephen High and Bernhard Dobberstein 105

1. Introduction 105
2. Types of membrane proteins and their topological signals 105
   2.1. Proteins with uncleaved signal sequences 106
   2.2. Proteins with cleavable signal sequences 107
   2.3. The loop model for protein insertion into the membrane 108
   2.4. Biosynthesis of multiple spanning membrane proteins 109
3. Components involved in the insertion of proteins into the ER membrane . . . 111
   3.1. Targeting 111
   3.2. Membrane insertion 111
   3.3. GTP requirement 114
4. Discussion 115
References 117

Chapter 10. Translocation of proteins through the endoplasmic reticulum membrane: investigation of their molecular environment by cross-linking
Enno Hartmann and Tom A. Rapoport 119

1. Introduction 119
2. Results 120
    2.1. Experimental strategies 120
    2.2. The SSR-complex 121
    2.3. The TRAM protein 122
    2.4. Other glycoproteins 123
    2.5. Unglycosylated proteins 123
    2.6. The Sec proteins of yeast microsomes 124
3. Discussion 124
References 126

Chapter 11. The role of GTP in protein targeting to the endoplasmic reticulum
Stephen C. Ogg, Jodi M. Nunnari, Joshua D. Miller and Peter Walter . . . 129

References 135

Chapter 12. Consecutive steps of nucleoside triphosphate hydrolysis are driving transport of precursor proteins into the endoplasmic reticulum
Peter Klappa, Günter Müller, Gabriel Schlenstedt, Hans Wiech and Richard Zimmermann 137

1. Introduction 137
2. Results 138
    2.1. Ribonucleoparticles versus molecular chaperones 139
    2.2. Translocase 141
3. Discussion 143
    3.1. Components involved in protein transport into yeast endoplasmic reticulum 143
    3.2. Model for ribonucleoparticle-independent transport 143
    3.3. Open questions 144
References 145

Part C   Vacuoles

Chapter 13. Mechanism and regulation of import and degradation of cytosolic proteins in the lysosome/vacuole
Hui-Ling Chiang and Randy Schekman 149

1. Introduction 150
2. Intracellular protein degradation in mammalian lysosomes 151
    2.1. Microautophagy 151
    2.2. Macroautophagy 152
    2.3. Lysosomal protein degradation in cultured fibroblasts 153
3. Protein degradation in the yeast vacuole 156

3.1.	Catabolite inactivation	156
3.2.	Vacuolar degradation of FBPase	158
3.3.	Mechanism of FBPase degradation	159
3.4.	Covalent modifications and FBPase degradation	160
	3.4.1. Phosphorylation	160
	3.4.2. Ubiquitination	161
References		162

Chapter 14. The sorting of soluble and integral membrane proteins to the yeast vacuole
Christopher K. Raymond, Carol A. Vater, Steven Nothwehr, Christopher J. Roberts and Tom H. Stevens 165

1. Introduction 165
2. Results 167
   2.1. No single domain of the vacuolar integral membrane protein DPAP B is required for vacuolar delivery 167
   2.2. The cytoplasmic domain of DPAP A is necessary and sufficient for its localization to a late Golgi compartment 169
   2.3. Vps1p, which is required for the sorting of soluble vacuolar glycoproteins, shares extensive similarity with a subfamily of GTP-binding proteins . . 170
   2.4. Vps1p binds and hydrolyzes GTP 172
   2.5. Mutational analysis suggests that Vps1p is composed of two functionally distinct domains 173
3. Discussion 177
   3.1. Targeting of integral membrane proteins in the secretory pathway of *Saccharomyces cerevisiae* 177
   3.2. Vps1p, which is a GTPase required for the sorting of soluble vacuolar proteins, is composed of two functionally distinct domains 179
References 181

Part D   Peroxisomes

Chapter 15. Defining components required for peroxisome assembly in Saccharomyces cerevisiae
Jörg Höhfeld, Daphne Mertens, Franziska F. Wiebel and Wolf-H. Kunau . . 185

1. Introduction 185
2. Results 187
   2.1. Peroxisomal mutants of *Saccharomyces cerevisiae* 187
      2.1.1. Defects in peroxisome formation (type I pas mutants: pas1, pas2, pas3, pas5) 189
      2.1.2. Defects in peroxisome proliferation (type II pas mutants: pas4 and pas6) 191
      2.1.3. Defects in import of 3-oxoacyl-CoA thiolase (type III pas mutant: pas7) 191
   2.2. Cloning of peroxisomal genes 191
   2.3. Sequence analysis 192

		2.3.1.	PAS1	192
		2.3.2.	PAS2	194
		2.3.3.	PAS4	194
	2.4.	Identification and characterization of the gene products		195
	2.5.	Analysis of the function of the cloned genes for peroxisome biogenesis		196
		2.5.1.	Overexpression of PAS4	196
		2.5.2.	Site directed mutagenesis	197
		2.5.3.	Conditional peroxisomal mutants	197
	2.6.	Fusion proteins as tools for further investigations		199
3.	Discussion			200
	3.1.	Peroxisomal mutants as a tool to dissect peroxisome biogenesis		200
	3.2.	Do peroxisomal prestructures exist in type I pas mutants?		202
	3.3.	Are type I pas mutants peroxisomal import mutants?		203
	3.4.	Do more peroxisomal import routes exist other than the SKL-mediated pathway?		203
	3.5.	Are type II pas mutants affected in peroxisome proliferation?		204
4.	Conclusions			205
References				205

Chapter 16. Structure and assembly of peroxisomal membrane proteins
Joel M. Goodman, Lisa J. Garrard and Mark T. McCammon 209

1.	Introduction		210
2.	Results		210
	2.1.	Assembly of peroxisomal proteins	212
	2.2.	Proliferation of peroxisomal components	213
	2.3.	Structure and composition of peroxisomal membranes	213
	2.4.	Cloning of genes encoding membrane proteins	214
	2.5.	Expression and sorting of PMP47	215
3.	Discussion		217
References			219

Chapter 17. Mechanisms of transport of proteins into microbodies
Suresh Subramani 221

1.	Introduction	221
2.	A C-terminal tripeptide is a major targeting signal for proteins of the microbody matrix	223
3.	Certain variants of the SKL tripeptide can also function as PTS	224
4.	Peroxisomal protein transport in microinjected mammalian cells	224
5.	Import deficiency in fibroblast cells from Zellweger's syndrome patients	225
6.	An amino-terminal PTS resides in the cleaved leader peptides of the peroxisomal thiolases	226
7.	Selective import deficiency in Zellweger cells	226
8.	Transport of membrane proteins into peroxisomes	227
9.	Summary	227
References		227

Chapter 18. Lessons for peroxisome biogenesis from fluorescence analyses of Zellweger syndrome fibroblasts
Paul B. Lazarow, Hugo W. Moser and Manuel J. Santos 231

1. Introduction 231
2. Results 232
3. Discussion 233
    3.1. Peroxisome membranes are always present: perhaps they are required for viability 233
    3.2. The peroxisome membranes are nearly empty ghosts: these are Peroxisome IMport (PIM) mutations 234
    3.3. Peroxisome membrane assembly has fewer requirements, or different requirements, from the packaging of peroxisome matrix proteins 234
    3.4. Empty peroxisome membrane ghosts divide 235
    3.5. Genetic complementation for peroxisome assembly is formally demonstrated 235
    3.6. Preliminary partial correlation of genotype and phenotype 236
    3.7. Future directions 236
References 236

Part E   Mitochondria

Chapter 19. The mitochondrial protein import machinery of Saccharomyces cerevisiae
Victoria Hines 241

1. Introduction 241
2. Components of the import machinery 242
    2.1. Import receptors 242
    2.2. The protein translocation channel 245
    2.3. Refolding and processing proteins 245
3. The mechanism of protein import 246
    3.1. Energy requirements 246
    3.2. Contact sites 247
    3.3. Protein sorting 248
4. Outlook 250
References 250

Chapter 20. Protein insertion into mitochondrial outer and inner membranes via the stop-transfer sorting pathway
Gordon C. Shore, Douglas G. Millar and Jian-Ming Li 253

1. Introduction 254
2. Results and discussion 256
    2.1. Mitochondrial topogenic sequences and stop-transfer sorting . . . 256
    2.2. The OMM signal-anchor sequence 256
    2.3. Stop-transfer sorting to the inner membrane 257

		2.4.	Polytopic proteins	259
		2.5.	Default sorting	259
	3.		Conclusions	261
References				262

Chapter 21. General and exceptional pathways of protein import into sub-mitochondrial compartments
Roland Lill, Christoph Hergersberg, Helmut Schneider, Thomas Söllner, Rosemary Stuart and Walter Neupert 265

1.	The general pathways for protein import into sub-mitochondrial compartments.	265
2.	Exceptional pathways of protein import	268
3.	MOM19 is imported into the OM without the aid of surface receptors	269
4.	Cytochrome *c* heme lyase is imported directly through the OM via a non-conservative sorting pathway	271
5.	Perspectives	274
References		275

Part F   Chloroplasts

Chapter 22. Targeting of proteins into and across the chloroplastic envelope
H.-M. Li, S.E. Perry and K. Keegstra 279

1.	Introduction		280
2.	Transport across the envelope membranes		281
	2.1.	Binding of precursors to the chloroplastic surface	281
	2.2.	Translocation of precursor across the envelope membranes	283
3.	Targeting of proteins into the envelope membranes		284
	3.1.	Targeting to the outer envelope membrane	284
	3.2.	Targeting to the inner envelope membrane	286
4.	Summary and future prospects		287
References			287

Chapter 23. Transport of proteins into the thylakoids of higher plant chloroplasts
Colin Robinson 289

1.	Introduction		289
2.	Results		291
	2.1.	Development of an in vitro assay for the import of proteins by isolated thylakoids	291
	2.2.	Energy requirements for the import of proteins into isolated thylakoids	292
	2.3.	Events in the stroma	292
	2.4.	Maturation of imported thylakoid lumen proteins	295
3.	Discussion		295
References			296

Chapter 24. Comparison of two different protein translocation mechanisms into chloroplasts
Jürgen Soll, Heike Alefsen, Birgit Böckler, Birgit Kerber, Michael Salomon and Karin Waegemann 299

1. Introduction 299
2. Results and discussion 299
   2.1. Import characteristics of pSSU and OEP 7 299
   2.2. Specificity and mechanism of OEP 7 insertion 301
References 306

Part G   Chaperones

Chapter 25. DnaJ homologs and protein transport
Takao Kurihara and Pamela A. Silver 309

1. Introduction 309
   1.1. Stimulation of protein transport by HSP70s and additional factors . . . 309
   1.2. E. coli DnaJ and GrpE function with and regulate bacterial HSP70 (DnaK) 310
      1.2.1. Bacteriophage λ and P1 replication and protein complex disassembly 311
      1.2.2. Refolding of thermally inactivated λcI857 repressor 312
      1.2.3. Proteolysis of puromycin-generated polypeptide fragments . . 312
      1.2.4. Stimulation of DnaK ATPase activity by DnaJ and GrpE . . . 313
2. Results 313
   2.1. DnaJ homologs 313
      2.1.1. Bacterial DnaJ homologs 316
      2.1.2. DnaJ homologs in the yeast *Saccharomyces cerevisiae* 316
      2.1.3. SCJ1 316
      2.1.4. YDJ1/MAS5 317
      2.1.5. SIS1 318
      2.1.6. NPL1/SEC63 319
   2.2. The J-region 321
      2.2.1. The NPL1/SEC63 J-region: localization to the ER lumen . . 321
      2.2.2. Genetic evidence for J-region role in KAR2 interaction and ER translocation 323
3. Discussion 323
   3.1. Model for NPL1/SEC63 function 323
   3.2. DnaJ homologs, J-regions and protein transport 324
References 325

Chapter 26. Chaperonin-mediated protein folding
Arthur L. Horwich, Shari Caplan, Joseph S. Wall and F.-Ulrich Hartl . . . 329

1. Introduction 329
2. In vivo analysis of chaperonin function 330
3. Role of hsp60 in biogenesis of mitochondrial-encoded proteins 331
4. Chaperonin-mediated folding reconstituted in vitro 333

5. Models for physical interactions of components in chaperonin-mediated folding   334
References   337

Index   339

Part A

Bacteria

CHAPTER 1

Where are we in the exploration of *Escherichia coli* translocation pathways?

BILL WICKNER

Molecular Biology Institute and Department of Biological Chemistry, University of California, Los Angeles, CA 90024-1570, USA

Abstract

The envelope of gram negative bacteria such as *Escherichia coli* is comprised of three layers, an inner or plasma membrane, an aqueous periplasmic space (with soluble proteins, membrane-derived oligosaccharide, and peptidoglycan), and an outer membrane. Considerable effort has been devoted to studying how the proteins, lipids, and carbohydrates that comprise each of these compartments are selectively transported to allow cell surface growth. A brief consideration of what we know of these processes, presented below, reveals that we have yet to answer, or even address, most of the significant questions in this field. There are several compelling reasons why the answers will be worth the considerable effort. The primary reason is surely our cultural drive to understand how nature functions. Such fundamental processes as selective targeting and transport of macromolecules across membranes must be conserved throughout evolution. What we learn from *E. coli*, with its unparalleled ease of combining biochemistry and genetics, will be applicable to other organelles and organisms, such as eukaryotic mitochondria, chloroplasts, peroxisomes, glyoxisomes, vacuoles, and endoplasmic reticulum. A second reason is that bacteria are major pathogens, and interruption of cell surface growth is a proven strategy in the pursuit of effective antibiotics. Recently, interest has grown in producing all manner of cloned eukaryotic proteins of medical and commercial importance in bacteria and exporting them to the cell surface to facilitate their correct folding and easy isolation. This review will briefly outline our current knowledge of protein targeting and translocation, list some of the major fundamental problems that remain unsolved, and sketch some unaddressed questions that may entice new investigators into this field.

1. Protein translocation pathways

Protein translocation across the plasma membrane is the most thoroughly studied aspect of bacterial cell surface growth. Indeed, while several eukaryotic and prokaryotic membranes have been studied for their protein translocation properties, only for the plasma membrane of *E. coli* have the genes for the proteins that catalyze translocation been cloned and sequenced, mutants described, and the proteins purified and reconstituted in fully functional form. Detailed reviews of the biochemistry and genetics of this system have appeared recently [1,2]. Virtually all periplasmic and outer membrane proteins are synthesized with an amino-terminal leader (signal) sequence [3,4]. Protein export is not coupled to ongoing protein synthesis in *E. coli* [5,6]. Indeed, elegant physiological studies have shown that much, or even most, export actually occurs post-translationally [5]. To avoid misfolding or misassociations, newly made presecretory proteins associate with chaperones, both the general cytosolic chaperones such as DnaK and GroEL and the export-specific chaperones such as SecB. SecB associates with the mature domain of preproteins [7]. This association is not a strict sequence recognition, as there is no sequence conserved among the many preproteins, but rather is a recognition of its unfolded character. It has been proposed that slow folding in the cytoplasm, and specifically inhibition of folding by the amino-terminal leader sequence [8], is a hallmark of exported proteins. The preprotein–SecB complex then binds to preprotein translocase, a complex, multisubunit membrane enzyme [9]. Translocase consists of two domains, a peripherally bound domain of SecA protein [10] and an integral, membrane-embedded domain of the SecY/E protein. The latter consists of three subunits, SecY protein, SecE protein, and a subunit which we term band 1 of undefined gene [9]. These proteins, and their binding relationships, are shown in Fig. 1.

The binding of the preprotein–SecB complex to translocase occurs at the SecA subunit. It is mediated by the specific affinities of SecA for SecB itself [10] and by recognition of the leader and mature domains of the preprotein [11,12]. As for the SecB recognition of preproteins, the SecA recognition must not rely on strict identification of a sequence. Rather, the basic and apolar character of leader regions, and the state of the intermediate folding characteristic of the mature domain, must govern SecA recognition. Correct targeting of the preprotein to the membrane is thus governed by a binding and recognition cascade. The availability of pure, functional preproteins, SecB, and SecA must allow the chemical basis of this recognition to be addressed. Satisfactory understanding will, of course, surely require a determination of the crystal structures of SecA, SecB, and of preprotein in complex with each.

The SecA protein, which is peripherally bound to the plasma membrane, can undergo fully functional dissociation and association from the membrane during in vitro manipulations. Its membrane binding is mediated by its specific associations with the membrane-embedded SecY/E protein [10] as well as its high affinity for the

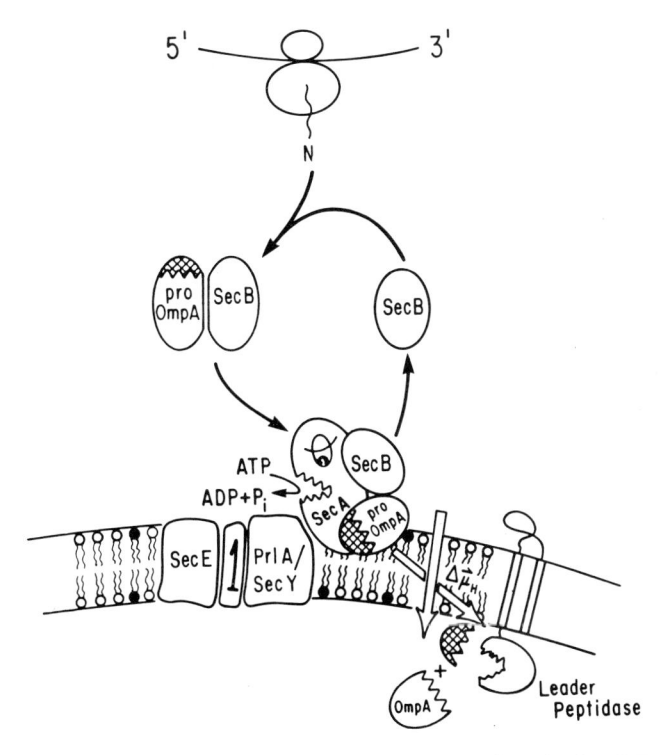

Fig. 1. A model of the binding relationships among the subunits of *E. coli* preprotein translocase. From [9].

acidic phospholipids phosphatidylglycerol and cardiolipin [12,13]. These associations of SecA have a dramatic effect on its function. SecA only shows a high affinity for binding precursor protein and SecB when it is correctly bound at SecY/E and acidic lipid. In turn, the binding of preprotein to such membrane-bound SecA activates SecA for the binding and hydrolysis of ATP, an activity we term the translocation ATPase [14]. Even prior to hydrolysis, the energy of ATP binding drives a small domain of approximately 20–30 residues across the membrane, initiating the translocation process [15]. How does this occur? One possibility is that SecA actually dips into the membrane, thereby stuffing a segment of preprotein across to the other side. A more likely possibility is that the energy of ATP binding drives the transfer of a segment of preprotein bound on one site of SecA (the mature binding site) to the other SecA site (the leader binding site), and that the displaced polypeptide from the leader site is thereby driven across the membrane. ATP hydrolysis then allows preprotein release from its association with SecA. Upon release, a rapid translocation of the polypeptide chain in the amino to carboxy direction occurs, driven largely by $\Delta\mu H^+$, the membrane electrochemical potential [15]. It is entirely unclear how the membrane potential functions to promote translocation, as well as how the membrane-transiting polypeptide chain actually crosses the plane of the bilayer. Does it cross through the lipid, through a proteinaceous tunnel formed by the membrane-spanning helices of the SecY/E subunits, or along a lipid-exposed surface of SecY/E? During translocation, the transiting preprotein may rebind to SecA and undergo additional cycles of ATP-driven translocation of small domains, especially when tightly folded subdomains are encountered. Late in translocation, or after translocation has completed, the preprotein remains tethered to the outer surface of the plasma membrane by its membrane-spanning leader peptide. Cleavage by leader peptidase releases the (now mature) protein for folding in the periplasm or to continue its journey to the outer membrane [16]. Folding of the translocated chain may be aided by catalysis, perhaps by SecD or SecF [17] or by periplasmic proteins.

Not all exported proteins employ translocase. M13 procoat [18] and MalF protein (K. McGovern and J. Beckwith, pers. commun.) assemble across the plasma membrane without the aid of translocase function. Recently, it has been found that one domain of leader peptidase assembles across the membrane without the aid of translocase while the other requires its function (R. Dalbey and A. Kuhn, pers. commun.). It is not known how translocase-independent proteins or domains manage the feat of membrane insertion, although procoat will itself assemble in vitro into liposomes which are devoid of integral membrane proteins [19]. Considerable study may be needed to understand the chaperones which facilitate these direct membrane integration events. Mutations introduced into the mature domains of procoat block its own mode of membrane assembly [20]. Why then does it not simply switch to use translocase? Understanding these questions may shed additional light on the recognition specificity of translocase itself.

2. Open questions

At least two other important areas of protein delivery to the cell surface are not well studied. One is the assembly of multispanning transport proteins into the plasma membrane. Most of these proteins are made without identifiable leader regions, and very little is known of their assembly pathways. A second area of almost total ignorance is how certain proteins are selectively incorporated into the outer membrane after they have first completed transfer across the plasma membrane. Is this selective incorporation based on the unfavorable electrochemical potential across the inner membrane, on recognition of lipopolysaccharide in the outer membrane, or on binding to other, pre-assembled outer membrane proteins? These problems are clearly accessible to investigation.

The substantial progress in molecular understanding of protein export stands in stark contrast to our ignorance of how lipids are exported. Biological membranes clearly have asymmetric lipid distribution, and the rates of spontaneous translocation (at least in model systems) are far too slow to account for transfer of newly made lipid from the inner leaflet to the outer leaflet. Almost nothing is known of how glycerolipids and lipopolysaccharide are transferred to the outer membrane.

Finally, in this litany of our ignorance, it should be noted that bacteria, in common with all other organisms, regulate intracellular metabolism according to a cell cycle. It will be intriguing to learn whether the cell cycle control enyzmes measure cell surface growth prior to initiating a new round of DNA synthesis, and how septation is coordinated with the processes of cell surface growth, macromolecular synthesis, and DNA replication.

References

1. Wickner, W., Driessen, A.J.M. and Hartl, F.-U. (1991) Annu. Rev. Biochem. 60, 101–124.
2. Bieker, K.L., Phillips, G.J. and Silhavy, T. (1990) J. Bioenerg. Biomembr. 22, 291–310.
3. Randall, L.L. and Hardy, S.J.S. (1989) Science 243, 1156–1159.
4. Gierasch, L.M. (1989) Biochemistry 28, 923–930.
5. Randall, L.L. (1983) Cell 33, 231–240.
6. Zimmermann, R. and Wickner, W. (1983) J. Biol. Chem. 258, 3920–3925.
7. Randall, L.L., Topping, T.B. and Hardy, S.J.S. (1990) Science 248, 860–863.
8. Liu, G., Topping, T.B. and Randall, L.L. (1989) Proc. Natl. Acad. Sci. USA 86, 9213–9217.
9. Brundage, L., Hendrick, J.P., Schiebel, E., Driessen, A.J.M. and Wickner, W. (1990) Cell 62, 649–657.
10. Hartl, F.-U., Lecker, S., Schiebel, E., Hendrick, J.P. and Wickner, W. (1990) Cell 63, 269–279.
11. Cunningham, K. and Wickner, W. (1989) Proc. Natl. Acad. Sci. USA 86, 8630–8634.
12. Lill, R., Dowhan, W. and Wickner, W. (1990) Cell 60, 271–280.
13. Hendrick, J. and Wickner, W. (1991) J. Biol. Chem. 266, 24596–24600.
14. Lill, R., Cunningham, K., Brundage, L., Ito, K., Oliver, D. and Wickner, W. (1989) EMBO J. 8, 961–966.
15. Schiebel, E., Driessen, A.J.M., Hartl, F.-U. and Wickner, W. (1991) Cell 64, 927–939.
16. Dalbey, R.E. and Wickner, W. (1986) J. Biol. Chem. 260, 15925–15931.
17. Gardel, C., Benson, S., Hunt, J., Michaelis, S. and Beckwith, J. (1987) J. Bacteriol. 169, 1286–1290.
18. Wolfe, P., Rice, M. and Wickner, W. (1985) J. Biol. Chem. 260, 1836–1841.
19. Geller, B.L. and Wickner, W. (1985) J. Biol. Chem. 260, 13281–13285.
20. Kuhn, A., Wickner, W. and Kreil, G. (1986) Nature 322, 335–339.
21. Schatz, P.J. and Beckwith, J. (1990) Annu. Rev. Genet. 24, 215–248.

W. Neupert and R. Lill (Eds.), *Membrane Biogenesis and Protein Targeting*
© 1992 Elsevier Science Publishers B.V. All rights reserved.

CHAPTER 2

Components involved in bacterial protein translocation*

CHRIS HARRIS[1] and PHANG C. TAI[2]

[1]*Department of Biochemistry and Molecular Biology, Harvard University, Cambridge, MA 02139, USA and* [2]*Department of Biology, Georgia State University, University Plaza, Atlanta, GA 30303, USA*

1. Introduction

The last few years have seen an explosive advance in the understanding of how proteins are translocated across the cytoplasmic membrane of *Escherichia coli*. It has been possible, through the use of site-directed mutagenesis, to better understand the sequence constraints upon the signal peptide as well as the intragenic information contained in the mature region of the precursors for protein export (reviewed by Gennity et al. [1] and Harris [2]). Thus it is now known that of the three elements which comprise a signal peptide, the hydrophobic core (the H element) is the only one essential for protein export, despite the observation that the length of the H element can be varied [3–5], and the elements are without strict sequence homologies (reviewed by von Heijne [6]). In addition, the insertion of basic amino acids into the N-terminal regions of the mature protein, the N(m) element, of several secretory proteins inhibits the export of these proteins [7–9].

The convergence of biochemical and genetic approaches in demonstrating the involvement of the products of *secA*, *secB*, *secE* (also known as *prlG*) and *secY* (also known as *prlA*) in protein translocation has been an important development and has underscored the advantage of microbial systems for studying a physiological process (reviewed by [10–13]). Mutations in the peripheral membrane protein SecA or the integral membrane proteins SecY/PrlA and SecE/PrlG can either cause pleiotropic defects in export or relieve the export defect of signal sequence mutations. The soluble protein SecB affects the export of a subset of proteins in vivo. The involvement of these Sec proteins in protein export has recently been demonstrated in the in vitro translocation system. This development has been particularly satisfying, since

---

* Dedicated to the memory of Philip Bassford, Jr. Parts have been presented at the Lunteren Lectures, September 24–27, 1991.

the previously independent biochemical and genetic approaches have now identified the same components involved in the process and have thus affirmed the validity of each approach for further analysis, e.g., the roles of other *sec* genes (*secD* and *secF*). The SecY/PrlA (and perhaps SecE/PrlG) is now believed to be the translocator, or part of it, where peptides are translocated [14,15] (but see later section). SecA is essential and is related to the requirement of ATP hydrolysis in protein translocation [16], while SecB is involved in the maintenance of the translocation-competent conformation of precursor molecules, and perhaps also in targeting to membrane translocation sites [17–19].

This chapter describes some of our recent work on determining the sequence constraints upon the signal peptide and N(m) element, and on the biochemical characterization of some components involved in protein translocation. The main focuses are to determine the minimal length of the signal peptide, to characterize further the nature of N(m) element mutations, to provide biochemical evidence on whether SecY/PrlA membrane protein is essential, and to examine how SecD and SecF may function in protein translocation.

2. The minimal length of a prokaryotic signal peptide

TEM β-lactamase, which is encoded by plasmids such as pBR322, is targeted to the periplasm by expressing *E. coli* cells. At the N-terminus of this protein, there is a 23 amino acid signal peptide, the sequence of which is displayed in Table I. The β-lactamase signal peptide possesses all three elements characteristic of prokaryotic signal peptides (see also [6]): an N-terminal or N element where a basic amino acid

TABLE I
Sequences of mutant β-lactamase signal peptides

Peptide	Predicted signal peptide elements[a]				
	N		H		C
Wild-type	MSIQHFR	–	VALIPFFAACL	–	PVFA
2-8-4	MR	–	L_8	–	PVFG
2-7-4	MR	–	L_7	–	PVFG
2-6-4	MR	–	L_6	–	PVFG
1-7-4	M	–	L_7	–	PVFG
2-7-1	MR	–	L_7	–	G
2-IL$_9$-1	MR	–	IL_9	–	G
2-IL$_{13}$-1	MR	–	IL_{13}	–	G

[a]The borders for the N, H, and C elements are estimated by looking for a polar residue to serve as the border between the N and H elements and by looking for a polar or turn-forming residue to demarcate the border between the H and C elements (see also [6]).

resides; a central hydrophobic, or H, element; and a C-terminal or C element that includes a turn-forming residue, proline, and residues with small side chains at positions 3 and 1. For β-lactamase, the predicted lengths of the natural N, H, and C elements are 7, 12, and 4 amino acids, respectively, and the signal peptide is designated as 7-12-4.

It was found that the 23 amino acid signal peptide of β-lactamase could be replaced by a 12 amino acid peptide, 1-7-4 (see Table I for the amino acid sequence of 1-7-4 which is designated as the amino acid residues of N, H, C elements, respectively, with the H element replaced by homologous leucine residues), without preventing export of β-lactamase in vivo. Whereas deletion of the β-lactamase signal peptide caused export to be blocked even 10 min after the initiation of protein labeling (data not shown), more than one-third of 1-7-4/β-lactamase was exported 30 s after labeling (Harris, unpublished data). Thus, the N element of the β-lactamase signal peptide could be reduced in size to a single amino acid without blocking protein export, albeit with a slower kinetics. The presence of a single arginine residue did, however, double the amount of export observed within 30 s. For processing to occur, the proline from the C element could not be deleted; four amino acids appeared to be the minimal length of the C element. Most naturally occurring C elements comprise 6 amino acids [6].

When the peptides 2-8-4, 2-7-4, and 2-6-4 were tested as β-lactamase signal peptides in vivo, very rapid, efficient export was effected by the 2-8-4 and 2-7-4 peptides; however, the 2-6-4 signal peptide displayed no ability to promote β-lactamase export (Harris, unpublished data). The predicted H element lengths of signal peptides 2-8-4, 2-7-4, and 2-6-4 are 8, 7, and 6 amino acids, respectively. From these data, it was apparent that the functional H element minimally requires the presence of 7 amino acids.

Arguing against a seven residue minimal length for the H element is the observation that point mutations and deletions within the C element of the 2-7-4 signal peptide were found to block β-lactamase export. For example, it was observed that removal of proline, valine, and phenylalanine from the 2-7-4 C element created a 2-7-1 signal peptide which did not support the export of any β-lactamase. On the other hand, signal peptides 2-IL$_9$-1 and 2-IL$_{13}$-1, lacking the proline, valine, and phenylalanine, were very active in promoting β-lactamase export even though signal peptide cleavage did not occur. The salient difference between the peptide sequences of 2-7-1, 2-IL$_9$-1, and 2-IL$_{13}$-1 is that the predicted length of the 2-7-1 H element is much shorter than the predicted H element length for the other two peptides. Therefore, the export defect brought on by deleting proline, valine, and phenylalanine was specific to a signal peptide with a shortened H element. The most likely explanation for these data is that the shortened H element recruits residues from the C element in order to elicit protein export. Since the H element of 2-7-4 extends into its C element, the minimal length of the signal peptide H element is likely to be longer than 7 amino acids.

The effective length of the 2-7-4 H element can be determined by noting the position of the C element mutations that block its function. In signal peptide 2-7-EVFG, only the proline residue has been replaced, yet the signal peptide is unable to support export. This supports the notion that the proline is conscripted by the shortened H element. In non-functional signal peptide 2-7-PGQG, the proline is unchanged, but the valine and phenylalanine are replaced. Thus one or both of these residues may also participate as part of the 2-7-4 H element. Therefore, it is likely that at least 9 amino acids, not 7, are required in order to comprise a functional H element in the signal peptide studied.

H elements with predicted lengths shorter than 9 amino acids are known, such as the 8 amino acid H element of the natural lipoprotein signal peptide [20], and the 6 amino acid H elements of engineered lysozyme [3] and carboxypeptidase [4] signal peptides. However, as the present study indicates, the H element length must be determined experimentally, by examining the export consequences of mutations in the C element (see also [21,22]). Instead of six residues, we suspect that nine or more hydrophobic residues from all signal peptides are minimally recognized by the cellular export machinery.

3. The N(m) element: an export requirement for low basicity as well as low polarity

All N(m) mutations in the N-terminal mature region of a secretory protein known to block export involve the introduction of basic amino acids into this region. Although the inserted basic amino acids raised both the hydrophilicity and the positive charge of the N(m) elements, Summers and colleagues have shown that the export of a β-lactamase/triosephosphate isomerase hybrid protein is blocked due to the hydrophilicity of the basic amino acid introduced into its N(m) element [23,24]. Specifically, the export of β-lactamase/triosephosphate isomerase was blocked by the presence of an arginine, but not by the presence of a lysine, within its N(m) element, in a TIM14 region. Since lysine and arginine each contributes a single positively charged moiety to the N(m) element, and arginine has a much more hydrophilic side chain than lysine, their differing export consequences are likely to derive from another property, the increased polarity.

However, it can now be reported that a β-lactamase mutant, Bla/TIM14/Bla, was blocked by the charge of its N(m) element, not its polarity. Bla/TIM14/Bla and β-lactamase/triosephosphate isomerase share the same amino acid sequence between residues +1 to +14 [2,23]. However, export of Bla/TIM14/Bla was blocked when either arginine +3 or lysine +4 were present within its N(m) element (Harris, unpublished data). Thus, arginine and lysine do not have differing effects on the export of β-lactamase, yet they do on the export of triosephosphate isomerase.

By itself, this result does not contradict the idea that the polarity of the Bla/TIM14/Bla N(m) element is too high to enable its export. This statement holds equally true for alkaline phosphatase and for lipoprotein, two other proteins for which the insertion of an arginine has the same export consequences as does the insertion of a lysine [7,8]. The polarity requirement on the β-lactamase, alkaline phosphatase, and lipoprotein N(m) elements could, for some reason, be less stringent than exists for triosephosphate isomerase export.

Thus, it was necessary to test directly whether the export defect conferred by the Bla/TIM14/Bla N(m) element is derived from its charge or from its polarity. If, as is the case for β-lactamase/triosephosphate isomerase, the export of Bla/TIM14/Bla is blocked by the polarity of its N(m) element, then mutations which increase the polarity of this element, even if they remove charged amino acids, should not suppress its export block. Conversely, if the export of Bla/TIM14/Bla is blocked by the basicity of its N(m) element, then mutations which increase the polarity of this N(m) element but remove basic amino acids should allow Bla/TIM14/Bla to be exported. Two mutants were made in which the Bla/TIM14/Bla N(m) element was made less charged but is predicted to be more polar. In Bla/TIM (R3P,F5Q)/Bla, arginine +3 is replaced with proline and phenylalanine +5 is replaced by glutamine. In Bla/TIM(R3P,K4N,F5P,F6S)/Bla, four N(m) mutations are made, including the replacement of both arginine +3 and lysine +4 with uncharged but polar amino acids. Despite the polarity of their N(m) elements, both Bla/TIM(R3P,F5Q)/Bla and Bla/TIM(R3P,K4N,F5P,F6S)/Bla were exported (Harris, unpublished data). Therefore, the export defect of Bla/TIM14/Bla appears to derive from the charge of its N(m) element residues.

Thus, although N(m) mutants have previously been thought to represent a single class of intragenic export mutants, these data suggest that there are at least two types of N(m) mutants: those blocked by polarity and those blocked by charge. It is likely that N(m) mutations previously reported to block the export of lipoprotein, alkaline phosphatase, and OmpA [7–9] may cause export defects by virtue of their charge since, like the β-lactamase protein that is shown to be blocked by the charge of its N(m) element, each is a true *E. coli* secretory protein. Conversely, the triosephosphate isomerase sequence which comprises the bulk of the β-lactamase/triosephosphate isomerase fusion protein is normally found in the cytoplasm of chicken cells [23]. It has been suggested that the low polarity threshold tolerated by the N(m) element of β-lactamase/triosephosphate isomerase will prove to be imposed upon *E. coli* export by any foreign, eukaryotic, and/or cytoplasmic protein, simply because such proteins are not designed to interact with the full repertoire of *E. coli* export factors [2]. Since nuclease A is also a foreign protein to the *E. coli* cytoplasm — this protein derives from *Staphylococcus aureus* — export of the nuclease A protein with the N(m) mutation [25] may also be blocked due to its N(m)'s polarity, not its positive charge. However, like β-lactamase, nuclease A is also prokaryotic and secretory. Thus, the way in which export of nuclease A is blocked by an N(m) mutation should prove to be interesting.

4. Is SecY/PrlA essential for protein translocation?

The *secY* gene encodes a 49 kDa integral membrane protein [26,27] with ten transmembrane domains [28]. Mutations in *secY* have been isolated which result in a pleiotropic defect in protein export [29,30]. Other mutations, the *prlA* alleles, were isolated even earlier as strong suppressors of signal sequence defects [31]. Because of these two mutant phenotypes, and its location in the cytoplasmic membrane, SecY/PrlA has long been considered to be a good candidate for a central component of the translocation machinery in the membrane.

In vitro demonstration of the temperature-sensitive *secY*24 mutation [29] came from the observations that membranes prepared from these cells grown at permissive temperatures were active in translocation of alkaline phosphatase and OmpA precursors, but were inactivated upon incubation of these membranes at 40°C, whereas $SecY^+$ membranes were unaffected [32]. These results demonstrate that the defect of the *secY*24 membranes is directly due to the *secY* mutation and not indirectly due to the growth defect of *secY*24 cells at 42°C. These studies mark the first biochemical demonstration of the defect of *sec* gene products in protein translocation in vitro. The insertion of major prolipoprotein (pLpp) into membrane vesicles occurs spontaneously even in the absence of functional involvement of SecY protein [33]. However, modification and processing of pLpp is blocked in the inactivated SecY24 membranes, suggesting that the functional involvement of SecY in the translocation process occurs after the initial interaction of the precursors with the membranes.

In order to investigate whether the function of SecY protein is essential for protein translocation in vitro, we prepared membranes that were greatly reduced in the amount of SecY protein. Membranes isolated from a temperature-sensitive upstream polar *secY* mutant (KI200; kindly provided by K. Ito) grown at a nonpermissive temperature contained less than 3% of SecY as determined immunologically, but possessed 50–60% translocation activity, both in extent and in kinetics, of proOmpA protein including signal peptide cleavage, as compared to control membranes isolated from cells grown at permissive conditions (J.P. Lian, unpublished data). To determine whether this translocation activity was due to the presence of a small amount of functional SecY, IgGs against synthetic peptides corresponding to N-terminal or C-terminal cytoplasmic domains of SecY were used to inhibit SecY-dependent translocation activity. The translocation activity of OmpA precursors, including processing of signal peptides, of these SecY-deficient membranes was only marginally affected by SecY antibodies, while that of control membranes containing normal amounts of SecY was inhibited about 50%, which differs from an earlier report of complete inhibition [34]. (In contrast, antibodies against SecD, SecE or SecF completely inhibited the translocation.) Furthermore, membrane vesicles reconstituted from solubilized membrane proteins under conditions that remove more than 99% of SecY [35] were still about 50% as active in protein translocation. To further reduce the residual SecY protein to a negligible amount, membrane vesicles were reconstituted from SecY-deficient mutant (KI200) membranes and were found to be able to efficiently translocate proteins. These SecY-deficient membranes were dependent on ATP and SecA protein for translocation, just as SecY-containing membranes. The translocation of prolipoprotein was affected even less than proOmpA in these SecY deficient membrane vesicles. We conclude, tentatively, that SecY protein contributes, but is not essential, for protein translocation, at least under the in vitro conditions used.

These findings do not necessarily contradict the observations that point mutations in SecY/PrlA result in pleiotropic secretion defects or in suppression of defective signal peptides. There are many examples of severe functional defects in processes caused by point mutations of a gene, yet no effect when the protein is removed. In this context, the contributory, but not obligatory, role of SecY/PrlA protein in protein translocation is reminiscent of the role of proton motive force in the process (see [36]).

This conclusion is somewhat surprising, considering the overwhelming genetic and biochemical evidence for the essential requirement of SecY in protein translocation. The question of limitation of an in vitro translocation system [37] cannot be ignored. However, it should be noted that the genetic manipulation of specific null mutation in *secY* has not yet been done and is difficult to do because of its location in the middle of the essential operon [26]. An in-frame deletion of the *secY* gene would be essential to resolve this paradox. On the other hand, genetic evidence suggests that SecD, SecE and SecF are also involved in protein translocation; the SecY-deficient membrane vesicles contained immunologically sufficient amounts of these Sec proteins to allow efficient translocation in the absence of SecY. Moreover, purified antibodies against peptides of SecD, SecE, or SecF completely inhibited the translocation, suggesting also that these Sec proteins are important in protein transport. Whether this interpretation contradicts the conclusion from other in vitro translocation systems [15,38,39] that SecY is essential, remains to be resolved. However, in these two systems, the translocation does not include cleavage of signal peptides and we have shown previously that aberrant reaction often allows translocation of precursor molecules, as determined by proteinase K resistance, but does not proceed to cleavage of signal peptides [40,41], and prolipoprotein can translocate into liposomes spontaneously [33].

5. In vitro suppression of defective signal peptides

Regardless of whether SecY is essential for protein translocation, the suppression of defective signal peptides in the cells is well documented [11,31,42]. The strong suppressor alleles *prlA*4 and *prlG*1 have been used to demonstrate genetically the functions of PrlA and PrlG, and the interaction of SecY/PrlA and SecE/PrlG [14,43]. However, there is no clear demonstration of the in vitro suppression of defective signal peptides by PrlA4 or PrlG1 membranes and a possible explanation has been offered [44].

We have recently been able to show that the translocation of OmpA protein with defective signal peptides can indeed be suppressed in vitro by membranes prepared from cells with various *prlA* and *prlG* alleles (J. Yu, unpublished data), thus strengthening the in vivo observations. The deletion of hydrophobic core residues Ala_7-Ile-Ala_9 ($\Delta 7$–9) and point mutation R14($Gly_{14} \rightarrow Arg_{14}$) in signal peptides of OmpA severely reduces the in vitro translocation of these precursors in the wild-type membranes, as has been observed in vivo [1]. The translocation of OmpA precursors carrying $\Delta 7$–9 mutation and R14 was greatly improved by membranes from cells carrying *prlA*4 mutation [31], and even more efficiently in a newly isolated *prlA* super suppressor mutant carrying a *prlA*665 mutation, surprisingly, in the first periplasmic domain of PrlA protein (P. Bassford, gift and pers. commun.). The translocation of wild-type OmpA precursors was equally efficient in these membranes. Moreover, the suppression was dependent on the presence of the mutated PrlA/SecY, since the SecY-deficient membranes were not active in allowing translocation of OmpA precursors with defective signal peptides.

Similarly, membranes prepared from cells containing suppressor *prlG*1 supported the translocation of OmpA precursors with mutant $\Delta 7$–9 and R14 signal peptides. The efficiency of the PrlG1 suppression that allowed the translocation of defective signal peptide precursors was near or greater than that of PrlA4 (but not as efficient as PrlA665).

These observations confirm and substantiate the in vivo conclusion that PrlA and PrlG are directly involved in the suppression of defective signal peptides by interacting with these precursors, and not by indirect, secondary effects. It has been 10 years since the original genetic identification of *prl* mutations!

6. Roles of SecD and SecF in protein translocation

Genetic studies have revealed that mutations in *secD* locus, containing *secD* and *secF* genes, cause pleiotropic secretion defects at non-permissive temperatures [45,46]. The *secD* and *secF* genes code for 65 kDa and 35 kDa membrane proteins, respectively, and cold-sensitive mutants in these alleles cause rapid accumulation of precursors at temperatures below 23°C [46]. Because of their large periplasmic domains in the predicted amino acid sequences of these two proteins, and because genetic selection for suppressing signal peptide defects has not picked up mutations in these genes, it has been suggested that SecD and SecF function late in protein translocation [46].

We sought to biochemically characterize the functions of SecD and SecF in protein translocation in vitro using cold-sensitive *secD* and *secF* mutants, in collaboration with J. Beckwith. The membranes prepared from *secDcs* and *secFcs* mutants and their wild-type strain grown at either 37°C or 18°C were all equally active in protein translocation with optimal amount of SecA. (In fact, the *secDcs* and *secFcs* mutant membranes prepared from cells grown at 18°C were more active without exogenous SecA because of overproduction of SecA in the membranes.) The same membranes from cold-sensitive mutants grown at 37°C or 18°C were slightly (about 30-50%) less active in translocation than wild-type membranes under identical conditions (J. Vigiduriene, unpublished data). The time course of the translocation suggested that the impairment was more pronounced at the later process, consistent with the notion that SecDcs and SecFcs interfere with the late stage of translocation. To determine whether the SecDcs and SecFcs may impair the late step, Na_2CO_3 treatment with translocated membranes was used to release the translocated, processed proteins from lumen- or peripheral-bound form inside the membranes. Indeed, the release of translocated OmpA and alkaline phosphatase was severely impaired in SecDcs and SecFcs membranes when translocation was carried out at 16°C under conditions that most translocated, processed proteins were released from wild-type membranes. Moreover, this impairment of release with SecDcs and SecFcs membranes is specific at low temperatures; the inhibition was not observed when translocation was carried out at 37°C with these same membranes. Furthermore, the impairment of release was not observed with SecYcs membranes that were partially defective at low temperatures. These observations suggest that cold-sensitive *secDcs* and *secFcs* mutations affect the release of translocated mature proteins, thus jamming the overall protein translocation, perhaps in the recycle process. Preliminary data indicate that this is indeed the case. These results further support the notion from genetic studies that SecD and SecF function in the later step of translocation. It should be noted, however, that IgGs against cytoplasmic domains of SecD and SecF completely inhibit protein translocation, suggesting that the release may not be the only function of SecD and SecF in the translocation process. Obviously, further studies are necessary to define the functions of SecD and SecF.

7. An inhibitor of protein translocation

We have previously shown that the translocation defect of temperature-sensitive SecY24 membranes in vitro can be compensated by the addition of purified SecA protein, and, to a lesser extent, by SecB protein [47,18]. However, an S300 extract from the *secB*⁻ strain was not active in suppressing the SecY24 defect, even though the extract had wild-type level of active SecA. We found that this extract contained an inhibitor (SecI, for secretion inhibitor) that interfered with the suppression of SecY24 defect by SecA protein, and this inhibitory activity could be neutralized by SecB protein (J. Fandl, H. Xu, unpublished data). Moreover, the inhibitory activity of purified SecI could also be observed with normal translocation of wild-type membranes, and has also been purified from wild-type cells. Immunological and structural studies revealed that SecI is identical to Wickner's trigger factor, which was reported to promote translocation by triggering the conformational change of precursor molecules necessary for traversing membranes [48], but is now believed to be a molecular chaperone for cell division [49]. However, we found that SecI was present in larger amounts in cells grown toward a stationary phase and there was more activity in *secB* mutants. Based on the inhibition of SecI on SecA activity and its interaction with SecB, we propose a regulatory function for SecI in protein translocation.

8. Perspective

Other than some minor sidetracks (e.g., *secC* and trigger factor), the studies of bacterial protein translocation by genetic and biochemical approaches have provided a remarkably consistent theme. All the important components are probably now identified; the detailed mechanism of the function of each component, sequential reaction, and the interaction of various components need to be worked out. Proton motive force clearly contributes to the physiological process, yet it is difficult to define its role since it can cause pleiotropic effects. Unanswered questions include the exact role of ATP hydrolysis in SecA function, the function of SecB and whether it interacts with signal peptide or mature region, or both [50] of a precursor, whether SecY is essential, how SecE and SecY interact, and how SecD and SecF may function in the release of mature proteins after translocation from the membranes. Whether there is a functional bacterial equivalent of the mammalian signal recognition particle remains unclear.

Complete reconstitution of functional membranes from purified components and ingenious genetic design should help to clarify the mechanism(s) of protein export in bacteria in the near future.

Acknowledgement

The work in the authors' laboratory has been supported by NIH grants GM 34766 and GM 41835.

References

1. Gennity, J., Goldstein, J. and Inouye, M. (1990) J. Bioenerg. Biomembr. 22, 233.
2. Harris, C. (1991) Ph.D. Thesis, Harvard University.
3. Yamamoto, Y., Taniyama, Y., Kikuchi, M. and Ikehara, M. (1987) Biochem. Biophys. Res. Commun. 149, 431.
4. Bird, P., Gething, M.-J. and Sambrook, J. (1990) J. Biol. Chem. 265, 8420.
5. Zhang, Y. and Broome-Smith, J.K. (1989) Mol. Microbiol. 3, 1361.

6. von Heijne, G. (1985) J. Mol. Biol. 184, 99.
7. Yamane, K. and Mizushima, S. (1988) J. Biol. Chem. 263, 19690.
8. Li, P., Beckwith, J. and Inouye, H. (1988) Proc. Natl. Acad. Sci. USA 85, 7685.
9. MacIntyre, S., Eschbach, M. and Mutschler, B. (1990) Mol. Gen. Genet. 221, 466.
10. Fandl, J.P. and Tai, P.C. (1990) J. Bioenerg. Biomembr. 22, 369.
11. Bieker, K.L., Phillips, G. and Silhavy, T. (1990) J. Bioenerg. Biomembr. 22, 369.
12. Schatz, P. and Beckwith, J. (1990) Annu. Rev. Genet. 24, 215.
13. Wickner, W., Driessen, A.J. and Hartl, F. (1991) Annu. Rev. Biochem. 60, 101.
14. Bieker, K.L. and Silhavy, T. (1990) Cell 61, 833.
15. Brundage, L., Hendrick, J.P., Schiebel, E., Driessen, A.J. and Wickner, W. (1990) Cell 62, 649.
16. Lill, R., Cunningham, K., Brundage, L., Ito, K., Oliver, D. and Wickner, W. (1989) EMBO J. 8, 961.
17. Weiss, J.B., Ray, P.H. and Bassford, P.J. (1988) Proc. Natl. Acad. Sci. USA 85, 8978.
18. Kumamoto, C.A., Chen, L.L., Fandl, J.P. and Tai, P.C. (1989) J. Biol. Chem. 246, 2242.
19. Watanabe, M. and Blobel, G. (1989) Cell 58, 695.
20. Inouye, S., Wang, S., Sekizawa, J., Halegoua, S. and Inouye, M. (1977) Proc. Natl. Acad. Sci. USA 74, 1004.
21. Fikes, J.D., Bankaitis, B.A., Ryan, J.P. and Bassford, P.J. (1987) J. Bacteriol. 169, 2345.
22. Fikes, J.D. and Bassford, P.J. (1987) J. Bacteriol. 169, 2352.
23. Summers, R.G., Harris, C.R. and Knowles, J.R. (1989) J. Biol. Chem. 264, 20082.
24. Summers, R.G. (1988) Ph.D. Thesis, Harvard University.
25. Liss, L.R., Johnson, B.L. and Oliver, D.B. (1985) J. Bacteriol. 164, 925.
26. Cerretti, D.P., Dean, D., Davis, G.R., Bedwell, D.M. and Nomura, M. F. (1983) Nucleic Acids Res. 11, 2599.
27. Akiyama, Y. and Ito, K. (1985) EMBO J. 4, 3351.
28. Akiyama, Y. and Ito, K. (1987) EMBO J. 6, 3465.
29. Shiba, K., Ito, K., Yura, T. and Cerretti, D. (1984) EMBO J. 3, 631.
30. Riggs, P.D., Derman, A.I. and Beckwith, J. (1988) Genetics, 118, 571.
31. Emr, S.D., Hanley-Way, S. and Silhavy, T.J. (1981) Cell 23, 79.
32. Fandl, J. and Tai, P.C. (1987) Proc. Natl. Acad. Sci. USA 84, 7448.
33. Tian, T., Wu, H.C., Ray, P.H. and Tai, P.C. (1989) J. Bacteriol. 171, 1987.
34. Watanabe, M. and Blobel, G. (1989) Proc. Natl. Acad. Sci. USA 86, 2248.
35. Watanabe, M., Nicchitta, C.V. and Blobel, G. (1990) Proc. Natl. Acad. Sci. USA 87, 1960.
36. Tai, P.C. (1990) in: T. Krulwich (Ed.), Bacterial Energetics, Academic Press, Orlando, FL, p. 393.
37. Tai, P.C. (1986) Curr. Top. Microbiol. Immunol. 125, 43.
38. Nishiyama, K., Kabuyama, Y., Akimaru, J, Matsuyama, S., Tokuda, H. and Mizushima, S. (1991) Biochim. Biophys. Acta 1065, 89.
39. Akimaru, J., Matsuyama, S., Tokuda, H. and Mizushima, S. (1991) Proc. Natl. Acad. Sci. USA 88, 6545.
40. Chen, L.L. and Tai, P.C. (1987) J. Bacteriol. 168, 2373.
41. Chen, L.L. and Tai, P.C. (1989) Arch. Microbiol. 153, 90.
42. Bankaitis, V.A. and Bassford, P.J., Jr. (1985) J. Bacteriol. 161, 169.
43. Bieker, K.L. and Silhavy, T.J. (1989) Proc. Natl. Acad. Sci. USA. 86, 968.
44. Weiss, J., MacGregor, C.H., Collier, D.N., Fibes, J.D., Ray, P. and Bassford, P. (1989) J. Biol. Chem. 264, 3021.
45. Gardel, C., Benson, S., Hunt, J., Michaelis, S. and Beckwith, J. (1987) J. Bacteriol. 169, 1286.
46. Gardel, C., Johnson, K, Jacq, A. and Beckwith, J. (1990) EMBO J. 9, 3209.
47. Fandl, J.P., Cabelli, R., Oliver, D. and Tai, P.C. (1988) Proc. Natl. Acad. Sci. USA 85, 8953.
48. Crooke, E. and Wickner, W. (1987) Proc. Natl. Acad. Sci. USA 84, 5216.
49. Guthrie, B. and Wickner, W. (1990) J. Bacteriol. 172, 5555.
50. Aberman, E., Emr, S. and Kumamoto, C. (1990) J. Biol. Chem. 265, 18154.

51. Boyd, D. and Beckwith, J. (1990) Cell 62, 1031.
52. Cabelli, R.J., Chen, L.L., Tai, P.C. and Oliver, D.B. (1988) Cell 55, 683.
53. Puziss, J.W., Fikes, J.D. and Bassford, P.J. (1989) J. Bacteriol. 171, 2303.

CHAPTER 3

Molecular characterization of Sec proteins comprising the protein secretory machinery of *Escherichia coli*

SHOJI MIZUSHIMA, HAJIME TOKUDA and
SHIN-ICHI MATSUYAMA

Institute of Applied Microbiology, University of Tokyo, Yayoi, Bunkyo-ku, Tokyo 113, Japan

Abstract

Genetic studies have suggested that SecA, SecD, SecE, SecF and SecY are general components of the protein secretory machinery of *Escherichia coli*. We have succeeded in overproducing and purifying all of these proteins. The function of SecA, a peripheral membrane protein, was studied using a conventional in vitro translocation system. Reconstitution of SecA analogs from its subfragments was performed to localize the sites on the SecA molecule which interact with ATP and a presecretory protein. Changes in the conformation of SecA upon interaction with ATP, a presecretory protein, phospholipids and membrane vesicles were also studied. Reconstitution of proteoliposomes from purified SecE and SecY was achieved. The proteoliposomes exhibited protein translocation activity in the presence of SecA and ATP, indicating that SecA, SecE and SecY are essential components of the secretory machinery. On the other hand, the involvement of SecD in the release of translocated proteins from the membrane was demonstrated with spheroplasts. The function of SecF remains unknown. Based on these observations, the molecular mechanism underlying the translocation of presecretory proteins across the cytoplasmic membrane is discussed.

1. Introduction

The translocation of proteins across the cytoplasmic membrane in prokaryotes and

---

Abbreviations: SDS, sodium dodecyl sulfate; Octylglucoside, n-octyl-β-D-glucopyranoside; AMP-PNP, adenosine 5'-(β,γ -imino)triphosphate.

that across the endoplasmic reticulum membrane in eukaryotes are similar in that both require the signal peptide that is attached to the amino terminus of the proteins. Several experiments have demonstrated that eukaryotic signal peptides are effective in prokaryotic cells and vice versa, suggesting that the basic mechanism underlying the translocation process is similar in all living things. Protein secretion is a typical example of such a translocation process.

Bacterial cells, especially *Escherichia coli* cells, are advantageous for the molecular analysis of such a translocation mechanism in that genetic and gene engineering methods can easily be applied to these organisms. Genetic studies on *E. coli* have revealed the involvement of the following gene products in the process of protein secretion; SecA, SecB, SecD, SecE, SecF and SecY [1]. The enzymes responsible for the cleavage of the signal peptide and those for the digestion of the cleaved signal peptide [2] are also essential for protein secretion. However, they do not seem to be essential for the transmembrane translocation itself [3].

Recently, cellular components of *E. coli* which are homologous to 7S RNA and the 54K protein of the eukaryotic signal recognition particle (SRP) were found, suggesting the presence of an SRP-like system in prokaryotes [4]. However, no direct evidence suggesting the general importance of these components in protein translocation in *E. coli* is available yet [5].

This review summarizes the results of biochemical analyses of Sec proteins, focusing on the results obtained in our own laboratory. Many of the results were obtained in reconstitution studies.

2. *Overproduction of Sec proteins*

For extensive biochemical studies, large quantities of purified components are usually essential. Recent progress in recombinant DNA technology has made the overproduction of a required protein possible. SecB overproduction has been achieved by Weiss et al. [6] and by Kumamoto et al. [7]. We have achieved the overproduction of SecA, SecD, SecE, SecF and SecY. The overproduction of SecA has also been reported by Oliver et al. [8]. A general way of constructing an overproducing strain is to place a cloned gene under the control of a controllable high-expression promoter on a plasmid. The *tac* promoter-operator is certainly one of the best for this purpose. The use of a run-away plasmid, whose replication is derepressed at a high temperature, is also advisable [9].

We have succeeded in cloning the *secA*, *secE*, *secY*, *secD* and *secF* genes downstream of the *tac* promoter-operator and in the overproduction of the Sec proteins encoded by these genes. SecA [10], SecE [11], and SecF [12] were overproduced independently. The overproduction of SecD alone may also be possible, although this has only been achieved so far with the operon comprising both the *secD* and *secF* genes [12]. The overproduction of SecY alone has been unsuccessful due to rapid breakdown of the overproduced SecY [13]. The overproduction was achieved with the simultaneous overproduction of SecE [11]. This suggests that there is a firm interaction between these two membrane proteins and that the interaction stabilizes SecY. The existence of such an interaction has also been suggested genetically [14] and biochemically [15].

3. Purification of Sec proteins

All the Sec proteins thus overproduced have been purified. Since the overproduced SecA is localized mainly in the cytosol, it was easily purified [10,16]. Very recently we realized that more than 50% of the SecA thus purified had lost eight amino acid residues from its amino terminus. When SecA was directly extracted from cells with sodium dodecyl sulfate (SDS) and isolated on an SDS-polyacrylamide gel, it was intact with respect to the amino terminus, indicating that proteolysis took place during the SecA purification. The proteolysis was almost completely prevented when the entire purification was performed in the presence of a mixture of the following protease inhibitors: benzamidine, aminobenzamidine, p-AMSF, antipain, leupeptin and phosphoramidon. The activity of in vitro translocation of the intact SecA purified in the presence of these inhibitors was the same as that in the case of the previously purified SecA (A. Shinkai, H.-M. Lu, S. Matsuyama and S. Mizushima, unpublished data).

For the purification of SecE and SecY, the membrane fraction containing overproduced amounts of these proteins was solubilized with 2.5% n-octyl-β-D-glucopyranoside (octylglucoside). These proteins were then separated and purified on FPLC columns of Mono Q or Mono S and Superose 12HR [17,18]. It should be noted that SecY thus overproduced was rapidly degraded in vitro, especially after solubilization of the membrane. The degradation was significantly suppressed when an *ompT* mutant lacking outer membrane protease OmpT was used as the host cell [19,20].

For the purification of SecD and SecF, differential solubilization of the membrane with various detergents was found to be useful [20a]. For SecD purification, the membrane fraction containing an overproduced amount of SecD was treated with 1.5% cholate and SecD was solubilized from the insoluble fraction with 2.5% octylglucoside. Further purification of SecD was carried out on FPLC columns of Mono P and Superose 6. For SecF, the membrane fraction was first treated with 6% cholate and then with 2% octylglucoside. The resultant insoluble fraction was further treated with 0.25% Sarcosyl to solubilize SecF, which was then purified by means of molecular sieving.

4. Estimation of the numbers of Sec proteins and of the secretory machinery in one E. coli cell

The overproduction of Sec proteins made it possible to estimate their quantities in a cell on an SDS-polyacrylamide gel after staining with Coomassie Brilliant Blue. The extent of overproduction of an individual Sec protein can be determined by means of immunoblotting with a specific antiserum. Thus we estimated the numbers of these Sec proteins in one normal cell (Table 1) (see also [11]). SecE, SecD, SecF and SecY were exclusively localized in the cytoplasmic membrane, whereas only 10% of SecA was found in the membrane fraction, the rest being in the cytosol. Assuming that one secretory machinery in the cytoplasmic membrane comprises on each of these proteins (SecA was found to exist as a dimer), the possible number of the machinery in one cell is, therefore, assumed to be around 500. SecF may not be the core constituent of the machinery. The number of ribosomes in one rapidly growing cell is in the order of 10^4. About 10% of the proteins synthesized on ribosomes have to be translocated across the cytoplasmic membrane to reach their final destinations, which are the periplasm, the outer membrane and extracellular medium. It is reasonable to assume, therefore, that one secretory machinery is responsible for the translocation of proteins synthesized on about ten ribosomes. This sounds reasonable since the time required for the translocation of one protein molecule is assumed to be much faster than that for the translation of one protein molecule.

Such a ten-to-one hypothesis is possible only when the translocation reaction is not coupled with the translation reaction (post-translational translocation). In the case of co-translational translocation, one secretory machinery has to be occupied by one ribosome synthesizing a presecretory protein, and the rate of translocation has to be the same as that of translation. The estimated number of the machinery, therefore,

TABLE 1

Numbers of Sec proteins in one E. coli cell [20a]

Protein	Protein content (%)[a]	Overproduction[b] (fold)	No. of molecules per normal cell
SecA	25/WC	50–100	2500–5000
SecE	3.4/M	80–160	300–600
SecY	5.3/M	50–100	200–400
SecD	6.9/M	20–40	450–900
SecF	16.5/M	1500–3000	30–60

[a] Contents of each Sec protein in the whole cell (WC), or the total membrane fraction (M) of Sec protein-overproducing cells were densitometrically determined on SDS-polyacrylamide gel stained with Coomassie Brilliant Blue.
[b] The degree of overproduction was determined by immunoblot analysis after SDS-polyacrylamide gel electrophoresis of overproducing and non-overproducing cell samples.

favors post-translational translocation. A large numer of in vitro studies have suggested that translocation is not necessarily coupled with translation. In vivo studies also support this view [21].

5. Functions of SecA in protein translocation

SecA is a homodimer of a 102 kDa subunit [22] and is essential for protein translocation both in vivo [23,24] and in vitro [8,10,25]. Our cross-linking studies revealed that SecA interacts with presecretory proteins by recognizing the amino terminal positive charge of the signal peptide [16]. The extent of cross-linking increased when the amino terminal positive charge was increased from zero to +2 and then +4. The increase paralleled the increase in the rate of translocation of these presecretory proteins. These results, together with the fact that SecA is a peripheral membrane protein, indicate that SecA plays a leading role in the initial step of the translocation process by recognizing presecretory proteins.

SecA exhibits ATPase activity [26]. None of the other Sec proteins has been demonstrated to do so. Thus, SecA is seemingly the only interaction site for ATP, which is essential for translocation. The ATPase activity is enhanced in the presence of presecretory proteins, membrane phospholipids and SecY [27,28]. Thus SecA is designated as the translocation ATPase that functions on the inner surface of the cytoplasmic membrane.

We studied the sites in the SecA molecule that interact with ATP and presecretory proteins. ProOmpF-Lpp, a model presecretory protein composed of proOmpF and the major lipoprotein, was used [29]. SecA denatured in 6 M guanidine-HCl can be completely renatured upon dilution and dialysis [30]. This technique was used to reconstitute SecA analogs from amino terminally truncated and carboxyl terminally truncated SecA fragments. When the reconstitution was performed with two truncated fragments that are large enough to complement each other structurally, the resultant SecA analogs were active as to cross-linking with ATP [31] and with the presecretory protein [32]. Thus, by carrying out reconstitution studies with a large variety of combinations of the truncated fragments, the loci in the SecA molecule that interact with these molecules were determined, as depicted in Fig. 1. The ATP-binding and the presecretory protein-binding sites were located in the amino terminal region of the SecA molecule. The $secA^{ts}$ mutations causing general protein export defects were localized within the ATP-binding region. Therefore, it is likely that the amino terminal region plays major roles in the function of SecA. The carboxyl terminal region can be removed by 70 amino acid residues without loss of activity [31].

Upon interaction with ATP and a presecretory protein (proOmpA), the SecA molecule underwent conformational changes. This was demonstrated by examining the changes in the sensitivity of SecA to staphylococcal protease V8 [33]. Such

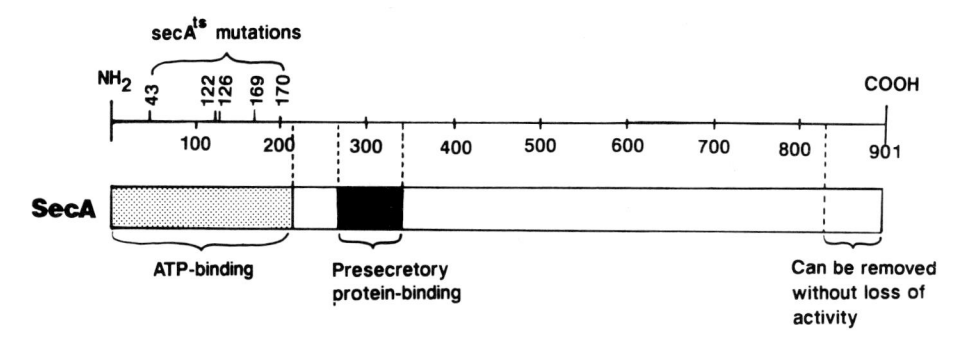

Fig. 1. A diagrammatic representation of functional domains in the SecA molecule. The numbers represent amino acid residues from the amino terminus of SecA. The shaded box indicates the ATP binding region, and the solid black box indicates the presecretory protein binding region. Positions of $secA^{ts}$ mutations that cause general protein export defect are also shown [55].

changes were also observed upon the addition of everted membrane vesicles or phospholipids. The V8 digestion profiles differed with the compounds or vesicles added, suggesting that SecA takes on different conformations upon interaction with them in the process of translocation. These conformation changes may play a role in leading presecretory proteins into the membrane.

The major phospholipids of the *E. coli* cytoplasmic membrane are phosphatidylglycerol, cardiolipin and phosphatidylethanolamine. We observed that phosphatidylglycerol and cardiolipin, acidic phospholipids, were effective in the conformation change of SecA, whereas phosphatidylethanolamine was not [33]. Another zwitterionic lipid, phosphatidylcholine, was also 'ineffective'. The same phospholipid specificities have been observed for stimulation of the ATPase activity of SecA, SecA- phospholipid binding [28] and in vitro translocation activity [34].

SecA exists both on the inner surface of the membrane and in the cytosol. After disruption of cells through a French pressure cell, about 10% of the SecA was recovered in the membrane fraction and 90% was found in the cytosol. About 50% and 75% of the membrane-bound SecA was solubilized on treatment with 2 M and 4 M urea, respectively [35]. Then the question arose as to whether a presecretory protein first binds to the soluble SecA to form a complex, which in turn moves onto the membrane, or a presecretory protein binds to the SecA preexisting on the secretory machinery. The binding of presecretory proteins to the soluble SecA in a manner dependent of the positive charge at the amino terminus of the signal peptide supports the former possibility [16]. The stimulation of in vitro translocation by the externally added SecA, in large excess of the amount that can be retained by the membrane, also support the importance of the soluble form of SecA [36]. Furthermore, we recently observed that the concentration of SecA required for in vitro translocation differed with the species of presecretory proteins [37]. This also favors the view that the formation of the SecA–presecretory protein complex is the first event in the protein translocation. On the other hand, a certain fraction of SecA is always found on the membrane, the amount being roughly equimolar to the amounts of SecE and SecY (S. Matsuyama and S. Mizushima, unpublished data), suggesting that the membrane-bound SecA constitutes a part of the functional secretory machinery. It is probable that SecA on the membrane is replaced by the cytosolic form which possesses a presecretory protein during the translocation process. Alternatively, presecretory proteins may first be recognized by the cytosolic SecA and then transferred to the membrane-bound SecA.

SecB, a chaperone protein in the cytosol, was shown to play a role in the first step of the translocation of some presecretory proteins, most likely by keeping them translocation competent [38,39]. The cytostolic SecA may serve as a chaperone as well.

6. Functions of SecE and SecY

SecE and SecY are integral membrane proteins, which possess three and ten membrane spanning domains, respectively [40,41]. Although accumulating genetic evidence indicates the participation of these proteins in the translocation reaction, no biochemical evidence demonstrating their essentiality was available until the functional reconstitution of the translocation system was achieved. We first reconstituted translocationally active proteoliposomes from purified SecE and a SecE-deficient membrane extract [17], and then from purified SecY and a SecY-deficient membrane extract [20]. Finally, we succeeded in reconstituting active proteoliposomes from purified preparations of SecE and SecY [18]. The reconstituted translocation activity was SecA- and ATP-dependent. It was concluded, therefore, that SecE, SecY and SecA are essential components of the protein secretory machinery, and that translocation activity can be reconstituted with only these three proteins and phospholipids. Brundage et al. [15] isolated a complex composed of SecE, SecY and another protein called 'band 1'. They reconstituted translocationally active proteoliposomes with this complex. Several lines of evidence suggest that the active species comprises a tightly associated complex of these three subunits [42]. It is not clear, however, whether the band 1 protein is a functional component of the machinery. Reconstitution with purified band 1 has not yet been successful.

SecE has three transmembrane stretches. Genetic analysis revealed that the third stretch is sufficient for the functioning of SecE [43]. We confirmed this and further found that the third stretch was functional in stabilizing SecY in the membrane, as intact SecE does, suggesting that this stretch is the site of interaction with SecY [44]. A reconstitution study also demonstrated that the carboxyl terminal segment possessing the third stretch was functionally active [44].

What are the functions of SecE and SecY in the translocation reaction? The interactions of SecA with both SecY [28,45] and SecE (K. Kimura, M. Akita, S. Matsuyama, H. Tokuda and S. Mizushima, unpublished data) have been demonstrated, indicating that SecE and SecY, and probably a complex comprising them, constitute the site of the binding of SecA. Presecretory proteins may then be transferred from SecA to SecE/SecY for transmembrane translocation. It is assumed, therefore, that SecE/SecY may form a channel through which presecretory proteins are translocated. It is interesting in this respect that the overproduction of both SecE and SecY in the membrane resulted in the collapse of ΔpH across the membrane [46]. The ΔpH collapse required a presecretory protein, SecA and ATP. The overproduction of SecY and SecE was also required. These facts suggest that SecY or SecE, or both are responsible for the protein translocation-coupled counterflux of protons. This may be the reason, or at least one of the reasons, for the ΔpH requirement for protein translocation.

7. Functions of SecD and SecE

The *secD* and *secF* genes, which constitute an operon, were first identified as cold-sensitive mutations for protein secretion [47]. SecD and SecF encoded by these genes each have multiple transmembrane stretches and a large periplasmic domain. Based on these facts, they are supposed to be involved in a rather late stage of the protein translocation.

We have purified these two proteins, which were then subjected to reconstitution experiments together with SecE and SecY [20a]. Despite extensive studies under different conditions, SecD or SecF, or both did not significantly enhance the translocation activity of reconstituted proteoliposomes. Very recently, the involvement of SecD in the release of translocated proteins from the membrane was demonstrated with the aid of anti-SecD IgG (S. Matsuyama, Y. Fujita and S. Mizushima, submitted). There is no biochemical evidence at present, therefore, to support the direct involvement of SecF in the translocation reaction.

8. Discussion

Based on the evidence described so far, together with other relevant evidence, we would like to discuss possible mechanisms of translocation of presecretory proteins across the cytoplasmic membrane of *E. coli*. Of course, some parts of the following discussion are still highly speculative and thus further critical experiments are needed. Our working model is illustrated in Fig. 2.

We assume that the first essential step of the translocation reaction is the binding of a presecretory protein to the SecA dimer in the cytosol. For some secretory proteins, the binding of SecB to the nascent polypeptide chain, to keep it in a translocation competent conformation, takes place before the SecA binding [38,39]. ATP binds to SecA [33]. This stimulates the binding of a presecretory protein to SecA [32], which in

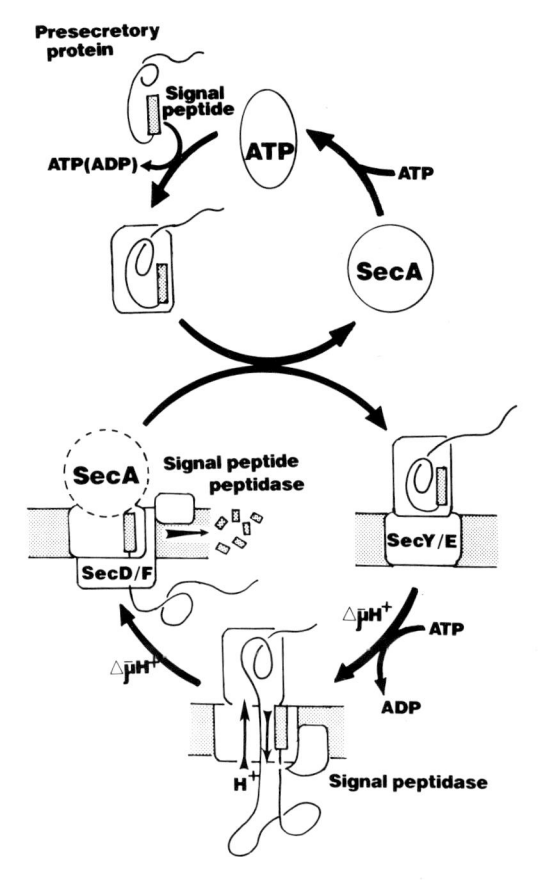

Fig. 2. A working model of translocation of presecretory proteins across the cytoplasmic membrane of *E. coli*. SecB is not depicted. For details, see text.

turn causes the release of the prebound ATP from SecA [33]. In each step, the SecA molecule undergoes a conformation change [33]. The hydrolysis of ATP may not be required up to this stage since adenosine 5'-(β,γ-imino)triphosphate (AMP) is as active as ATP. The presecretory protein is then transferred onto the secretory machinery in the membrane. This may be achieved by the binding of the SecA-presecretory protein complex to the machinery or through the exchange of the SecA-presecretory protein complex with SecA pre-existing on the membrane. Alternatively, the transfer of the presecrtory protein from the cytosolic SecA to the membrane-bound form may occur. Thus a translocation complex composed of the presecretory protein, SecA, SecY, SecE and membrane phospholipids is formed. The formation induces translocation ATPase activity [26–28], which is essential for at least the early stage of the translocation reaction [48–51]. Proton motive force, $\Delta\tilde{\mu}H^+$, is also required, but is not essential, up to this stage. The translocation of a polypeptide chain in the subsequent stage of the translocation reaction, on the other hand, does not require ATP hydrolysis, whereas $\Delta\tilde{\mu}H^+$ is still required [49,50]. The protein translocation as a whole may be coupled with the counterflux of protons [46]. This may be the reason for the requirement of membrane potential, $\Delta\psi$, and ΔpH, components of $\Delta\tilde{\mu}H^+$, for protein translocation. Thus, ATP and $\Delta\tilde{\mu}H^+$ function differently at different steps of the translocation reaction. This was also discussed recently in detail by Wickner and associates [51,52]. It is still unclear whether ATP and $\Delta\tilde{\mu}H^+$ are required repeatedly in a stepwise fashion in the translocation of one protein molecule.

The possibility that the contribution of $\Delta\tilde{\mu}H^+$ might include $\Delta\psi$-driven electrophoresis of acidic regions and ΔpH-driven deprotonation of basic residues in the cytoplasm and reprotonation in the periplasm has been discussed [51,53]. These, however, should not be the major roles of $\Delta\psi$ and ΔpH, since the translocation of a model presecretory protein completely devoid of charged residues in the mature domain still depends on both $\Delta\psi$ and ΔpH [54].

The mechanisms by which polypeptide chains complete their translocation are still rather unclear. Although proteoliposomes reconstituted from purified SecE and SecY are active as to translocation in the presence of SecA and ATP, the reconstituted activity was very low compared with that of intact membrane vesicles containing the same amounts of SecE and SecY [18]. Furthermore, the results obtained with the current method of translocation measurement (resistance to externally added proteinase K) do not necessarily reflect the complete translocation of the polypeptide chain. Genetic studies have demonstrated the involvement of SecD and SecF in protein secretion. The reconstitution of proteoliposomes in the presence of purified SecD and SecF did not result in enhancement of the translocation activity, however. Since the reconstituted activity was merely judged by proteinase K resistance, it is possible that these proteins play a role in a late stage of the secretion process [47]; for example, the stage of the polypeptide release from the membrane. Our recent observation as the role of SecD in the release of translocated proteins supports this view.

Acknowledgements

The work performed in our laboratory was supported by grants from the Ministry of Education, Science and Culture of Japan, the Nisshin-Seifun Foundation and the Naito Foundation. We thank Miss Iyoko Sugihara for secretarial support.

References

1. Bieker, K.L., Phillips, G.J. and Silhavy, T.J. (1990) J. Bioenerg. Biomembr. 22, 291–310.
2. Dev, I.K. and Ray, P.H. (1990) J. Bioenerg. Biomembr. 22, 271–290.
3. Ichihara, S., Hussain, M. and Mizushima, S. (1982) J. Biol. Chem. 257, 495–500.
4. Rapoport, T. (1991) Nature 349, 107–108.

5. Bassford Jr., P.J., Beckwith, J., Ito, K., Kumamoto, C., Mizushima, S., Oliver, D., Randall, L., Silhavy, T., Tai, P.C. and Wickner, W. (1991) Cell 65, 367–368.
6. Weiss, J.B., Ray, P.H. and Bassford Jr., P.J. (1988) Proc. Natl. Acad. Sci. USA 85, 8978–8982.
7. Kumamoto, C.A., Chen, L., Fandl, J. and Tai, P.C. (1989) J. Biol. Cham. 264, 2242–2249.
8. Cunningham, K., Lill, R., Crooke, E., Rice, M., Moore, K., Wickner, W. and Oliver, D. (1989) EMBO J. 8, 955–959.
9. Yasuda, S. and Takagi, T. (1983) J. Bacteriol. 154, 1153–1161.
10. Kawasaki, H., Matsuyama, S., Sasaki, S., Akita, M. and Mizushima, S. (1989) FEBS Lett. 242, 431–434.
11. Matsuyama, S., Akimaru, J. and Mizushima, S. (1990) FEBS Lett. 269, 96–100.
12. Mizushima, S., Tokuda, H. and Matsuyama, S. (1991) Methods Cell Biol. 34, 107–146.
13. Akiyama, Y. and Ito, K. (1986) Eur. J. Biochem. 159, 263–266.
14. Bieker, K.L. and Silhavy, T.J. (1990) Cell 61, 833–842.
15. Brundage, L., Hendrick, J.P., Schiebel, E., Driessen, A.J.M. and Wickner, W. (1990) Cell 62, 649–657.
16. Akita, M., Sasaki, S., Matsuyama, S. and Mizushima, S. (1990) J. Biol. Chem. 265, 8164–8169.
17. Tokuda, H., Akimaru, J., Matsuyama, S., Nishiyama, K. and Mizushima, S. (1991) FEBS Lett. 279, 233–236.
18. Akimaru, J., Matsuyama, S., Tokuda, H. and Mizushima, S. (1991) Proc. Natl. Acad. Sci. USA 88, 6545–6549.
19. Akiyama, Y. and Ito, K. (1990) Biochem. Biophys. Res. Commun. 167, 711–715.
20. Nishiyama, K., Kabuyama, Y., Akimaru, J., Matsuyama, S., Tokuda, H. and Mizushima, S. (1991) Biochim. Biophys. Acta 1065, 89–97.
20a. Matsuyama, S., Fujita, Y., Sagara, K. and Mizuchima, S. (1992) Biochim. Biophys. Acta, in press.
21. Randall, L.L. (1983) Cell 33, 231–240.
22. Akita, M., Shinkai, A., Matsuyama, S. and Mizushima, S. (1991) Biochem. Biophys. Res. Commun. 174, 211–216.
23. Oliver, D.B. and Beckwith, J. (1981) Cell 25, 765–772.
24. Oliver, D.B. and Beckwith, J. (1982) Cell 30, 311–319.
25. Cabelli, R.J., Chen, L., Tai, P.C. and Oliver, D.B. (1988) Cell 55, 683–692.
26. Lill, R., Cunningham, K., Brundage, L.A., Ito, K., Oliver, D. and Wickner, W. (1989) EMBO J. 8, 961–966.
27. Cunningham, K. and Wickner, W. (1989) Proc. Natl. Acad. Sci. USA 86, 8630–8634.
28. Lill, R., Dowhan, W. and Wickner, W. (1990) Cell 60, 259–269.
29. Yamane, K., Matsuyama, S. and Mizushima, S. (1988) J. Biol. Cham. 263, 5368–5372.
30. Shinkai, A., Akita, M., Matsuyama, S. and Mizushima, S. (1990) Biochem. Biophys. Res. Commun. 172, 1217–1223.
31. Matsuyama, S., Kimura, E. and Mizushima, S. (1990) J. Biol. Chem. 265, 8760–8765.
32. Kimura, E., Akita, M., Matsuyama, S. and Mizushima, S. (1991) J. Biol. Chem. 266, 6600–6606.
33. Shinkai, A., Lu, H.-M., Tokuda, H. and Mizushima, S. (1991) J. Biol. Chem. 266, 5827–5833.
34. Kusters, R., Dowhan, W. and de Kruijff, B. (1991) J. Biol. Chem. 266, 8659–8662.
35. Lu, H.-M., Yamada, H. and Mizushima, S. (1991) J. Biol. Chem. 266, 9977–9982.
36. Yamada, H., Tokuda, H. and Mizushima, S. (1989) J. Biol. Chem. 264, 1723–1728.
37. Hikita, C. and Mizushima, S. (1992) J. Biol. Chem. 267, 12375–12379.
38. Kumamoto, C.A. (1990) J. Bioenerg. Biomembr. 22, 337–351.
39. Hartl, F.-U., Lecker, S., Schiebel, E., Hendrick, P. and Wickner, W. (1990) Cell 63, 269–279.
40. Schatz, P.J., Riggs, P.D., Jacq, A., Fath, M.J. and Beckwith, J. (1989) Genes Dev. 3, 1035–1044.
41. Akiyama, Y. and Ito, K. (1987) EMBO J. 6, 3465–3470.
42. Brundage, L., Fimmel, C.J., Mizushima, S. and Wickner, W. (1992) J. Biol. Chem. 267, 4166–4170.
43. Schatz, P.J., Bieker, K.L., Ottemann, K.M., Silhavy, T.J. and Beckwith, J. (1991) EMBO J. 10, 1749–1757.

44. Nishiyama, K., Mizushima, S. and Tokuda, H., J. Biol. Chem. 267, 7170–7176.
45. Fandl, J.P., Cabelli, R., Oliver, D. and Tai, P.C. (1988) Proc. Natl. Acad. Sci. USA 85, 8953–8957.
46. Kawasaki, S., Mizushima, S. and Tokuda, H., submitted.
47. Gardel, C., Johnson, K., Jacq, A. and Beckwith, J. (1990) EMBO J. 9, 3209–3216.
48. Tani, K., Shiozuka, K., Tokuda, H. and Mizushima, S. (1989) J. Biol. Chem. 264, 18582–18588.
49. Tani, K., Tokuda, H. and Mizushima, S. (1990) J. Biol. chem. 265, 17341–17347.
50. Geller, B.L. and Green, H.M. (1989) J. Biol. Chem. 264, 16465–16469.
51. Schiebel, E., Driessen, A.J.M., Hartl, F.-U., and Wickner, W. (1991) Cell 64, 927–939.
52. Wickner, W., Driessen, A.J.M. and Hartl, F.-U. (1991) Annu. Rev. Biochem. 60, 101–124.
53. Driessen, A.J.M. and Wickner, W. (1991) Proc. Natl. Acad. Sci. USA 88, 2471–2475.
54. Kato, M., Tokuda, H. and Mizushima, S. (1991) J. Biol. Chem. 267, 413–418.
55. Schmidt, M.G., Rollo, E.E., Grodberg, J. and Oliver, D.B. (1988) J. Bacteriol. 170, 3404–3414.

CHAPTER 4

Distinct steps in the insertion pathway of bacteriophage coat proteins

ANDREAS KUHN and DOROTHEE TROSCHEL

Department of Applied Microbiology, University of Karlsruhe, W-75 Karlsruhe, Germany

Abstract

The 50 amino acid long bacteriophage M13 coat protein is one of the most extensively studied inner membrane proteins of *Escherichia coli*. It is synthesized as a precursor protein with a cleavable leader peptide that is 23 amino acids in length. The mechanism by which the M13 procoat protein inserts into the membrane was investigated by biochemical as well as genetic methods. The most striking result was that the bacterial protein secretion machinery is not required for the membrane transport of this particular protein. Using a variety of procoat mutant proteins, the insertion pathway was dissected into specific steps, each of them requiring certain features of the precursor protein.

1. Introduction

One intriguing aspect of protein transport across biological membranes is that the complex translocation machineries that have evolved in prokaryotic and eukaryotic systems are not indispensable for the transport process in general. Particular small proteins insert into membranes independently of many of the proteinaceous components such as SecA, SecB and SecY/E in *Escherichia coli* [1–3], or SRP, DP and SSR of mammalian cells [4]. This suggests that the complex translocation machineries have not been created to perform protein transport per se but rather to extend the range of proteins able to assume a transmembrane configuration. In bacteria, the more complex proteins have first to interact with the soluble Sec components, SecB and SecA, to bind to the membrane surface (targeting) and subsequently to translocate across the membrane with the help of SecE and SecY (insertion). Small and simple proteins that do not require these components must have the structural properties necessary for targeting and translocation within their primary amino

M13   **PF3**

periplasm

cytoplasm

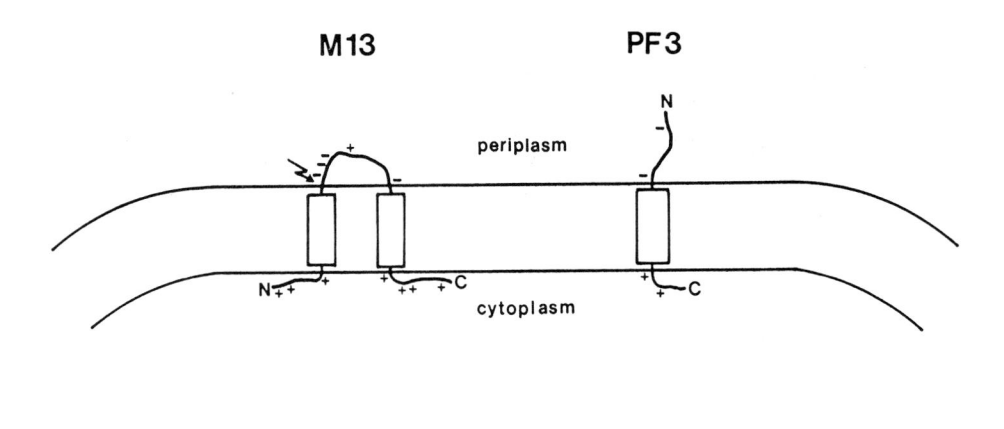

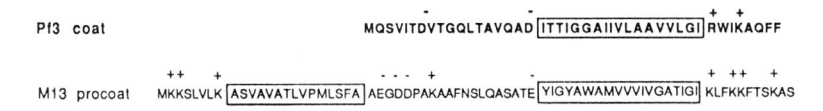

```
                                              -        -              + +
Pf3 coat                        MQSVITDVTGQLTAVQAD ITTIGGAIIVLAAVVLGI RWIKAQFF

                   ++    +              ---  +          -             +  ++   +
M13 procoat     MKKSLVLK ASVAVATLVPMLSFA AEGDDPAKAAFNSLQASATE YIGYAWAMVVVIVGATIGI KLFKKFTSKAS
```

Fig. 1. Insertion mechanisms of the M13 and Pf3 coat proteins into the *E. coli* inner membrane. Membrane insertion of the M13 procoat protein occurs via a loop leaving the N and the C terminus in the cytoplasm. After insertion, the procoat protein is processed by leader peptidase (arrow). The Pf3 coat protein, in contrast, is made without a leader sequence. The N terminus is transported to the periplasmic face of the membrane, whereas the C terminus stays in the cytoplasm. The charged residues are indicated by symbols and the hydrophobic domains by rectangles.

acid sequence. Possibly, to insert into the membrane, they use an ancestral translocation pathway before translocase evolved.

To investigate how these Sec-independent proteins insert into the membrane, we have focused our studies on two coat proteins of the filamentous bacteriophages M13 and Pf3. *E. coli* phage M13 and *Pseudomonas aeruginosa* phage Pf3 insert into the bacterial cytoplasmic membrane during their assembly into viral particles. Both proteins are similar in length (50 and 44 residues, respectively), and do not share any sequence similarity [5]. As schematically drawn in Fig. 1, both coat proteins contain a hydrophobic region of about 20 amino acids flanked by an acidic amino terminal region and a basic region at the carboxy terminus. The acidic regions of both proteins are located in the periplasm, whereas the basic regions remain in the cytoplasm [6,7].

2. Results and discussion

2.1. Pf3 coat protein requires no leader sequence for membrane insertion

In contrast to the M13 coat protein, Pf3 coat protein is synthesized without a leader (signal) sequence at its amino terminus. To investigate the mechanism of how Pf3 coat protein inserts into the bacterial membrane, the coding gene was cloned into an expression vector and transformed into *E. coli*. Protein synthesis and membrane insertion were followed in pulse-chase experiments using [^{35}S]-methionine [7]. Membrane insertion of Pf3 coat protein occurred very rapidly, and after a pulse time of 3 min, essentially all Pf3 coat protein was found translocated. The membrane-inserted state of the coat protein was proven by its accessibility to externally added proteinase K and its membrane association, as determined by fractionation experiments. In addition, the transmembrane form proved resistant to trypsin added to the outside. This is consistent with the cytoplasmic location of the carboxy terminus, since the only arginyl and lysyl residues are found in the carboxyl terminal region of the protein.

2.2. Hybrid coat proteins of M13 and Pf3

The absence of a leader sequence in the Pf3 coat protein and its small size show that little sequence information is required to achieve membrane insertion. Possibly, the presence of a short hydrophobic stretch with an adjacent basic region suffices for the protein to enter a membrane. These features are also the characteristic properties of a signal peptide [8]. In the Pf3 coat protein, the basic region is located carboxyl terminal to the hydrophobic stretch, whereas in signal peptides the basic region is at the amino terminus. Correspondingly, the signal peptides precede the translocated protein region, while in the Pf3 coat protein, the translocated region is followed by the hydrophobic and basic elements. Taken together, it is likely that the Pf3 coat protein contains a 'reverse signal peptide' [9].

To further understand the function of a leader sequence, we have made a number of hybrid proteins between the Pf3 coat protein and the M13 coat protein [7]. A mutant protein, named MPF1, in which the periplasmic region of the Pf3 coat was fused in front of the hydrophobic and basic region of the M13 coat normally inserted into the membrane of *E. coli* (Fig. 2). However, another mutant, named MPF4, consisting of the entire M13 coat protein and only three additional amino acids at the amino terminus was incapable of membrane insertion. This suggests that the properties of the periplasmic region are crucial for determining whether translocation occurs without a leader sequence. One major difference between the two coat proteins is the number of charged residues in that region. M13 coat protein has five charged amino acids (four acidic, one basic), whereas the Pf3 coat has only two (both acidic). The idea that the different amount of charges is critical in this respect is consistent with the view that the transport of charged residues requires more energy, since the water shell surrounding these residues either has to be removed or co-transported [10]. In accordance with this idea, substitution of the two aspartyl residues with asparagyl residues in the MPF4 protein, resulted in slow membrane insertion. Moreover, membrane translocation was fully restored when the glutamic acid

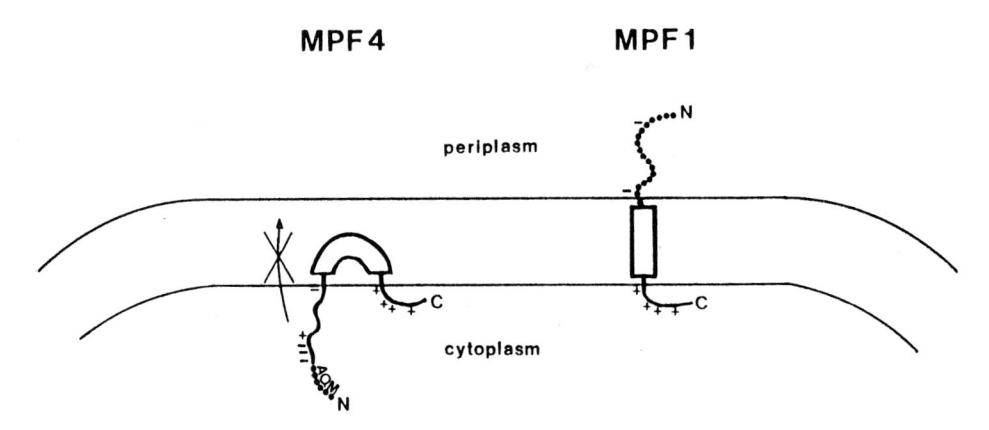

Fig. 2. The capability of the membrane anchoring region of the M13 coat protein to translocate N terminal extensions is limited to small, weakly charged regions. Fusion proteins between the M13 coat protein (full lines) and the Pf3 coat protein (dotted lines) were constructed. MPF1 contains the first 18 amino acyl residues of Pf3 fused to the membrane anchoring region of M13. The N terminal region of Pf3 was translocated to the periplasm. In the case of the MPF4 protein, however, which contains only the first three amino acids of the Pf3 protein fused to the M13 coat protein, the N terminal region was not transported.

residue at the amino terminus was changed to a glutamine [7].

The periplasmic part of the Pf3 coat protein seems capable of translocation only when it is short and uncharged. For the translocation of more charged regions, a leader sequence is necessary, as is the case for the M13 procoat protein.

2.3. M13 procoat protein first binds electrostatically to the membrane surface

Bacteriophage M13 procoat protein is synthesized on free polysomes prior to its assembly into the inner membrane of *E. coli* [11]. The newly synthesized protein has to be correctly targeted to the membrane without the aid of SecB and SecA [3]. It seems that, distinct from other translocated proteins, procoat binds to the membrane surface by a direct interaction and is not mediated by chaperone molecules. This initial step has been thoroughly studied in a series of mutants in the amino and carboxyl terminal region of the procoat protein (Fig. 3). Procoat mutants with neutral or negatively charged residues accumulated in the non-processed form [12,13]. Resistance to proteinase K added to the outside of the cell showed that the mutant proteins were not translocated across the membrane. In addition, cell fractionation experiments suggested that these precursor proteins were mainly located in the cytoplasm. The procoat mutants with either the amino or the carboxyl terminal region changed, showed intermediate behaviour; they were found as precursors in the cytoplasmic fraction and as mature proteins in the membrane fraction.

To investigate membrane binding, an in vitro assay was developed, in which various mutant proteins were synthesised in the presence of artificial liposomes

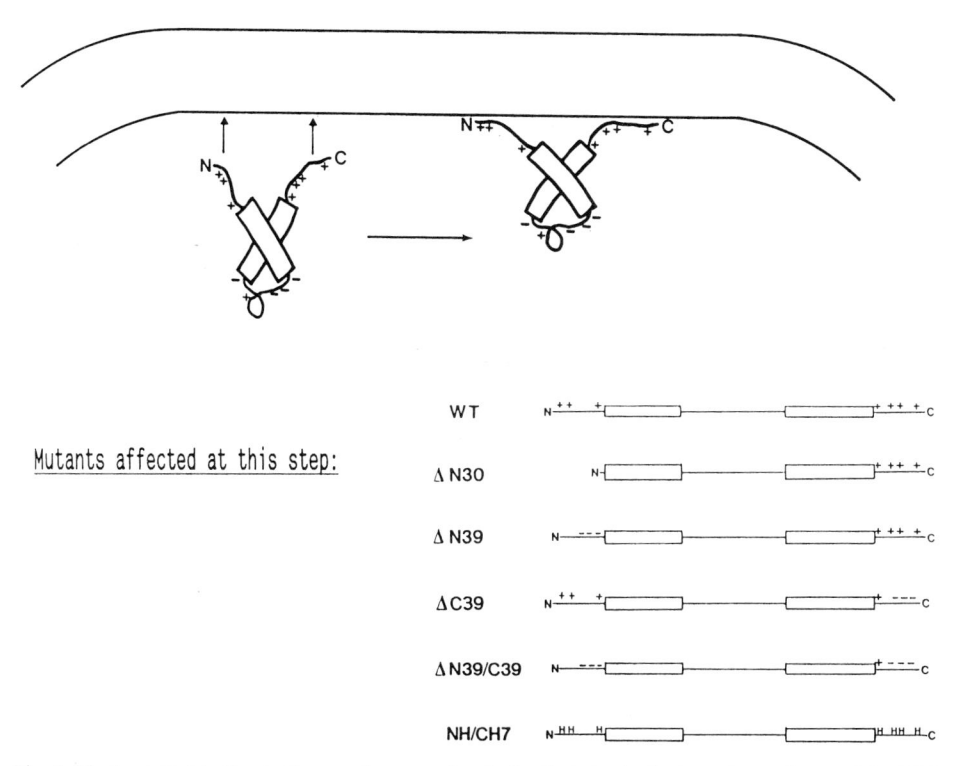

Fig. 3. Electrostatic binding to the membrane surface is the first step in the insertion pathway of the M13 coat protein. Both, the N and C terminal ends of the protein are positively charged and can therefore interact with the negatively charged membrane surface. These positive charges are important for membrane targeting since mutants lacking positive charges at either end of the protein are severely inhibited in membrane binding. The mutant proteins are shown schematically; only the N and C terminal charges are indicated; the central domain of the protein remained unchanged. The mutant NH/CH7 has histidyl residues at both ends; the degree of protonation depends on the actual pH.

[13]. Binding of the M13 procoat protein to these liposomes was assayed by flotation experiments in sucrose step gradients. Wild-type procoat bound efficiently (>95%) to liposomes consisting of 25% phosphatidylglycerol (negatively charged) and 75% phosphatidylcholine (zwitterionic), whereas a procoat mutant with six negatively charged residues in the terminal regions, named ΔN39/C39, showed very weak binding (10%). The binding of this mutant protein was improved when the liposomes used were prepared exclusively from phosphatidylcholine. These results support the view that the initial step in the membrane biogenesis of the M13 procoat protein is mainly an electrostatic binding to the membrane.

The electrostatic nature of the targeting step was further analyzed with another procoat mutant, NH/CH7, which substitutes the seven lysyl residues for histidyl residues in both terminal regions [14]. Under physiological conditions (pH 7.6–7.8), these histidyl residues occur in the unprotonated state. At this pH, the mutant precursor accumulated in the cytoplasm. This was shown in vivo by pulse-chase experiments and by cell fractionation studies. Interestingly, after shifting the intracellular pH to acidic conditions, a small portion of the mutant protein was translocated and processed to the mature form. Acidification of the intracellular pH was performed by transferring the pulse-labelled cells to a K^+-free minimal medium (pH 5.0) containing the non-bactericidal polymyxin B derived nonapeptide PMBN [15]. This made the cells sensitive to the subsequent treatment with nigericin, a K^+/H^+ antiporter. As a consequence, the intracellular pH adjusted to the external value resulting in a breakdown of ΔpH. The now protonated histidyl residues led to a positive charge at the N and C termini in the mutant protein. The appearance of the processed form under these conditions demonstrated that the procoat protein was successfully inserted into the membrane. This again supports the hypothesis that an electrostatic interaction of the precursor with the negatively charged membrane surface is a crucial step in protein translocation. Moreover, these results show that the cytoplasmically accumulated precursor remains translocation-competent and is transported post-translationally.

2.4. Both hydrophobic regions are required for the partitioning of the M13 procoat protein into the membrane

The M13 procoat protein has two hydrophobic regions, the leader sequence and the membrane anchor sequence of the mature part. Introduction of strongly charged residues into these regions disrupts their hydrophobicity (Fig. 4). Procoat mutants with arginyl residues in the centre of either hydrophobic region (OL8 and OM30, respectively) showed a strong inhibition of membrane insertion and processing [16]. As indicated by the resistance of the precursor to proteinase K digestion, the hydrophilic central region containing the cleavage site was not exposed to the periplasm. The mutant precursors were efficiently cleaved in vitro by purified leader peptidase. This further suggests that the lack of cleavage in vivo is indeed due to the fact that the periplasmic region was not translocated to the outer face of the inner membrane. That a direct hydrophobic interaction of the precursor with the membrane is involved in the translocation event is supported by the following observations. (1) The precursor proteins with an arginine substitution were not extracted under alkaline conditions [13]. Conclusively, the mutant proteins were still able to interact with the membrane by hydrophobic forces in addition to the electrostatic interactions mentioned above. (2) Binding experiments with liposomes, comprising 25% phosphatidylglycerol and 75% phosphatidylcholine, showed that even the double mutant protein (OL8-OM30) was as efficiently bound as the wild-type protein [9]. (3) The mutant precursors were translocated with altered kinetics in vivo. However, this effect was less pronounced for mutants carrying the arginyl residues not in the centre but at more peripheral positions of the hydrophobic stretches (OL14, OM26).

Additional experiments showed, however, that it is not simply the hydrophobicity

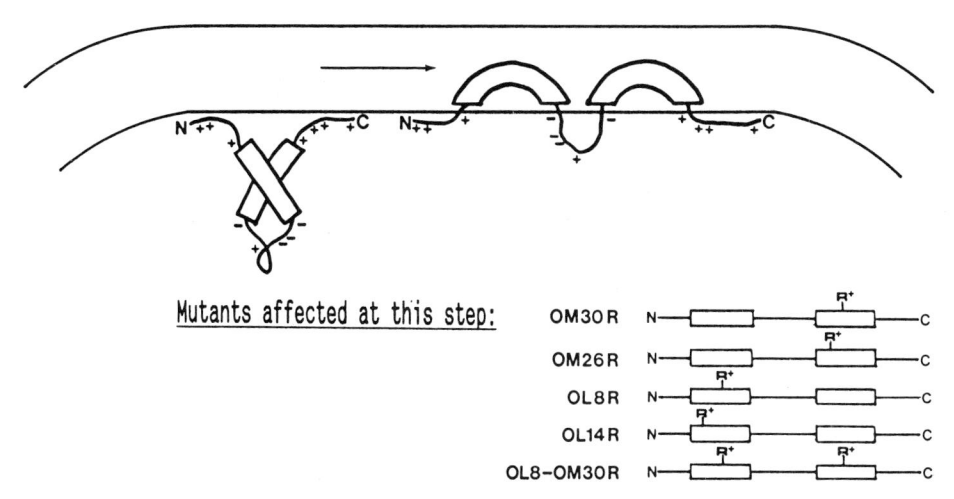

Fig. 4. Hydrophobic partitioning involves both hydrophobic regions of the M13 procoat protein. Hydrophobic interactions between the two membrane spanning regions of M13 start the translocation event. Both hydrophobic regions are required for efficient membrane insertion since mutations in either of the two segments lead to a strongly reduced translocation. The double mutant OL8-OM30R is not translocated at all.

of the two regions that is crucial for the membrane insertion process. Mutants carrying prolyl residues in either of the two hydrophobic regions infer that the conformation and three-dimensional structure of the protein regions also play an important role in membrane translocation (Fig. 5). The more prolines that were introduced into the hydrophobic stretches, the more drastic was the reduction of the processing and translocation rate. The prolines were placed at positions -12 and -9 in the leader and at $+26$, $+29$ and $+32$ in the mature region (A. Kuhn, unpublished data). A mutant carrying two additional prolines in each of the two regions (LM4P) showed a total block in translocation. Very likely, the hydrophobic regions assume an α-helical conformation during translocation [17,18]. Hence, the formation of the two α-helices, which might be the energy basis for the insertion process, is probably distorted by the additional prolyl residues.

2.5. Translocation of the negatively charged periplasmic region of the M13 procoat protein is not primarily an electrophoretic event

The insertion of the M13 procoat protein into the inner membrane of *E. coli* requires the electrochemical potential which is more positively charged on the outer surface. The translocated periplasmic region of the wild-type protein contains four negatively charged residues and one positively charged. By this reasoning, a charge transfer process could be the driving force for the translocation event [19]. To test this idea, mutants were constructed that substitute the negatively charged amino

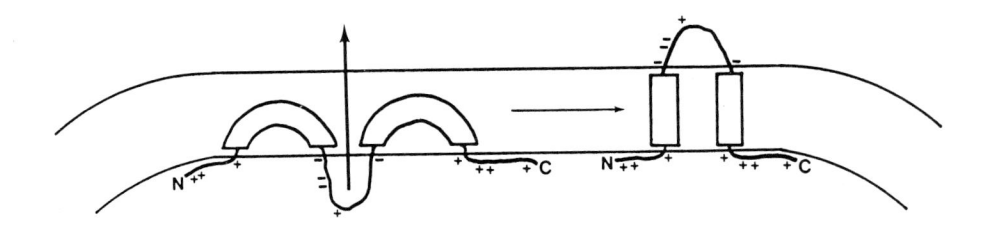

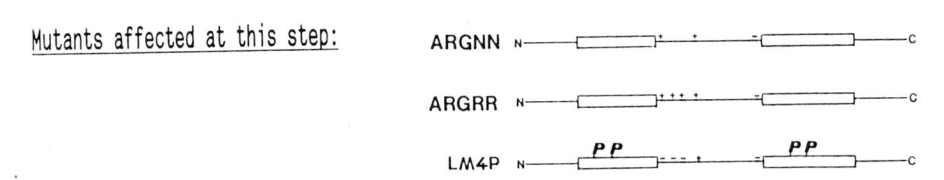

Fig. 5. The translocation of the periplasmic region is not driven mainly by electrophoretic forces, but rather by hydrophobic forces via the formation of transmembrane α-helices. The negatively charged central region of the M13 procoat protein is transported to the positively charged periplasmic side of the membrane. When the charge of the central region is changed to a net charge of +3 as in the case of the mutant ARGRR, translocation still occurs, although with slower kinetics.

acids with neutral or even positively charged residues [20]. Surprisingly, these mutants were only slightly affected in their translocation efficiency; only a mutant carrying four positively charged residues in the periplasmic region, named ARGRR, showed a more pronounced reduction of the translocation rate. Thus, an electrophoretic mechanism is not the major driving force for the membrane insertion of the M13 procoat protein. As mentioned in the previous section, the energy necessary for membrane insertion of this particular protein probably arises mainly from conformational changes of the hydrophobic regions during the translocation process.

In earlier experiments, procoat mutants at position +2 were selected that were dramatically less dependent on a membrane potential [21]. In other proteins, this mechanism may differ, since a membrane potential-dependent charge effect has been observed for pro-OmpA [22]. In addition, pre-alkaline phosphatase and pro-OmpA are extremely sensitive to the introduction of a positively charged residue close to the cleavage site [22,23]. In these proteins, the introduction of even one positive charge in the first part of the mature sequence blocks translocation completely. Possibly, the positively charged residues prevent these precursors from interacting with the *sec* gene products. Since the M13 procoat protein inserts into the membrane without the aid of any Sec proteins, it is logical that in this case the positive charges close to the cleavage site do not have such an inhibitory effect.

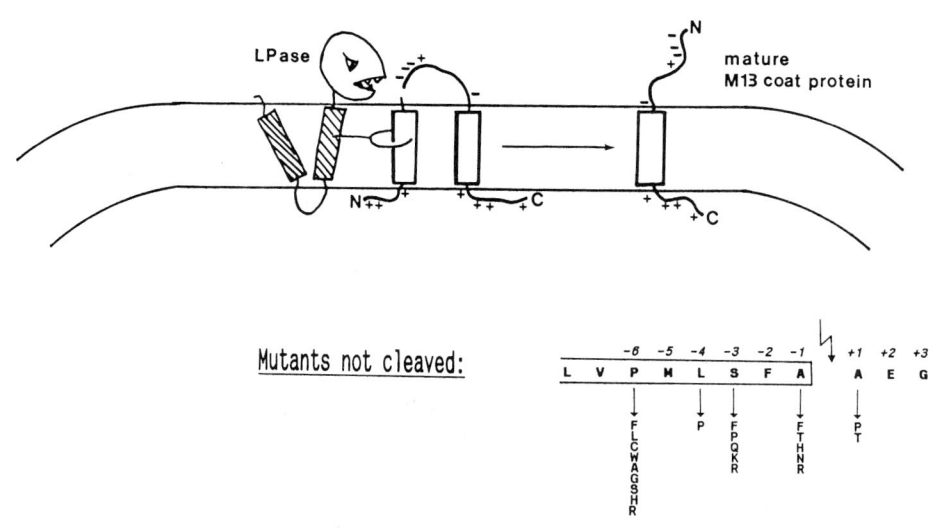

Fig. 6. Cleavage by leader peptidase completes the translocation process. The precursor region between the positions −6 and +1 is crucial for efficient processing. The proline residue at position −6 as well as the small amino acids at −1 and −3 are critical for processing.

2.6. M13 procoat protein as a model substrate for leader peptidase

The amino terminal leader peptides of most *E. coli* transported proteins are cleaved off by leader peptidase I (LPase) during membrane translocation (Fig. 6). To characterize the substrate specificity of this interesting enzyme, the sequences of a large number of precursor proteins have been systematically compared and analyzed [24,25]. These studies pointed out that the part of the leader region from −6 to −1 of the precursor protein is crucial for substrate recognition by the peptidase. Although the various leader peptides differ largely in their amino acid sequences, they have characteristic patterns in the carboxyl terminal region, namely small amino acids at positions −1 and −3 and a helix-breaking residue near positions −4 to −6. To study the leader peptidase specificity in more detail, the residues were systematically changed at positions −6 to +1 of the procoat protein into a number of other residues by site-directed mutagenesis [25]. The mutants were then analyzed for membrane insertion and cleavage by in vivo pulse-chase experiments or in vitro using purified leader peptidase. With the analysis of over 60 procoat mutants, we found that at positions −1 and −3, only a substitution of the alanine at position −1 by serine and of the serine at position −3 by a glycine did not affect the efficiency of cleavage. The proline at position −6 is indispensable since nine different mutations at this position, including substitution by glycine, led to uncleavable precursor proteins. At the other positions of the recognition region (−5, −4, −2 and +1), only the change to proline at positions −4 and +1 seriously affected the cleavage reaction, probably by distorting the molecular architecture of the substrate.

2.7. From membrane to phage

The insertion of the M13 coat protein into the inner membrane of *E. coli* is an important step in the phage assembly pathway since the encapsidation of filamentous phages takes place at the bacterial membrane [26]. The most intriguing feature in the pathway is the incorporation of the viral DNA into the phage coat. It is difficult to exactly characterize the regions of the major coat protein that are important for DNA packaging because most mutations in this protein might already affect the membrane insertion itself. For this reason, only mutations keeping the hydrophobicity of the membrane anchor and the positively charged residues at both ends of the protein intact can be studied. This was done in *E. coli* cells infected with M13 phages carrying an amber mutation (am8) in the coding region for the major coat protein [27]. Plasmids encoding various mutant coat proteins were transformed into these cells, and assembly of intact phage particles was analyzed in a plaque assay. Few mutations showed complementation of the am-phenotype. These were mainly rather conservative amino acid substitutions, such as the change of an alanyl residue to a glycyl residue at position $+1$ and an isoleucyl residue to a valyl residue at position $+22$. But on the other hand, the change of a lysyl residue to a seryl residue at position $+8$ also rescued the am8-defect (A. Kuhn, unpublished data).

Interestingly, the mutations affecting phage assembly, but not membrane insertion, of the coat protein are clustered in the membrane-anchoring region from residues $+25$ to $+35$. Therefore, the second function of this region may be to provide hydrophobic protein–protein interactions important for the formation of the phage coat. Protein–protein interactions are known to influence the specific arrangement of the α-helices of the coat proteins of various filamentous phages [28]. In addition, interactions between the viral DNA and the coat protein also seem to be important. For the filamentous bacteriophage fd, which is closely related to M13, the change of one lysyl residue to a glutamyl residue in the carboxyl terminal region (then carrying three positive charges) resulted in elongated phage particles. Possibly, the encapsidated DNA is more stretched and binds more coat protein molecules [29]. Another fd mutant was constructed that carried five positively charged amino acids in the carboxyl terminal region. When coat proteins of the two mutants were expressed simultaneously, the phage lengths were again shortened [30]. These results suggest that the packing of the DNA in the phage particle is determined by an electrostatic interaction between the positively charged carboxyl terminal region of the coat protein and the negatively charged phosphodiester links of the DNA (Fig. 7).

2.8. The Sec-independent insertion pathway is limited to simple membrane translocation domains

Two components of the bacterial protein secretion machinery essential for the trans-

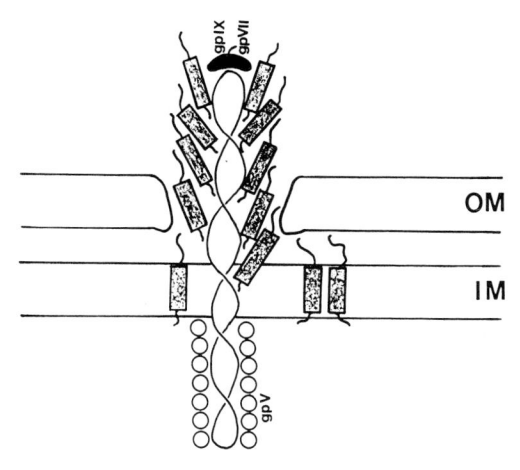

Fig. 7. Phage assembly in the bacterial membrane. There are several bacterial and phage-derived proteins involved in the assembly process. Only the most important structural phage proteins are shown (gene-products gpV, gpVII and gpIX). These proteins are packed around the helical phage-DNA. The grey boxes symbolize the mature M13 major coat protein, which is the product of gene VIII.

port of the majority of membrane and periplasmic proteins are the *secA* and the *secY* gene products. Very few translocated proteins do not need the aid of SecA or SecY for their membrane transport. Among them are the two unusually small coat proteins of the bacteriophages M13 and Pf3 [3,7]. To investigate whether it is simply the size of the proteins that makes their membrane insertion *sec*-independent, in vivo translocation experiments were performed using a fusion protein between the M13 procoat protein and the *sec*-dependent leader peptidase [12,31]. This fusion protein, consisting of all but the last residue of procoat and a carboxyl terminal fragment (103 amino acids) of leader peptidase, retained its *secY*-independence, as demonstrated in pulse-chase experiments in a *secY* temperature-sensitive strain (CJ107). The procoat leader peptidase fusion protein did not show a reduced translocation rate at the non-permissive temperature. In contrast, the processing of the *secY*-dependent OmpA wild-type protein was strongly inhibited as expected under these conditions. A similar fusion protein of procoat and the mature part of β-lactamase proved to insert independently of *secY* (N. Berger and A. Kuhn, unpublished data). These results clearly show that it is not simply the length of the M13 procoat protein that makes its membrane insertion *secY*-independent.

The properties that make a transport protein *sec*-independent can either reside in the leader sequence or in the mature part. To determine this, experiments were performed where the leader sequences of the pro-OmpA protein and the M13 procoat protein were exchanged [31]. Two fusion proteins were constructed, one consisting of the M13 procoat leader sequence and the mature part of OmpA (L(coat)-M(OmpA)), and the other one of the pro-OmpA leader sequence joined to the mature part of the M13 coat protein (L(OmpA)-M(coat)). Both hybrid proteins were processed to the mature form much more slowly than the wild-type proteins. This shows that in principle leader peptides of various proteins are interchangeable, but that individual leaders are optimally tuned to their particular mature proteins for efficient translocation. SecY-dependence of the two hybrid proteins examined in the CJ107 strain showed that the L(OmpA)-M(coat) fusion protein did not require SecY for membrane transport. In contrast, the L(coat)-M(OmpA) hybrid still needed the intact SecY for translocation. Therefore, the SecY-dependence or independence of transported proteins is not simply determined by different leader peptides but resides in the mature part of the protein.

To investigate which features of a mature region allow the *sec*-independent membrane insertion, another hybrid protein between M13 procoat and OmpA was constructed, in which 174 amino acids of the OmpA protein (residues 67–240) were positioned between the leader and the mature part of the procoat protein [32]. This hybrid protein required SecA and SecY for membrane insertion, as did the smaller versions of the hybrid proteins with internal deletions (with 98 and 106 OmpA-derived amino acids). By site-directed deletion, additional fusion proteins were constructed having only 80, 60, 40 or 20 residues of OmpA left. Membrane translocation of the 80-hybrid protein was still dependent on SecA and SecY (Hugenschmidt and Kuhn, unpublished data). When the deletions exceeded this certain length, the hybrid proteins did not insert into the membrane at all. The reason for this behaviour is not clear, and it remains to be shown whether these short fusion proteins are still able to interact with the Sec components. To further understand how the size of a protein influences its mode of membrane insertion, the above-mentioned Sec-independent procoat leader peptidase fusion protein was changed. The procoat residues 27–48, which include the membrane anchor region, were deleted. Since the region which acts as a stop-transfer signal was now missing, the carboxyl terminal leader peptidase fragment was translocated. Interestingly, this hybrid protein needs the Sec components for membrane insertion even though its length is smaller than that of the original Sec-independent fusion protein. The conclusion from these results is that it is not the total size of a protein that is the determinant for Sec-dependence or independence but the size of the region translocated across the membrane. Strikingly, the two different modes of membrane insertion can reside in the same protein. This was shown for a fusion protein consisting of the 18 residue amino terminal periplasmic domain of Pf3 phage coat protein joined to the amino terminus of leader peptidase [33]. The amino terminus of this hybrid protein was translocated independently of the SecA and SecY proteins, whereas the carboxyl terminal region of leader peptidase required these components for insertion, as shown by experiments in SecA and SecY temperature-sensitive strains, respectively. Proteinase K digestion and immunoprecipitation with Pf3 coat antibodies of pulse-labelled cells showed that the Pf3 part of the fusion protein was sensitive to the protease treatment, whereas the leader peptidase part was resistant. This was verified by the loss of antigenicity against the Pf3 antibodies after protease digestion.

In conclusion, the different regions of a protein can be translocated via different mechanisms, depending on the molecular structure of their distinct elements. This is possibly also of general interest for deciphering the mode of membrane insertion of multispanning membrane proteins.

3. Conclusions

Membrane translocation of proteins is preconditioned by their correct targeting to the membrane surface. Two mechanisms have been characterized so far: the involvement of (1) electrostatic and hydrophobic protein–lipid interactions [13,34] and (2) protein–protein interactions [35]. Possibly, these two mechanisms are complementing each other such that the interaction with the Sec proteins of the translocation machinery supports the protein-lipid interaction of the protein precursors. Alternatively, the two mechanisms are distinct and some precursors are capable of interacting with the Sec components while others do not. This would mean that Sec-dependent and Sec-independent proteins translocate through biological membranes by two distinct pathways. In this context, one has to consider that an interaction with a Sec component does not always implicate membrane translocation. The non-translocated gene product 37 of bacteriophage T4 forms a complex with SecB during assembly of the tail fibres which takes place in the cytoplasm of E. coli [36]. Similar results have been achieved for GroEL and its interaction with cytoplasmically located proteins [37]. This suggests that the molecular chaperonins, even when they are bound to a substrate, do not contain sufficient information for membrane targeting. Rather, the precursor protein must exhibit additional features to yield membrane interaction. For example, electrostatic protein–lipid interactions are required for the membrane transport of prePhoE in addition to SecB [38].

Signal peptides presumably initiate the translocation process across the membrane resulting in the formation of a transmembrane loop [39]. The orientation of a hydrophobic region in the membrane is likely to be determined by the distribution of the positively charged residues flanking the hydrophobic sequence [40,41]. Signal peptides have positively charged residues at their amino terminus and direct the carboxyl terminal region across the membrane. However, the Pf3 coat protein lacks positively charged residues at the amino terminus and its amino terminal region is therefore placed into the periplasm. In addition, such a reverse signal sequence was also found in the M13 coat protein since the hydrophobic membrane-anchoring sequence also has the capability to translocate an amino terminal protein region. The reverse signal sequences may be widespread, acting also in the insertion of multispanning membrane proteins. It has been shown for a mutant of the asialoglycoprotein receptor that only the insertion of the signal sequence requires the signal recognition particle, whereas the other transmembrane regions insert in an SRP-independent mode [42]. Hence, insertion of such stop-transfer regions could follow a similar mechanism to that of the Pf3 coat protein, i.e. that they directly insert into the bilayer. However, the action of the reverse signal of Pf3 is limited to the translocation of short and only weakly charged sequences. Highly charged regions can only be translocated by the cooperative action of two membrane spanning regions in close neighbourhood as demonstrated for the M13 procoat protein. We propose that the specific features of each translocated segment determine the type of insertion signals and export mode. Small, weakly charged segments require only one hydrophobic region, whereas highly charged segments necessitate additional N terminal or internal signal sequences. Still longer hydrophilic regions would recruit cellular components to catalyze translocation through the membrane. Possibly one has to assume a more differentiated view of how various proteins cross biological membranes. Especially, the role of the Sec proteins has to be reconsidered; they may be specialized catalysts required only for the translocation of particular protein segments.

References

1. Collier, D.N., Strobel, S.M. and Bassford Jr., P.J. (1990) J. Bacteriol. 172, 6875–6884.
2. Cobet, W.W.E., Mollay, C., Müller, G. and Zimmermann, R. (1989) J. Biol. Chem. 264, 10169–10176.
3. Wolfe, P.B., Rice, M. and Wickner, W. (1985) J. Biol. Chem. 260, 1836–1841.
4. Wiech, H., Sagstetter, M., Müller, G. and Zimmermann, R. (1987) EMBO J. 6, 1011–1016.
5. Luiten, R.G.M., Puttermann, D.G., Schoenmakers, J.G.G., Konings, R.N.H. and Day, L.A. (1985) J. Virol. 56, 268–276.
6. Kuhn, A. (1987) Science 238, 1413–1415.
7. Rohrer J. and Kuhn, A. (1990) Science 250, 1418–1421.
8. von Heijne, G. and Gavel, Y. (1988) Eur. J. Biochem. 174, 671–678.
9. Kuhn, A., Rohrer, J. and Gallusser, A., (1990) J. Struct. Biol. 104, 38–43.
10. Engelman, D.M. and Steitz, T.A. (1981) Cell 23, 411–422.
11. Ito, K., Mandel, G. and Wickner, W. (1979) Proc. Natl. Acad. Sci. USA 76, 1199–1203.
12. Kuhn, A., Wickner, W. and Kreil, G. (1986) Nature 322, 335–339.
13. Gallusser, A. and Kuhn, A. (1990) EMBO J. 9, 2723–2729.
14. Gallusser, A., Troschel, D. and Kuhn, A., unpublished data.
15. Alatossava, T., Jütte, H., Kuhn, A. and Kellenberger, E. (1985) J. Bacteriol. 162, 413–419.
16. Kuhn, A., Kreil, G. and Wickner, W. (1986) EMBO J. 5, 3681–3685.
17. Nambudripad, R., Stark, W., Opella, S.J. and Makowski, L. (1991) Science 252, 1305–1308.
18. Briggs, M.S., Cornell, D.G., Dluhy, R.A. and Gierasch, L.M. (1986) Science 233, 206–208.
19. Enequist, H.G., Hirst, T.R., Harayama, S., Hardy, S.J.S. and Randall, L.L. (1981) Eur. J. Biochem. 116, 227–233.
20. Kuhn, A., Zhu, H.-Y. and Dalbey, R.E. (1990) EMBO J. 9, 2385–2388, 2429.
21. Zimmermann, R., Watts, C. and Wickner, W. (1982) J. Biol. Chem. 257, 6529–6536.
22. Zhu, H.-Y., Cheng, S., Kuhn, A., Geller, B. and Dalbey, R.E. (1992) J. Biol. Chem., submitted.
23. Li, P., Beckwith, J. and Inouye, H. (1988) Proc. Natl. Acad. Sci. USA 85, 7685–7689.
24. Kuhn, A. and Wickner, W. (1985) J. Biol. Chem. 260, 15914–15918.
25. Shen, L.M., Lee, J.-I., Cheng, S., Jütte, H., Kuhn, A. and Dalbey, R.E. (1991) Biochemistry 30, 11775–11781.
26. Russel, M. (1991) Mol. Microbiol. 5, 1607–1613.
27. Kuhn, A. and Wickner, W. (1985) J. Biol. Chem. 260, 15907–15913.

28. Day, L.A. (1988) Annu. Rev. Biophys. Chem. 17, 509–539.
29. Rowitch, D.H., Hunter, G.J. and Perham, R.N. (1988) J. Mol. Biol. 204, 663–674.
30. Greenwood, J., Hunter, G.J. and Perham, R.N. (1991) J. Mol. Biol. 217, 223–227.
31. Kuhn, A., Kreil, G. and Wickner, W. (1987) EMBO J. 6, 501–505.
32. Kuhn, A. (1988) Eur. J. Biochem. 177, 267–271.
33. Lee, J.-I., Kuhn, A. and Dalbey, R.E. (1992) J. Biol. Chem. 267, 938–943.
34. De Vrije, G.J., Batenburg, A., Killian, J.A. and de Kruijff, B. (1990) Mol. Microbiol. 4, 143–150.
35. Wickner, W., Driessen, A.J.M. and Hartl, F.U. (1991) Annu. Rev. Biochem. 60, 101–124.
36. MacIntyre, S., Mutschler, B. and Henning, U. (1991) Mol. Gen. Genet. 227, 224–228.
37. Hohn, T., Hohn, B., Engel, A., Wurtz, M. and Smith, P.R. (1979) J. Mol. Biol. 129, 359–373.
38. Kusters, R., de Vrije, T., Breukink, E. and de Kruijff, B. (1991) J. Biol. Chem. 264, 20827–20830.
39. Inouye, M. and Halegoua, S. (1980) CRC Crit. Rev. Biochem. 7, 339–371.
40. Dalbey, R.E. (1990) Trends Biochem. Sci. 15, 253–257.
41. Nilsson, I. and von Heijne, G. (1990) Cell 62, 1135–1141.
42. Wessels and Spiess, M. (1988) Cell 55, 61–70.

CHAPTER 5

Steps in the assembly of a cytoplasmic membrane protein: the MalF component of the maltose transport complex

BETH TRAXLER and JON BECKWITH

Department of Microbiology and Molecular Genetics, Harvard Medical School, 200 Longwood Avenue, Boston, MA 02115, USA

Abstract

We discuss the utility of MalF as a model system for the study of protein assembly into the cytoplasmic membrane of *Escherichia coli*. Our studies have demonstrated the importance of various sequences within MalF for membrane insertion and topology. The sequences that play the preeminent role in determining topology are the positively charged hydrophilic cytoplasmic domains. The position of these domains relative to hydrophobic membrane spanning segments (MSS) determines which MSS act as protein export signals and which act as stop-transfer sequences. Using alkaline phosphatase fusions, we have shown that both the cytoplasmic domains and the membrane spanning segments vary in the strength with which they promote proper topology. The process of membrane insertion of the MalF protein is independent of the normal secretion apparatus for protein export in *E. coli*. Finally, the influence of the other membrane components of the maltose transport complex, MalG and MalK, on MalF assembly is examined. While these proteins have no detectable effect on either the membrane insertion or topology of MalF, they do influence the final structure that MalF assumes in the membrane.

1. Introduction

1.1. Issues in the study of membrane protein assembly and structure

The protein folding problem has usually been framed in the context of those proteins that assemble in a hydrophilic environment. However, many important cellular proteins must assume their final configuration in a hydrophobic milieu, a lipid bilayer. The study of the assembly of membrane proteins is both a more complex and a simpler problem than the study of assembly of water-soluble proteins. It is more complex because most such proteins have domains that are in three different cellular compartments: the membrane, the cytoplasm, and the aqueous compartment on the other side of the membrane from the cytoplasm. Furthermore, the organization of the membranes into which membrane proteins insert is not well understood, even though much is known about the structure of model lipid bilayers. Clearly cellular membranes are not simple lipid bilayers, as they contain many proteins and may have lipid arranged in a variety of different ways. Thus, models for insertion and structure of membrane proteins may be based on simplifying assumptions that are not correct.

The problem of membrane protein assembly has its simpler aspects, because proteins within membranes may be very restricted in the types of secondary structure that they can assume. It is often easy to identify those components of proteins that are integral to a membrane by inspection of the amino acid sequence. Typical membrane spanning segments (MSS), thought to be α-helical, are sequences of about 20 amino acids, and are rich in hydrophobic amino acids [1].

Intimately connected with the problem of membrane protein assembly is the question of whether the insertion into the membrane is facilitated by any cellular factors. In particular, it has been suggested that most membrane proteins require for their insertion the same cellular machinery that is used for secretion of proteins *through* membranes. The in vitro study of membrane insertion of a number of eukaryotic membrane proteins has shown a dependence on a signal recognition particle, a central component of the secretion machinery [2]. In vivo studies in bacteria with mutants defective in proteins that make up the prokaryotic secretion apparatus suggest that the assembly of certain membrane proteins requires the *sec* gene products [3–5]. However, there are exceptions in both prokaryotes and eukaryotes [2,6].

Crucial to studies on membrane protein structure and function is the determination of the topological arrangement of such proteins within the membrane. Establishing a topology requires ascertaining which of the hydrophilic domains of the protein are localized to the cytoplasm and which to the extracytoplasmic face of the membrane. Most topologies have been determined either by proteolytic studies or by gene fusion analysis [7].

The study of topology leads to a study of the signals within proteins that assure their proper topology. Blobel first proposed that the term *topogenic* be used to describe those amino acid sequences in membrane proteins that contain information determining the topology [8]. At that time, he proposed that the hydrophobic membrane spanning segments could be separated into two classes: signal sequences which promote export of the external hydrophilic domains, and stop-transfer sequences which prevent translocation of cytoplasmic hydrophilic domains. More recently it has become clear that cytoplasmic domains of membrane proteins themselves, contain topogenic signals. In particular, the presence of an excess of basic amino acids in these domains insures a cytoplasmic location for them, and thus determines the orientation of the surrounding membrane spanning segments [9,10].

Finally, many of the issues involved in the study of membrane proteins can be divided into the analysis of the nature and formation of tertiary and quaternary structure (in the case of proteins in multisubunit complexes). Very little is known about the structure of membrane proteins; only a few such proteins or protein complexes have had their structure determined [11,12]. However, the interactions between different domains of a membrane protein that direct the final assembled structure are more easily examined. For instance, evidence is beginning to emerge that specific MSS can interact to influence the assembly of membrane proteins [13,14] for both glycophorins and T-cell antigen receptor. In other systems, such as the acetylcholine receptor, membrane protein assembly depends, at least in part, on interactions between hydrophilic domains [15].

1.2. The MalF protein as a model system

We have focussed on a membrane protein of *E. coli* which is part of a multicomponent complex. The maltose transport system of *E. coli* is dependent on the periplasmic maltose binding protein and a membrane complex that includes MalF, MalG and MalK [16]. MalF and MalG are integral to the cytoplasmic membrane, while MalK is a peripheral membrane protein associated with MalF–MalG at the cytoplasmic face of the membrane. The maltose binding protein apparently brings maltose to the membrane proteins which then transport the sugar into the cytoplasm. The energy for the transport process is thought to derive from ATP hydrolysis by MalK [17].

The MalF protein has a molecular weight of 57 kDa [18]. Inspection of the sequence suggests that it may have eight membrane spanning segments. The topology of this protein in the membrane has been studied using both alkaline phosphatase (PhoA) and β-galactosidase fusions [19,20]. These gene fusion approaches for analyzing membrane protein topology are based on a variation in the enzymatic activity of the reporter proteins that depends on their cellular location. Alkaline phosphatase is inactive when localized to the cytoplasm and active in the periplasm. The lack of activity in the cytoplasm probably results from the failure to form essential disulfide bonds in this compartment [21,22]. β-Galactosidase is active in the cytoplasm. However, when an export signal is attached to the enzyme, it becomes imbedded in the membrane where it is enzymatically inactive. By fusing either of these two reporter proteins to hydrophilic domains of membrane proteins, it is possible to deduce whether these domains are located in the cytoplasm or in the extracytoplasmic space. This deduction is based on the amount of enzymatic activity that the reporter proteins exhibit. Results of studies with fusions of either alkaline phosphatase or β-galactosidase to MalF suggest the structure shown in Fig. 1. According to this model,

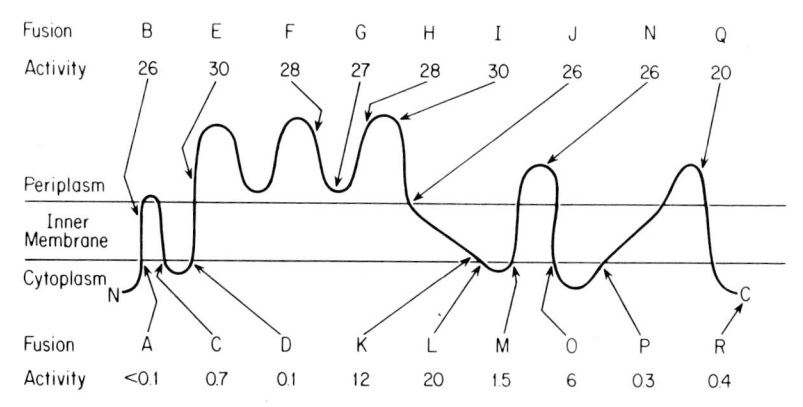

Fusion	B	E	F	G	H	I	J	N	Q
Activity	26	30	28	27	28	30	26	26	20

Fusion	A	C	D	K	L	M	O	P	R
Activity	<0.1	0.7	0.1	12	20	15	6	0.3	0.4

Fig. 1. Topological model of MalF. The model is based on the deduced amino acid sequence of the protein [18]. The MSS of MalF correspond to hydrophobic sequences of 20 or more amino acids. The fourth and seventh MSS are represented by slanted lines to reflect the longer length (approximately 30 amino acids) of these hydrophobic stretches. The positions of the fusion joints for MalF–PhoA fusions are indicated [19]. The different fusions are referred to throughout the text by the one letter designations shown. The alkaline phosphatase activity of each fusion is indicated by the number below each letter. Activity measurements were made on strains containing a single copy of each fusion construct integrated onto the chromosome.

the protein does have eight membrane spanning segments and one large periplasmic domain of about 180 amino acids.

1.3. The mechanism of insertion into the membrane of MalF

For assembly of MalF into the membrane, it must both incorporate its membrane spanning segments into the lipid bilayer and transfer several of its hydrophilic domains across the membrane into the periplasm. At first glance, the translocation of the 180 amino acid large periplasmic domain would appear to represent the same problem to the cell as the export of a periplasmic protein. A common solution would be to use the bacterial Sec machinery for both processes. Indeed, Wickner's group has clearly shown that the transfer across the membrane of the large periplasmic domain of *E. coli* leader peptidase is defective in certain *sec* mutants [5]. Other less direct experiments suggest the same conclusion with the SecY and Tsr proteins [3,4].

We have analyzed the dependence of MalF membrane insertion on the Sec machinery by several approaches [23]. Using either *sec* mutants (*secA* and *secD*) or sodium azide to inhibit SecA, we examined the translocation of periplasmic domains of MalF using alkaline phosphatase or β-galactosidase fusions. The translocation of these domains is assayed by determining their sensitivity to protease added to bacterial spheroplasts. Only those portions of proteins that are on the periplasmic face of the cytoplasmic membrane should be sensitive to the protease. If the interference with the Sec machinery were to prevent the insertion of the fusion proteins into the membrane, they should no longer be sensitive to protease. We found no difference in the kinetics of appearance of protease-sensitive fusion protein on the extracytoplasmic face of the membrane between wild-type conditions and conditions where the Sec machinery is impaired. We also have done these same experiments with native MalF protein, with the same results.

These results suggest that the MalF protein does not depend on the *sec* gene products for its insertion into the membrane. We have not totally ruled out Sec-dependence, however, since it may be that signals in the MalF protein have such a high affinity for the secretion apparatus that our means of inhibiting the pathway leave sufficient activity for the incorporation of this protein.

If the membrane insertion of MalF is independent of the *sec* gene products, then either there is a separate apparatus for the incorporation of this protein into the membrane or it can assemble without any accessory factors. If the latter were the case, it would mean that a large hydrophilic domain could be transferred unassisted across the lipid bilayer when attached to a potent enough signal [24,25]. Given that most MSS are considerably longer than the hydrophobic core of signal sequences for secreted proteins, it is possible that these MSS can act as potent export signals on their own.

2. The nature of topogenic signals in MalF

In 1986, von Heijne noted that in a large proportion of membrane proteins the cytoplasmic domains are enriched for basic amino acids when compared to periplasmic domains [26]. He proposed that these amino acids may be important in insuring the proper topology of the protein.

We have isolated a set of MalF–alkaline phosphatase fusion strains in which each hydrophilic domain of the protein has at least one alkaline phosphatase fusion joint [19] (Fig. 1). In addition to suggesting a topology for MalF, these fusions have allowed us to study individual topogenic signals within the protein. In particular, this analysis has verified the role of basic amino acids in establishing cytoplasmic domains as important determinants of MalF topology.

In most of the cytoplasmic domains of MalF, we have obtained fusions in which alkaline phosphatase is fused to both the beginning and the end of the hydrophilic loop. We find consistently that fusions to the amino terminal ends of these loops exhibit considerably more alkaline phosphatase activity than fusions to the carboxy terminal ends of these loops. (In all cases examined, we showed directly that any alkaline phosphatase activity found was localized to the periplasm.) These results are consistent with the proposal of von Heijne that basic amino acids anchor cytoplasmic domains. The amino terminal fusions are missing the basic amino acids, and thus the alkaline phosphatase is less stably localized to the cytoplasm.

We tested the role of the basic amino acids by mutagenizing a fusion (M) in which alkaline phosphatase is fused to the carboxy end of a cytoplasmic domain. We showed that reducing the amount of positive charge between the MSS and the alkaline phosphatase moiety results in elevated export of alkaline phosphatase [27]. In addition, the position of the positive charges in the loop was important. Neutralization of the positive charges closest to the MSS had the most profound effect on the fusion's activity. These results defined the third cytoplasmic domain as a topogenic element and indicated that it was the charge nature of the amino acid sequences in this loop, rather than the length or the specific sequence of the loop, that was important in the stable anchoring of alkaline phosphatase to the cytoplasm in this fusion.

Further studies revealed that there are multiple topological signals defining MalF's overall two-dimensional structure. A derivative of MalF in which the first cytoplasmic domain and MSS were deleted was still highly active for maltose transport. This indicated that in spite of the loss of normal amino terminal topogenic signals, the carboxy terminal portion of the protein retained the topological information necessary for proper membrane insertion [28].

To analyze individual topogenic signals within MalF, we introduced deletions of various MSS into several MalF–PhoA fusions. It seemed possible that the removal of a MSS could cause the inversion of the subsequent portions of the molecule. Since such an inversion would alter the location of the alkaline phosphatase moiety, the activity of the fusion protein should be altered. Thus, by assaying alkaline phosphatase activity of fusions carrying deletions, we determined whether portions of the molecule could be inverted. (In our studies, we also looked directly at the location of the alkaline phosphatase by using proteases to determine whether the protein is inside or outside.) Whether altering the orientation of the amino terminal portion of the membrane protein alters the topology downstream depends on: (1) whether there are topogenic signals carboxy terminal to the deletion and (2) how strong these signals are.

We found that a deletion of the first MSS and periplasmic domain of MalF converted the low activity MalF–alkaline phosphatase C fusion into a high specific activity protein (Fig. 2A) [28]. However, this same deletion did not change the specific activity of several later MalF–alkaline phosphatase fusion proteins (F, J, M, and R; Table I). Therefore, a deletion that could invert the orientation of the protein in the membrane on a short derivative of MalF did not effect this inversion on the longer, more complete derivatives of MalF. This work directly tested the proposal that the integration of a protein into the membrane occurs in a directed sequential manner, with carboxy terminal domains passively following a pattern determined by the orientation of the most N-terminal signal [29]. Our results indicated that the arrangement of a protein in the membrane is not fully determined by the N-terminal topological signal but rather that topological information is distributed throughout the MalF primary sequence.

A further detailed analysis showed that the cytoplasmic domains of MalF are the most important topogenic signals [30]. For example, as indicated above, deletion of

A *malF-TnphoA #C*   B *malF-TnphoA #D*

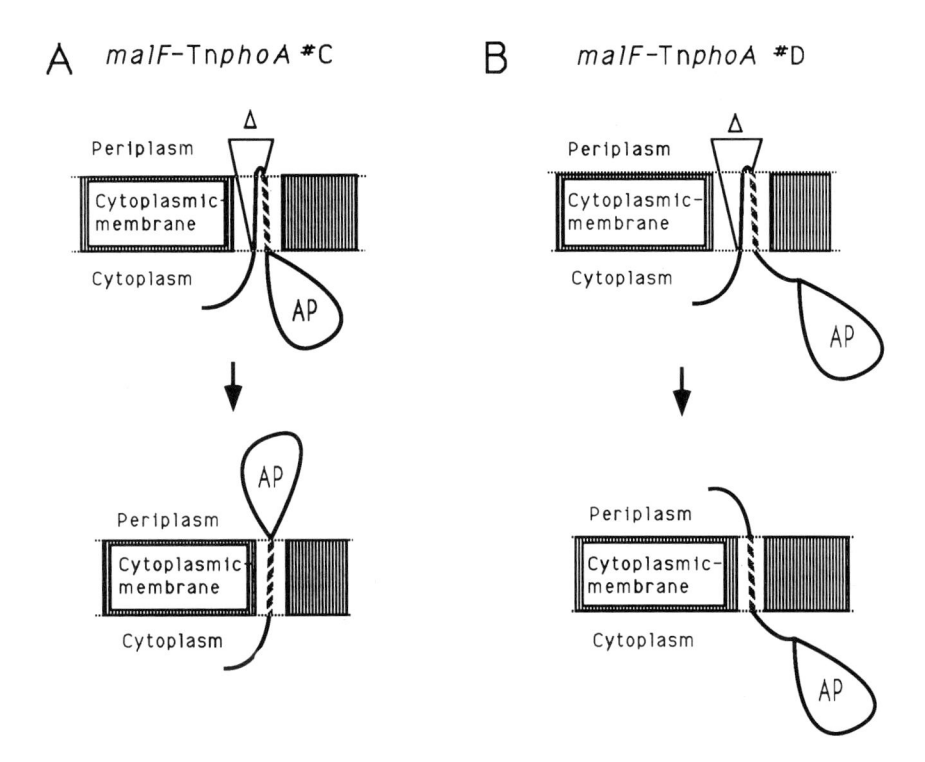

Fig. 2. Deletion analysis of MalF–PhoA fusions. We show schematic representations of the effect of deleting various segments of MalF on the topology of the rest of the protein, as assayed with fusions. In both cases shown here, MalF amino acids 17–39 are removed by oligonucleotide-directed mutagenesis. The topology of the protein produced by the deletion constructs is deduced from alkaline phosphatase activity measurements and proteolysis treatment of spheroplasts.

the first MSS and the periplasmic domain of MalF converted the MalF–alkaline phosphatase C fusion into a high specific activity protein. Yet, this same deletion had little effect on the specific activity of the MalF–alkaline phosphatase D fusion (Fig. 2B and Table I). The only difference between the C and D fusions was the presence of the second cytoplasmic domain of MalF in the D fusion. Cytoplasmic domains must therefore carry the information that determines their cytoplasmic localization, presumably in the form of the positively charged amino acids. This information thereby orients the MSS surrounding the cytoplasmic loops. The periplasmic domains, regardless of size or charge density, did not appear to contain localization signals. Furthermore, cytoplasmic domains varied in the strength of the signals they contain. As mentioned above, the second cytoplasmic domain of MalF (between MSS II and III) seemed to contain a strong topogenic signal whereas the third cytoplasmic domain (between MSS IV and V) was much weaker in comparison. The charge density of these cytoplasmic domains correlated well with the strength of the signals.

TABLE I

Alkaline phosphatase activity measurements of MalF–PhoA constructs

Fusion	Units of alkaline phosphatase activity[a]	
	Original MalF–PhoA	MalF–PhoA Δ aa 17–39
C	57	1180
D	40	78
F	943	1054
J	700	772
M	77	76
R	13	16

[a] Units of activity are assayed as described [19] from cells containing plasmids coding for the fusion proteins. The units of activity shown here for MalF–PhoA fusions are higher than those in Fig. 1 due to the high copy number of the plasmids. The deletion of MalF amino acids 17–39 (Δ aa 17–39) removes the first MSS and the first periplasmic domain of MalF. Alkaline phosphatase activity measurements are taken from [28,30].

3. Kinetics of assembly of MalF in the cytoplasmic membrane

We have also studied the relationship between the kinetics of membrane insertion and the topology of complex membrane proteins [31]. Newly synthesized MalF–alkaline phosphatase fusion proteins were examined by treatment of spheroplasts with protease. We observed a significant difference in the rate of accumulation of the protease-resistant alkaline phosphatase moiety in the periplasm among different MalF–PhoA fusions. We interpret these results to reflect a difference in export kinetics for the alkaline phosphatase portion of the fusion. Alkaline phosphatase is exported rapidly with most of the fusions in which alkaline phosphatase is joined to MalF in periplasmic domains (such as J or N). Alkaline phosphatase is exported slowly in those fusions in which alkaline phosphatase is joined to cytoplasmic domains of MalF (K, L, M, and O). Interestingly, alkaline phosphatase is also exported slowly in one fusion in which alkaline phosphatase is joined to a periplasmic domain of MalF (Q, following MSS VII).

It appears that the MSS preceding normally translocated domains of a membrane protein (such as MalF MSS III and V) are usually efficient export signals. However, MSS VII, which contains a lysine within the presumed MSS, is less efficient than the others. Calamia and Manoil [36] have observed a similar inefficiency of a MSS containing an arginine in the LacY protein. The efficiency was increased when this arginine was altered by mutation to an alanine. These findings raise the possibility that different extracytoplasmic domains of membrane proteins are exported at different rates, depending on the nature of the preceding export signal. Alternatively, since these fusion proteins are missing carboxy terminal portions of the membrane proteins, it may be that the participation of downstream sequences can enhance the export rate in the native protein.

The alkaline phosphatase moiety of cytoplasmic fusions (such as K, L, or O) can become periplasmically localized in the absence of the positive charges normally stabilizing the cytoplasmic localization. Although it occurs with moderate efficiency, this export is at a rate that is detectably slower than that for most periplasmic domains.

4. Assembly of MalF into the quaternary MalF–MalG–MalK structure

The functioning of MalF relies on its integration into the oligomeric maltose transport complex in the cytoplasmic membrane. This complex consists of one MalF, one MalG, and two MalK molecules [32]. We can follow the assembly of this complex by monitoring the proteolytic sensitivity of MalF in spheroplasts of *E. coli* cells.

We have found that newly synthesized MalF in pulse-labelled wild-type cells, is incorporated rapidly into the membrane, where it is protease-sensitive in spheroplasts (Fig. 3). This labelled membrane protein subsequently chases into a protease-resistant state after further growth of the culture. However, in spheroplasts from strains that are deficient for the production of either MalG or MalK, MalF is sensitive to

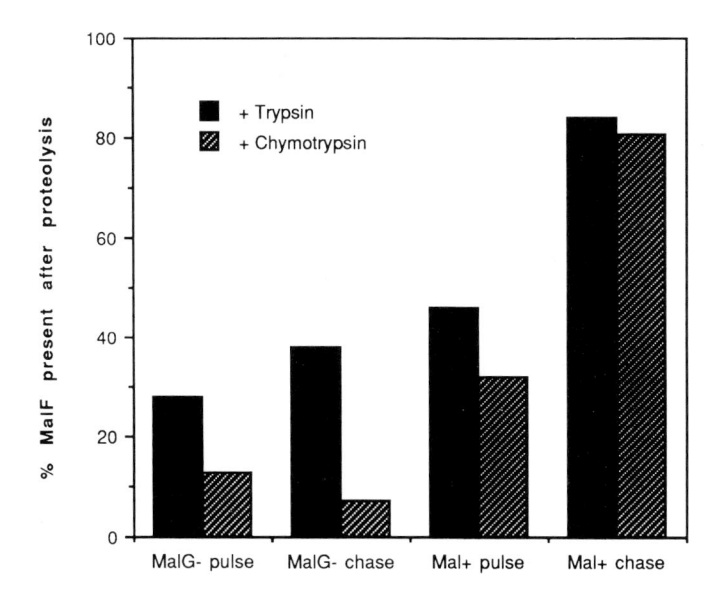

Fig. 3. The acquisition of protease resistance for MalF. Isogenic Mal$^+$ and MalG$^-$ strains were grown in minimal medium and pulse-labelled for 1 min with [^{35}S]methionine. One-half of each culture was chased with cold methionine for 30 min. Spheroplasts were treated with 10 μg/ml trypsin or chymotrypsin, and proteins were immunoprecipitated with antiserum against MalF. Proteins were separated by SDS-PAGE, and the amount of intact labelled MalF in the proteolysed samples was compared to the unproteolysed control.

proteases, even after a chase. The pattern of proteolytic peptides from MalF was the same in *malG⁻*, *malK⁻* and *malG⁻K⁻* strains. In contrast, cells missing the maltose binding protein, product of the *malE* gene, still produced protease-resistant MalF.

We studied the kinetics of interaction between MalF and MalG by differentially regulating their synthesis. This was done by using a strain which contained a *malG* amber mutation and a tightly regulated amber suppressor. Cells constitutively producing MalF, but not producing MalG, were pulse-labelled with [^{35}S]methionine. The labelled MalF protein incorporated into the membrane was protease-sensitive. Subsequently, expression of the amber suppressor was induced, resulting in the production of MalG, and the protease-resistance of the previously incorporated MalF was examined. We found that pulse-labelled MalF could be chased into the protease resistant state after induction of the amber suppressor (and therefore, of MalG synthesis). In these experiments, the MalF protein was in place in the membrane (as evidenced by its protease sensitivity in spheroplasts) before MalG was synthesized. When MalG was produced at a later time, it inserted into the membrane and eventually associated with MalF and MalK. Therefore, the different subunits of the complex do not necessarily have to interact before or immediately after membrane insertion.

These results have led us to propose the following:

(1) MalF ordinarily inserts into the membrane independently of MalG and MalK. The complex does not assemble in the cytoplasm before insertion and MalF itself does not require the other components for its insertion.

(2) Following insertion, MalF and MalG can diffuse in the membrane subsequent to insertion to assemble, along with MalK, into the transport complex. Another example of a complex in which assembly can occur post-translationally in the membrane is the influenza hemagglutinin (HA) [33]. With HA, newly synthesized monomers enter a common pool in the endoplasmic reticulum membrane and diffuse freely before trimerization and subsequent transport to the plasma membrane. This independent assembly of membrane components and subsequent diffusion is also analogous to the assembly of fragments of bacterial rhodopsin after incorporation into membrane vesicles [34].

(3) In the assembled MalFGK complex, protease-sensitive regions of MalF become inaccessible to protease either because they are masked or because of a conformational change. A third possibility, that the topology of the already membrane-assembled MalF is changed by its interaction with MalG and MalK, appears highly unlikely for energetic reasons. Furthermore, we have yet to see any dependence of MalF–AP fusion protein topology on other components of the system (Traxler, McGovern, Ehrmann, Boyd and Beckwith, unpublished results).

Our results suggest that by studying the proteolytic sensitivity of MalF, one can examine the assembly of a hetero-oligomeric membrane protein complex. We should be able to use this technique to estimate the rapidity of this process and evaluate the strength of various factors that may influence it, such as subunit concentration or temperature. This system should be a valuable tool for examining the process of subunit diffusion and intersubunit recognition as part of the assembly process.

5. Summary

We have focussed on the MalF protein as a model system for analyzing the membrane assembly of a complex membrane protein. We have examined the initial stages of this assembly in terms of: (1) the involvement of the Sec apparatus in MalF's insertion into the membrane and (2) the topological signals defining MalF's two-dimensional arrangement in the membrane. Our evidence suggests that MalF, unlike certain other cytoplasmic membrane proteins, does not require the Sec machinery for membrane incorporation. This observation is somewhat surprising given the need to translocate large hydrophilic domains of MalF across the membrane. In the absence of evidence for the existence of an alternative Sec apparatus, the results suggest that the multiple signals of MalF that define the protein's topology contain sufficient information by themselves, for the transport of hydrophilic regions to the extracytoplasmic compartment.

These topological signals consist of both the hydrophilic regions on the cytoplasmic face of the membrane and the hydrophobic MSS which enable the interactions with the membrane. We and others have demonstrated that the positively charged amino acids in hydrophilic domains are responsible for maintenance of certain regions on the cytoplasmic face of the membrane. There are five such hydrophilic regions in MalF, and they vary in the relative strength of the cytoplasmic localization signal that they carry. We propose that the position and the density of the positively charged amino acids in these cytoplasmic loops orients the MSS surrounding them and determines the strength of their localization signal. We have also observed that MSS acting as signal sequences vary in the strength of the export signal they contain. We have shown that both MSS III and V are potent export signals in the context of MalF–alkaline phosphatase fusions. In contrast, MSS VII is much less potent in this regard. We are beginning to examine later stages of MalF's membrane assembly, represented by its acquisition of tertiary and quaternary structure. We can study MalF's association with the other two subunits of the transport complex, MalG and MalK, by assaying the sensitivity of MalF to exogenous proteases. MalF is converted from a protease-sensitive to a protease-resistant form when it associates with the other proteins of the transport complex. These findings are consistent with the two-stage model [35] of membrane protein folding and oligomerization in which individual domains of proteins and individual proteins are first independently integrated into the membrane and subsequently interact to form higher order structures.

While membrane protein assembly has been examined in other systems, there are many advantages to studying this process in *E. coli* with the maltose transport complex. The *mal* genes are easily manipulated genetically. The Mal proteins are stably expressed by themselves or in combination with the other Mal proteins, and MalF and MalG are properly localized to the cytoplasmic membrane in the presence or absence of complex assembly. Finally, as a hetero-oligomeric complex, the maltose transport system is not too large to be conveniently studied, either in terms of the total number of subunits or the number of different subunits in the complex. Future experiments should allow us to more precisely examine the interactions, both hydrophobic and hydrophilic, that govern the architecture of complex membrane proteins and oligomeric membrane protein complexes.

Acknowledgments

This work was supported by a grant to J.B. from the National Insitutes of Health. J.B. is an American Cancer Society Research Professor.

References

1. Henderson, R. (1975) J. Mol. Biol. 93, 123–138.
2. Anderson, D.J., Mostov, K.E. and Blobel, G. (1983) Proc. Natl. Acad. Sci. USA 80, 7249–7253.
3. Gebert, J., Overhoff, F.B., Manson, M.D. and Boos, W. (1988) J. Biol. Chem. 263, 16652–16660.
4. Akiyama, Y. and Ito, K. (1989) J. Biol. Chem. 264, 437–442.
5. Wolfe, P.B. and Wickner, W. (1984) Cell 36, 1067–1072.
6. Wolfe, P.B., Rice, M. and Wickner, W.. (1985) J. Biol. Chem. 260, 1836–1841.
7. Jennings, M.L. (1989) Annu. Rev. Biochem. 58, 999–1027.
8. Blobel, G. (1980) Proc. Natl. Acad. Sci. USA 77, 1496–1500.
9. Nilsson, I. and von Heijne, G. (1990) Cell 62, 1135–1141.
10. Boyd, D. and Beckwith, J. (1990) Cell 62, 1031–1033.
11. Deisenhofer, J., Epp, O., Miki, K., Huber, R. and Michel, H. (1985) Nature 318, 618–624.
12. Henderson, R., Baldwin, J.M., Ceska, T.A., Zemlin, F., Beckmann, E. and Downing, K.H. (1990) J. Mol. Biol. 213, 899–929.
13. Bormann, B.J., Knowles, W.J. and Marchesi, V.T. (1991) J. Biol. Chem. 264, 4033–4037.
14. Cosson, P., Lankford, S.P., Bonifacino, J.S. and Klausner, R.D. (1991) Nature 351, 414–416.
15. Yu, X.-M. and Hall, Z.W. (1991) Nature 352, 64–67.
16. Shuman, H.A. (1987) Annu. Rev. Genet. 21, 155–177.
17. Dean, D.A., Davidson, A.L. and Nikaido, H. (1989) Proc. Natl. Acad. Sci. USA 86, 9134–9138.
18. Froshauer, S. and Beckwith, J. (1984) J. Biol. Chem. 259, 10896–10903.
19. Boyd, D., Manoil, C. and Beckwith, J. (1987) Proc. Natl. Acad. Sci. USA 84, 8525–8529.
20. Froshauer, S., Green, G.N., Boyd, D., McGovern, K. and Beckwith, J. (1988) J. Mol. Biol. 200, 501–511.
21. Derman, A.I. and Beckwith, J. (1991) J. Bacteriol. 173, 7719–7722.
22. Bardwell, J.C.A., McGovern, K. and Beckwith, J. (1991) Cell 67, 581–589.
23. McGovern, K. and Beckwith, J. (1991) J. Biol. Chem. 266, 20870–20876.
24. von Heijne, G. and Blomberg, C. (1979) Eur. J. Biochem. 97, 175–181.
25. Engelman, D.M. and Steitz, T.A. (1981) Cell 23, 411–422.
26. von Heijne, G. (1986) EMBO J. 5, 3021–3027.
27. Boyd, D. and Beckwith, J. (1989) Proc. Natl. Acad. Sci. USA 86, 9446–9450.

28. Ehrmann, M. and Beckwith, J. (1991) J. Biol. Chem. 266, 16530–16533.
29. Hartmann, E., Rapoport, T.A. and Lodish, H.F. (1989) Proc. Natl. Acad. Sci. USA 86, 5786–5790.
30. McGovern, K., Ehrmann, M. and Beckwith, J. (1991) EMBO J. 10, 2773–2782.
31. Traxler, B., Lee, C., Boyd, D. and Beckwith, J. (1991) J. Biol. Chem. 267, 5339–5345.
32. Davidson, A.L. and Nikaido, H. (1991) J. Biol. Chem. 266, 8946–8951.
33. Boulay, F., Doms, R.W., Webster, R. and Helenius, A. (1988) J. Cell Biol. 106, 629–639.
34. Popot, J.-L., Gerchman, S.-E. and Engelman, D.M. (1987) J. Mol. Biol. 198, 655–676.
35. Popot, J.-L. and Engleman, D.M. (1990) Biochemistry 29, 4031–4037.
36. Calamia, J. and Manoil, C. (1992) J. Mol. Biol. 224, 539–543.

CHAPTER 6

Structural characteristics of presecretory proteins: their implication as to translocation competency

SHOJI MIZUSHIMA, KATSUKO TANI, CHINAMI HIKITA and MASASHI KATO

Institute of Applied Microbiology, University of Tokyo, Yayoi, Bunkyo-ku, Tokyo 113, Japan

Abstract

Structural characteristics of the signal peptide domains and the mature domain of presecretory proteins are discussed in relation to the translocation competence of these proteins. The signal peptide domain is characterized by an amino terminal positive charge and a central hydrophobic stretch, both of which play important roles in protein translocation. Functions of the two parts are interrelated. Evidence suggesting that the two regions are recognized specifically by components of the secretory machinery is presented. The chemical structure of the signal peptide cleavage site and its possible implication in translocation competence are also discussed. Although translocation of the mature region is a rather passive process, the structure of this domain affects the translocation competence. A long hydrophobic stretch blocks ongoing translocation. The presence of positively charged residues near the amino terminus of the mature domain also inhibits translocation. Evidence has accumulated indicating that the mature region does not necessarily have to be an expanded linear polypeptide chain; polypeptide chains intramolecularly cross-linked to form a loop of a certain size or a mature domain composed of two polypeptide chains, which are cross-linked with relatively large cross-linkers, are also translocation competent. Based on these facts, the mechanism underlying the translocation of presecretory proteins is discussed.

Abbreviations: $\Delta\tilde{\mu}H^+$, proton motive force; $\Delta\psi$, membrane potential.

1. Introduction

Presecretory proteins are generally composed of an amino terminal signal peptide domain and a mature domain. How the individual domains are recognized by the secretory machinery for their translocation across membranes is discussed in this review.

The signal peptide domain usually comprises an amino terminal positively charged region and a central hydrophobic region. The former region is very necessary for the translocation of presecretory proteins across the cytoplasmic membrane in prokaryotic cells, whereas the requirement is not that strict for the translocation across the endoplasmic reticulum membrane of eukaryotic cells; about one-third of the signal peptides so far reported do not possess the positively charged region (NBRF Protein Data Base: National Biomedical Research Foundation, USA). On the other hand, a long hydrophobic stretch of amino acid residues at the central region is indispensable for the translocation in both prokaryotic and eukaryotic cells. Despite such a difference, the functions of signal peptides in both types of cells seem to be essentially the same, since signal peptides were shown to be interchangeable between some prokaryotic and eukaryotic systems. In the first part of this review, we discuss the roles of the individual regions of the signal peptide in protein translocation, mostly based on our own recent in vitro studies. The functions of signal peptides have been extensively reviewed with emphasis on mutations and genetics [1,2], biophysics [3,4] and statistics [5,6].

The addition of a signal peptide to some cytosolic proteins by means of gene engineering makes the protein translocation competent. This is not always true, however, indicating that proteins must meet certain structural requirements to be translocated. It is also unclear what the basic structure that is recognized or tolerated by the secretory machinery is. This is discussed in the latter part of this review.

2. Amino terminal positive charge of the signal peptide

In prokaryotic cells, presecretory proteins usually possess one to three positively charges amino acid residues at the amino terminus of their signal peptide. A number of signal peptide mutants with respect to the positive charge have been constructed. Replacement of the positively charged amino acid residues with negatively charged ones resulted in the in vivo accumulation of presecretory proteins, suggesting the importance of the positive charge for protein secretion [7-10]. The positive charge is thought to be involved in interaction with acidic phospholipids on the inner surface of membranes [11].

In order to gain a deeper insight into the functional role of the positively charged residues, we have carried out quantitative studies on the effects of the number and species of the positively charged residues on the rate of in vitro translocation [12].

ProOmpF-Lpp, a model presecretory protein, and mutant proteins derived from it were used. When the positive charge, which is +2 (LysArg) for the wild-type, was changed to 0, −1 or −2, little or no translocation was observed, indicating the essential requirement of the positive charge for translocation. The number of the positive charge was then altered by introducing different numbers of Lys or Arg. The rate of in vitro translocation was roughly proportional to this number, irrespective of whether the charged amino acid residues were Arg or Lys. Furthermore, His was also active when the pH was low enough to allow ionization of this residue, and the activity paralleled the number of His. We conclude, therefore, that the signal peptide requires a positive charge at its amino terminal region to function in the translocation reaction and that the rate of translocation is roughly proportional to the number of positively charged groups, irrespective of the amino acid species that donate the charges.

How are the positive charges recognized by the secretory machinery? We demonstrated the signal peptide-dependent interaction of presecretory proteins with SecA, a peripheral membrane protein involved in the translocation reaction, by means of chemical cross-linking [13]. Interestingly, the interaction became increasingly enhanced as the number of positively charged amino acid residues at the amino terminal region of the signal peptide was increased, irrespective of the species of amino acid residues donating the charges, as in the case of the translocation reaction. These results strongly suggest that the amino terminal positive charges of the signal peptide interact with SecA to initiate the translocation reaction. Inukai et al. [14] demonstrated that the amino terminal positively charged region stays on the inner surface of the cytoplasmic membrane of *E. coli* throughout the translocation reaction.

As mentioned earlier, the requirement of the positive charge was rather low in the eukaryotic translocation system. The eukaryotic counterpart of SecA, which is unknown, may not require such positive charges for recognition of the signal peptide. The possible absence of the membrane potential ($\Delta\psi$) in the endoplasmic reticulum membrane may have something to do with the low requirement. The $\Delta\psi$ of the bacterial membrane, which is inside-negative, is suggested to contribute to the stable localization of the positive charge on the inner surface of the membrane [15].

3. Central hydrophobic stretch of the signal peptide

Signal peptides possess a stretch of hydrophobic amino acid residues in their central region. The importance of the hydrophobic region for protein translocation was first demonstrated in vivo; deletion of this hydrophobic region or insertion of a charged amino acid residue into this region resulted in severe inhibition of protein secretion in *E. coli* [16–18]. In spite of the accumulation of such evidence, the basic principles of the hydrophobic region have remained unclear, most likely due to the fact that this region consists of a variety of amino acid residues. To overcome this problem,

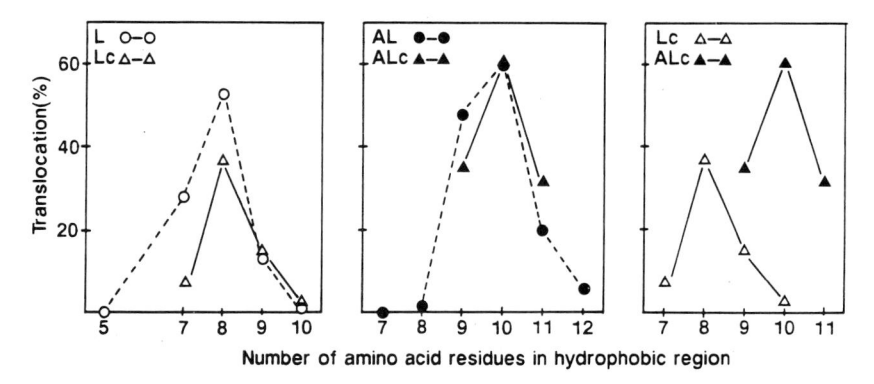

Fig. 1. Effect of the numbers of Leu (L) and Ala/Leu (AL) residues comprising the hydrophobic region of the signal peptide on the rate of in vitro translocation. Derivatives of proOmpF-Lpp possessing uncleavable (L, ○- - -○; AL, ●—●) or cleavable (Lc, △—△; ALc, ▲—▲) signal peptides were used as substrates. They were subjected to in vitro translocation. The translocation in the first 1 min was expressed as a percentage of the total input proOmpF-Lpp. From [23].

Kendall et al. [19–21] and Yamamoto et al. [22] systematically engineered the hydrophobic region on a DNA level to be composed of one species of amino acid residue, such as Leu, Val, Ala or Ile, in *E. coli* and yeast, respectively. Both groups showed in vivo that such artificial signal peptides were active as to secretion when the length of the hydrophobic stretch was appropriate. Kendall et al. [20] also showed that a longer hydrophobic stretch was required when the hydrophobicity per residue was lower, suggesting the importance of the total hydrophobicity.

We have examined in vitro the effect of the length of the hydrophobic stretch, composed of Leu alone or alternate Leu and Ala, on the rate of translocation of proOmpF-Lpps, model presecretory proteins [23]. The results also suggest the importance of the total hydrophobicity. In addition, the in vitro translocation exhibited high substrate specificity with respect to the number of hydrophobic residues; the maximum translocation activity was observed for the octamer in the case of Leu polymers and for the decamer in the case of polymers comprising Ala and Leu, with a sharp decrease on both sides (Fig. 1). The results suggest that the hydrophobic region is specifically recognized by a component(s) of the secretory machinery. This does not exclude the possible interaction of signal peptides with the hydrophobic core of the phospholipid bilayer. The implication of such an interaction in protein secretion has been discussed [4]. At present, it is unclear which components of the secretory machinery are involved in such an interaction in *E. coli*. In a eukaryotic system, a methionine-rich domain of the 54K subunit of the signal recognition particle (SRP) has been suggested to interact with a signal peptide [24,25].

4. Function of the positive charge can be compensated for by a longer hydrophobic stretch

So far, most studies on the roles of the amino terminal positive charge and the central hydrophobic stretch have been carried out independently. Some of the evidence, however, suggests that the functions of the two regions may be somehow interrelated. Bassford et al. [26] suggested on the basis of an in vivo experiment, that the defect caused by the loss of the positive charge may be compensated for by a longer hydrophobic stretch. Some of the eukaryotic signal peptides do not possess an amino terminal positive charge. A statistical study indicated that the total hydrophobicity of the first eight amino acid residues of the hydrophobic stretch was seemingly higher in the signal peptides without a positive charge at the amino terminus than in those with it [27]. This suggests that the strong hydrophobicity of the central region may substitute for the function of the positive charge.

We have constructed, on a DNA level, a large variety of model presecretory proteins, proOmpF-Lpps, possessing different numbers of Lys as the positively charged amino acid residues and different numbers of Leu as the hydrophobic amino acide residues (Fig. 2A). The effect of the positive charge on the in vitro translocation efficiency differed with the length of the hydrophobic stretch (Fig. 2B). The positive charge was strongly required for translocation when the hydrophobic region comprised eight Leu, whereas the translocation was independent of positive charge when the hydrophobic region comprised nine Leu [28]. Thus it is likely that the role of the positive charge can be compensated for by a longer hydrophobic stretch.

How are the functions of the two regions interrelated? The positively charged region is recognized by SecA and this recognition is seemingly the first step of the protein translocation reaction [13]. Since the translocation of a presecretory protein possessing no positive charge at the amino terminus of the signal peptide is SecA-dependent [28], it is probable that SecA also recognizes the hydrophobic region. A stronger hydrophobic interaction may compensate for the ionic interaction between the signal peptide and SecA.

Alternatively, the membrane may play some roles in the interactions with these two regions of the signal peptide. The interaction of the amino terminal positive charge with the acidic residues of phospholipids on the inner surface of the membrane has been suggested [11]. The involvement of $\Delta\psi$, which is inside-negative, in protein translocation also favors such a membrane-signal peptide interaction. The hydrophobic interaction between signal peptides and the phospholipid layer has been studied physiologically [4]. The signal peptide may first electrostatically interact with the membrane through the amino terminal positive charge, which in turn induces a hydrophobic interaction between the two to permit transmembrane orientation of the peptide. For a signal peptide possessing stronger hydrophobicity, the electrostatic interaction may not be needed.

The mechanisms by which the total hydrophobicity is recognized by the secretory

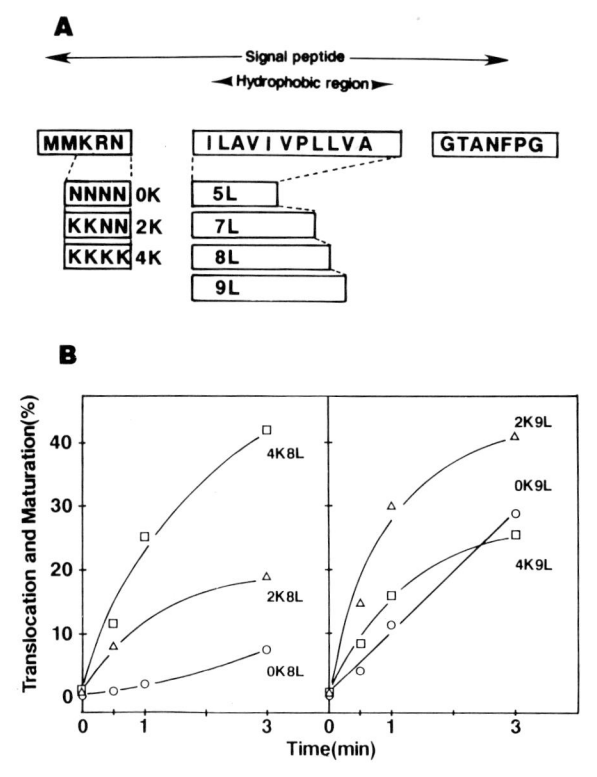

Fig. 2. Functions of N-terminal positive charge and the hydrophobic region are interrelated. ProOmpF-Lpps possession different numbers of N-terminal positively charged Lys (K) and different numbers of Leu (L) comprising the hydrophobic region were constructed on a DNA level (A) and their translocation competency was examined in vitro (B). For example, 0K9L possesses no Lys and nine Leu. Data were taken from [28].

machinery is unclear. A preliminary cross-linking experiment did not support the view that SecA is responsible for this specific recognition. The total hydrophobicity may be recognized by other components of the secretory machinery including the adjacent phospholipid layer.

5. *Carboxyl terminal region of the signal peptide including the cleavage site*

This region, which follows the hydrophobic region, is usually composed of several amino acid residues, on average around five. This number is, however, quite variable. This region is rather enriched with Ser and Thr [29]. This prompted us to examine the possible phosphorylation of these residues during secretory translocation. No positive evidence was obtained, however.

A statistical study demonstrated that small, neutral residues are abundant at positions −1 and −3 (the first and third residues from the carboxyl terminus of the signal peptide) [5]. In *E. coli*, position −1 is mostly occupied by Ala. On the contrary, the amino acid residues of the amino terminus of the mature domain are quite heterogenous. Almost any amino acid residue can occupy this position. In *E. coli*, however, the introduction of Pro to position +1 resulted in blocking of the signal peptide cleavage [30]. This may be due to the substrate specificity of signal peptidase. Pro at this position (+1) did not inhibit in vitro translocation [30]. Some of the eukaryotic presecretory proteins possess Pro at position +1. The amino acid sequence of the signal peptide cleavage site is partly determined by the substrate specificity of signal peptidases. For example, *E. coli* signal peptidase II (lipoprotein signal peptidase) cleaves X-Cys (X is Gly or Ala) only after Cys is modified by diacylglycerol through the thiol group. An in vitro enzymatic study demonstrated that the region from −4 to +5 is involved in the recognition by signal peptidase I of *E. coli* [31].

Our recent work suggests that the conformation determined by the amino acid composition of the region around the signal cleavage site has something to do with the proton motive force ($\Delta\tilde{\mu}H^+$) requirement for protein translocation [30]. We found that in vitro translocation into *E. coli* everted membrane vesicles of presecretory proteins, proOmpA and proOmpF-Lpp, possessing Pro at position +1 or +2, took place even in the absence of $\Delta\tilde{\mu}H^+$, although at a lower rate, whereas these proteins became translocation incompetent in the absence of $\Delta\tilde{\mu}H^+$ upon removal of the Pro. The position of Pro was then moved stepwise away from the amino terminus of the mature domain. The further the position was moved away, the slower was the rate of translocation in the absence of $\Delta\tilde{\mu}H^+$ (Fig. 3). It is suggested, therefore, that the conformational flexibility endowed by Pro on the junction region between the signal peptide and the mature domain allows the translocation in the absence of $\Delta\tilde{\mu}H^+$ and that this junction region must take on a particular conformation for initiation of the translocation reaction but not for the signal peptide cleavage. Gly could not replace the Pro residue for such an effect. Perlman and Halvorson [32] have discussed the possible requirement of a β-turn for recognition of the cleavage site. Yamamoto et al. [33] suggested that an α-helical structure should not be formed around the cleavage site for proper in vivo secretion in yeast. One can imagine then that Pro in the carboxyl terminal region of the signal peptide may have a similar function. No Pro was found at positions −1, −2 and −3 in the signal peptide. This is probably because of the requirement for signal peptide cleavage. Interestingly, Pro exists rather frequently at positions −4, −5 and −6 [34]. Furthermore, Pro at positions −4, −5 and −6 in an engineered signal sequence was shown to be important for the secretion of human lysozyme in yeast [33].

At present, it is unclear why the junction region takes on a particular conformation during the translocation reaction. In the initial step of the loop model for protein translocation [11], the amino terminal region including the signal peptide starts to make a loop, or a bent structure in the membrane, which grows until the signal cleavage site reaches the external surface of the membrane. It is probable that such a

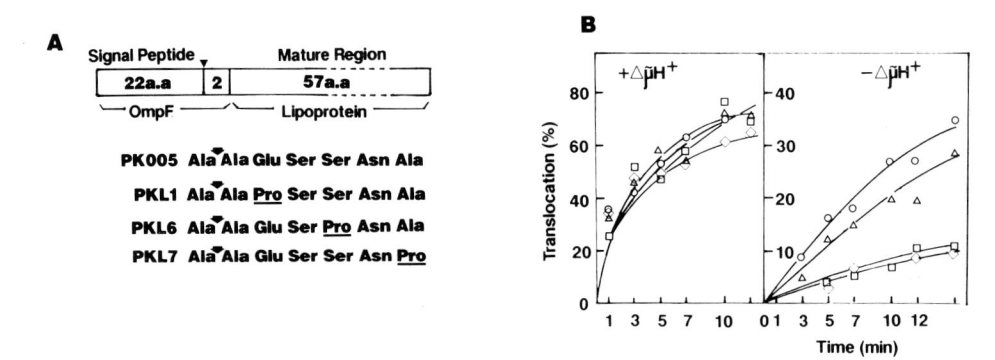

Fig. 3. The effect of Pro on the $\Delta\tilde{\mu}H^+$ requirement is specific to the signal peptide cleavage site. ProOmpF-Lpp possessing Pro at different positions near the signal peptide cleavage site were used as substrate (A) for in vitro translocation in the presence and absence of $\Delta\tilde{\mu}H^+$ (NADH) (B). ProOmpF-Lpps used were : PKL1, (○); PKL6, (△), PKL7, (□), PK005 (◇). Data were taken from [30].

conformation, in which the bent region is rather flexible and hence should not take on an α-helical conformation, is a key one in the initial stage of the translocation. $\Delta\tilde{\mu}H^+$ may be required for this conformation to be taken on when Pro is not present in this region. If this is the case, the process before this stage would not involve such a loop (bend) within the signal peptide region. Although some of the signal peptides possess Pro in the hydrophobic stretch, which usually exhibits a strong tendency to form an α-helix, the majority of them do not. Replacement of such a Pro with a hydrophobic amino acid residue did not result in detectable functional changes [19,23].

6. Roles of charged amino acid residues in the mature domain in protein translocation

Although the translocation of the mature domain is assumed to be a rather passive process, it is not true that any polypeptide chains can be translocated when it is equipped with a signal peptide. The mature domain has to satisfy some structural requirements to be translocated. In general, the mature domain does not contain an extensive stretch of hydrophobic residues [35] and the insertion of a long hydrophobic stretch suffices to stop the transfer of a protein across the membrane [36], suggesting that the secretory machinery does not tolerate the translocation of such a hydrophobic stretch.

Another factor that may affect the translocation competency of the mature domain is the charge distribution. The importance of the positive charge preceding the hydrophobic stretch of the signal peptide for protein translocation has been discussed previously. On the other hand, the amino terminal region of the mature domain, which follows the signal peptide, is generally neutral or negatively charges [15,37]. Li et al.

[38] and Yamane and Mizushima [15] have demonstrated that the introduction of basic amino acid residues into this region results in strong inhibition of their translocation across the cytoplasmic membrane of *E. coli*. More recent studies have further demonstrated that the blocking effect of a positive charge depends on its distance from the amino terminus of the mature domain; the further from the amino terminus the positive charges are, the less blocking effect on translocation they have [39,40].

It should be noted in this connection that the stop-transfer hydrophobic domains of membrane proteins are often followed by positively charged residues [15,41]. Topological studies on membrane proteins suggested that some hydrophobic regions with amino termini directed toward the cytoplasm have a signal peptide-like function, whereas those with the opposite orientation function as a stop-transfer sequence (for example, see [42]). It should also be noted that in integral membrane proteins the hydrophilic segments facing the cytoplasm are generally positively charged [41]. Taking all the evidence together, we assume that the positively charged residues, which are somehow stabilized on the cytosolic surface of membranes, have a positive effect on the following hydrophobic stretch as a signal peptide and on the preceding hydrophobic domain as a stop-transfer sequence [15]. Such dual effects of the positive charges may account for the orientation of hydrophobic domains in the membrane [15,43,44]. Nilsson and von Heijne [45] proposed that it is not the net positive charge after the hydrophobic stretch but the charge balance around it that is important in determining the orientation of polypeptide chains. The bacterial cytoplasmic membrane usually has a membrane potential ($\Delta\psi$), which is cytosolic side-negative. This must contribute to the stabilization of the positive charges on the cytosolic surface of the membrane. On the other hand, no evidence for the presence of $\Delta\psi$ on the endoplasmic reticulum membrane has been presented.

Proteins are usually highly charged. All the secretory proteins and probably other proteins as well, are charged. The question then arises as to whether charges on the mature domain are essential for the transmembrane movement of proteins, or how charged amino acid residues are translocated across the hydrophobic core of membranes. Since the translocation reaction in *E. coli* requires $\Delta\tilde{\mu}H^+$, which is composed of $\Delta\psi$ and ΔpH, the possibility that the contribution of $\Delta\tilde{\mu}H^+$ might include $\Delta\psi$-driven electrophoresis of acidic regions and ΔpH-driven deprotonation of basic residues in the cytoplasm and reprotonation in the periplasm has been discussed [46,47]. These, however, should not be the major roles of $\Delta\psi$ and ΔpH, since model presecretory proteins completely devoid of charges, both negative and positive, on the mature domain can be translocated in a $\Delta\psi/\Delta pH$-dependent manner [48]. This also indicates that the mature domain does not have to be charged for translocation.

7. Chemical structure of the mature domain tolerated by the secretory machinery

The following evidence suggests that presecretory proteins are translocated across membranes as an unfolded polypeptide chain: (1) SecB [49–51] and GroEL [52] of *E. coli*, and Hsp70 family proteins [53,54] of eukaryotic cells help to maintain prosecretory proteins in their translocation-competent conformations, which is most likely an expanded polypeptide chain; (2) a stretch of hydrophobic amino acid residues blocks ongoing translocation [36]; (3) presecretory proteins do not generally possess a stretch of hydrophobic amino acid residues except in the region of the signal peptide [35]; and (4) the formation of intramolecular disulfide bridges in a preprotein blocks translocation [55].

We recently demonstrated, however, that the polypeptide chain does not necessarily have to be linear and fully expanded for translocation. ProOmpA, the precursor of an outer membrane protein of *E. coli*, has Cys at positions +290 and +302. ProOmpA possessing a disulfide bridge between them was translocated in vitro without cleavage of the bridge [56]. The translocation was $\Delta\tilde{\mu}H^+$-dependent. In the absence of $\Delta\tilde{\mu}H^+$, translocation was blocked at the position of the disulfide bridge. When the two Cys were irreversibly cross-linked with a chemical reagent, $\Delta\tilde{\mu}H^+$-dependent translocation was also observed [57]. The total molecular mass of the loop formed upon the cross-linking was 1841 Da, suggesting that the secretory machinery tolerates the translocation of such a non-linear polypeptide loop. This, in turn, suggests that a channel or pore for the passage of secretory proteins, if one exists, would not be one that just fits a linear polypeptide chain. The $\Delta\tilde{\mu}H^+$-dependent translocation of a polypeptide chain possessing a cross-linked side chain comprising 20 amino acid residues has also been observed (K. Tani and S. Mizushima, unpublished data). The import into mitochondria of precursor proteins coupled to either a stilbene disulfonate derivative or cytochrome *c* has also been reported [58].

The tolerance of the protein translocation machinery may not be limitless, however. Maher and Singer [55] showed that the in vitro translocation across the endoplasmic reticulum membrane of human preprolactin did not take place when the protein possessed disulfide bonds at Cys^{53}-Cys^{164} and Cys^{181}-Cys^{188}. In order to determine the limit permitted by the secretory machinery, proOmpA was engineered to be able to form a disulfide bridge, which resulted in the formation of a loop composed of 5, 13, 16, 20 or 30 amino acid residues. The loops composed of less than 16 amino acid residues were translocated into everted membrane vesicles of *E. coli* in the presence of $\Delta\tilde{\mu}H^+$, whereas those larger than that were not (K. Uchida, K. Tani and S. Mizushima, unpublished data).

Since the secretory machinery is undoubtedly for proteins, the polypeptide bonds are assumed to be the principal structure that is recognized by the machinery. The passage of the cross-linked proOmpA, however, suggests the possibility that the translocation reaction advances through the cross-linked region, recognizing the polypeptide loop as a large side chain. If this is the case, substances to be translocated through the secretory machinery need not necessarily be solely held together by polypeptide bonds.

This view was recently proven to be true, since the following substance was found to be translocation competent, in terms of both proteinase K-resistance and signal peptide cleavage, in the presence of $\Delta\tilde{\mu}H^+$, when it was equipped with a signal peptide; two polypeptide chains chemically cross-linked with N,N'-bis(3-maleimidopropionyl)-2-hydroxy-1,3-propanediamine (MPHP) between the carboxyl terminal and amino terminal Cys (M. Kato and S. Mizushima, submitted). Furthermore, when a polypeptide chain comprising 15 amino acid residues was cross-linked to the carboxyl terminus of a presecretory protein in the reverse orientation, i.e., the carboxyl terminus in front, the reversed polypeptide region as well as the genuine mature region became proteinase K-resistant after in vitro translocation in the presence of $\Delta\tilde{\mu}H^+$, with the concomitant cleavage of the signal peptide (M. Kato and S. Mizushima, unpublished data). It is highly likely, therefore, that substrates for the secretory machinery are not necessarily solely composed of polypeptide bonds. The machinery seemingly tolerates the translocation of a variety of molecules provided that they are preceded by a signal peptide.

Acknowledgements

The work performed in our laboratory was supported by grants from the Ministry of Education, Science and Culture of Japan, the Nisshin-Seifun Foundation and the Naito Foundation. We thank Miss Iyoko Sugihara for secretarial support.

References

1. Bieker, K.L., Philips, G.J. and Silhavy, T.J. (1990) J. Bioenerg. Biomembr. 22, 291–310.
2. Gennity, J., Goldstein, J. and Inouye, M. (1990) J. Bioenerg. Biomembr. 22, 233–269.
3. Briggs, M.S. and Gierasch, L.M. (1986) Adv. Protein Chem. 38, 109–180.
4. Jones, J.D., McKnight, C.J. and Gierasch, L.M. (1990) J. Bioenerg. Biomembr. 22, 213–232.
5. von Heijne, G. (1985) Curr. Top. Membr. Transp. 24, 151–179.
6. von Heijne, G. (1985) J. Mol. Biol. 184, 99–105.
7. Inouye, S., Soberon, X., Franceschini, T., Nakamura, K., Itakura, K. and Inouye, M. (1982) Proc. Natl. Acad. Sci USA 79, 3438–3441.
8. Vlasuk, G.P., Inouye, S., Ito, H., Itakura, K. and Inouye, M. (1983) J. Biol. Chem. 258, 7141–7148.
9. Iino, T., Takahashi, M. and Sako, T. (1987) J. Biol. Chem. 262, 7412–7417.
10. Bosch, D., De Boer, P., Bitter, W. and Tommassen, J. (1989) Biochim. Biophys. Acta 979, 69–76.
11. Inouye, M. and Halegoua, S. (1980) CRC Crit. Rev. Biochem. 7, 339–371.
12. Sasaki, S., Matsuyama, S. and Mizushima, S. (1990) J. Biol. Chem. 265, 4358–4363.
13. Akita, M., Sasaki, S., Matsuyama, S. and Mizushima, S. (1990) J. Biol. Chem. 265, 8164–8169.
14. Inukai, M. and Inouye, M. (1983) Eur. J. Biochem. 130, 27–32.
15. Yamane, K. and Mizushima, S. (1988) J. Biol. Chem. 263, 19690–19696.

16. Emr, S.D., Hedgpeth, J., Clement, J.-M., Silhavy, T.J. and Hofnung, M. (1980) Nature 285, 82–85.
17. Emr, S.D. and Silhavy, T.J. (1983) Proc. Natl. Acad. Sci. USA 80, 4599–4603.
18. Bankaitis, V.A., Rasmussen, B.A. and Bassford Jr., P.J. (1984) Cell 37, 243–252.
19. Kendall, D.A., Bock, S.C. and Kaiser, E.T. (1986) Nature 321, 706–708.
20. Chou, M.M. and Kendall, D.A. (1990) J. Biol. Chem. 265, 2873–2880.
21. Kendall, D.A. and Kaiser, E.T. (1988) J. Biol. Chem. 263, 7261–7265.
22. Yamamoto, Y., Taniyama, Y., Kikuchi, M. and Ikehara, M. (1987) Biochem. Biophys. Res. Commun. 149, 431–436.
23. Hikita, C. and Mizushima, S. (1992) J. Biol. Chem. 267, 4882–4888.
24. Bernstein, H.D., Poritz, M.A., Strub, K., Hoben, P.J., Brenner, S. and Walter, P. (1989) Nature 340, 482–486.
25. Zopf, D., Bernstein, H.D., Johnson, A.E. and Walter, P. (1990) EMBO J. 9, 4511–4517.
26. Puziss, J.W., Fikes, J.D. and Bassford Jr., P.J. (1989) J. Bacteriol. 171, 2303–2311.
27. Takagi, Y., Morohashi, K., Kawabata, S., Go, M. and Omura, T. (1988) J. Biochem. 104, 801–806.
28. Hikita, C. and Mizushima, S. (1992) J. Biol. Chem. 267, 12375–12379.
29. Mutoh, N., Inokuchi, K. and Mizushima, S. (1982) FEBS Lett. 137, 171–174.
30. Lu, H.-M., Yamada, H. and Mizushima, S. (1991) J. Biol. Chem. 266, 9977–9982.
31. Dierstein, R. and Wickner, W. (1986) EMBO J. 5, 427–431.
32. Perlman, D. and Halvorson, H.O. (1983) J. Mol. Biol. 167, 392–409.
33. Yamamoto, Y., Taniyama, Y. and Kikuchi, M. (1989) Biochemistry 28, 2728–2732.
34. von Heijne, G. (1986) Nucleic Acids Res. 14, 4683–4690.
35. Mizushima, S. (1987) in: M. Inouye (Ed.), Bacterial Outer Membranes as Model Systems, Wiley, New York, 163–185.
36. Davis, N.G. and Model, P. (1985) Cell 41, 607–614.
37. von Heijne, G. (1986) J. Mol. Biol. 192, 287–290.
38. Li, P., Beckwith, J. and Inouye, H. (1988) Proc. Natl. Acad. Sci. USA 85, 7685–7689.
39. Summers, R.G. and Knowles, J.R. (1989) J. Biol. Chem. 264, 20074–20081.
40. Summers, R.G., Harris, C.R. and Knowles, J.R. (1989) J. Biol. Chem. 264, 20082–20088.
41. von Heijne, G. (1986) EMBO J. 5, 3021–3027.
42. Akiyama, Y. and Ito, K. (1987) EMBO J. 6, 3465–3470.
43. Yamane, K., Akiyama, Y., Ito, K. and Mizushima, S. (1990) J. Biol. Chem. 265, 21166–21171.
44. Boyd, D. and Beckwith, J. (1989) Proc. Natl. Acad. Sci. USA 86, 9446–9550.
45. Nilsson, I. and von Heijne, G. (1990) Cell 62, 1135–1141.
46. Schiebel, E., Driessen, A.J.M., Hartl, F.-U. and Wickner, W. (1991) Cell 64, 927–939.
47. Driessen, A.J.M. and Wickner, W. (1991) Proc. Natl. Acad. Sci. USA 88, 2471–2475.
48. Kato, M., Tokuda, H. and Mizushima, S. (1992) J. Biol. Chem. 267, 413–418
49. Collier, D.N., Bankaitis, V.A., Weiss, J.B. and Bassford Jr., P.J. (1988) Cell 53, 273–283.
50. Kumamoto, C.A. and Gannon, P.M. (1988) J. Biol. Chem. 263, 11554–11558.
51. Weiss, J.B., Ray, P.H. and Bassford Jr., P.J. (1988) Proc. Natl. Acad. Sci. USA 85, 8978–8982.
52. Bochkareva, E., Lissin, N.M. and Girshovich, A.S. (1988) Nature 336, 254–257.
53. Deshaies, R.J., Koch, B.D., Werner-Washurne, M., Craig, E.A. and Schekman, R. (1988) Nature 332, 800–805.
54. Chirico, W.J., Waters, M.G. and Blobel, G. (1988) Nature 322, 805–810.
55. Maher, P.A. and Singer, S.J. (1986) Proc. Natl. Acad. Sci. USA 83, 9001–9005.
56. Tani, K., Tokuda, H. and Mizushima, S. (1990) J. Biol. Chem. 265, 17341–17347.
57. Tani, K. and Mizushima, S. (1991) FEBS Lett. 285, 127–131.
58. Vestwever, D. and Schatz, G. (1988) J. Cell Biol. 107, 2045–2049.

CHAPTER 7

Sequence determinants of membrane protein topology

GUNNAR VON HEIJNE

Department of Molecular Biology and Karolinska Institute Center for Structural Biochemistry, NOVUM, S-141 57 Huddinge, Sweden

Abstract

Most integral membrane proteins are built according to a helical bundle-design with one or more apolar α-helices spanning the lipid bilayer. The membrane assembly of this class of proteins is guided by the apolar segments and their polar flanking regions. In particular, membrane proteins from a variety of sources have been shown to follow the positive inside-rule: positively charged residues abound in the non-translocated parts of the protein, but are scarce in the translocated parts. Recent work in a number of labs, including our own, has shown that addition or removal of positively charged amino acids from critical locations in the chain can have dramatic consequences for the topology of a membrane protein.

1. Introduction

Integral membrane proteins come in two basic structural families: helical bundles and β-barrels. In the first group, the membrane-spanning segments of the polypeptide chain form a tightly packed bundle of apolar α-helices oriented more or less perpendicularly to the membrane [1,2]; in the second group, a large number of β-strands form a closed barrel with an apolar, lipid-facing external surface and a more polar interior surface lining a central aqueous pore [3]. Helical bundle proteins are found in the plasma (inner) membrane of both prokaryotic and eukaryotic cells, as well as in the inner membrane of mitochondria and the thylakoid membrane of chloroplasts. β-Barrel proteins seem to be restricted to the outer membrane of Gram-negative bacteria, and may also possibly be found in the outer mitochondrial membrane.

In analogy with globular proteins, the three-dimensional structure of a membrane protein is the result of a folding process dependent on the primary amino acid sequence. For membrane proteins, folding appears to be intimately coupled to the membrane insertion event and it is unlikely that anything even vaguely similar to the final structure will be formed except in a lipid or other non-polar environment. In addition, every membrane protein has a unique topology that defines how it weaves back and forth across the membrane. Thus, topological information must also be encoded within the polypeptide chain.

Most helical bundle proteins seem to utilize one or other of the intra-cellular systems that are normally used to translocate proteins across membranes for their membrane assembly. In this sense, they may be regarded as incompletely translocated proteins, endowed with hydrophobic stop-transfer sequences that get stuck in the translocation machinery and prevent the translocation of downstream parts. Much of what is known about the signals and mechanisms of protein translocation can thus be immediately applied to the problem of membrane protein assembly, although, as will be discussed below, some bacterial proteins apparently do not require a functional translocation machinery for their membrane integration.

The β-barrel proteins of the outer bacterial membrane, on the other hand, must first negotiate the inner membrane without getting stuck, i.e. they must not contain stop-transfer sequences, yet be able to form a sufficiently hydrophobic structure to allow integration into the outer membrane. Their response to these competing requirements has been to evolve the amphipathic β-strand as their basic structural unit; since this is built upon an alternating pattern of polar/apolar residues, its local hydrophobicity is low and allows the chain to be translocated across the inner membrane, and only when the individual strands come together to form the barrel-structure do they reveal their true, globally hydrophobic nature.

In this chapter, our work on the assembly of inner membrane (i.e. helical bundle) proteins in *Escherichia coli* is discussed. Other recent reviews of this subject can be found in, for example, refs. 4 and 5.

2. Results

2.1. Signals and topologies

Most proteins destined for the secretory pathway of both prokaryotic and eukaryotic cells are initially made with a cleavable amino-terminal signal peptide. The primary amino acid sequence of signal peptides is only loosely conserved between different proteins, and all that is required for a functional signal peptide is an N-terminal positively charged segment (n-region), a central hydrophobic stretch (h-region), and a C-terminal cleavage cassette (c-region) that contains the processing site [6] (Fig. 1).

Provided that the nascent protein does not fold or aggregate too rapidly [7], the

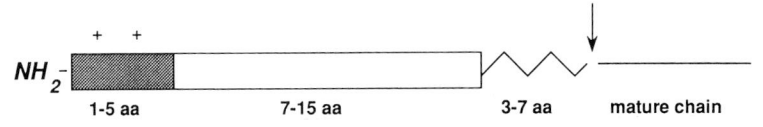

Fig. 1. Basic design of a signal peptide. The n-, h-, and c-regions are indicated.

presence of an N-terminal signal peptide will in most cases be sufficient to ensure translocation across the bacterial inner membrane or the endoplasmic reticular membrane of eukaryotic cells. In bacteria, an additional requirement is that a short region (up to ~30 residues) immediately downstream of the signal peptide must not contain too many positively charged residues [8]; it is at present unclear if there is a similar constraint in eukaryotic cells.

Conceptually, single-spanning membrane proteins can be constructed from a secretory protein in at least three different ways [9,10], all of which can be found in Nature (Fig. 2). Class I proteins have, in addition to their signal peptide, a stop-transfer sequence that will block further translocation and ultimately anchor the protein in the membrane. Almost any string of 15 or more apolar amino acids will have stop-transfer function [10], and the required minimum length seems to be a function of the overall hydrophobicity of the segment [11]. Since the signal peptide is removed from the class I proteins, their final topology will be N_{out}–C_{in}.

Class II proteins can be thought of as secretory proteins with an uncleaved signal peptide, and can indeed be constructed by changing the signal peptide c-region in such a way that it is no longer processed [12]. The uncleaved signal peptide will permanently anchor the protein in the membrane, and the topology will thus be N_{in}–C_{out}.

The class III proteins also lack a cleavable signal peptide, and typically have an N-

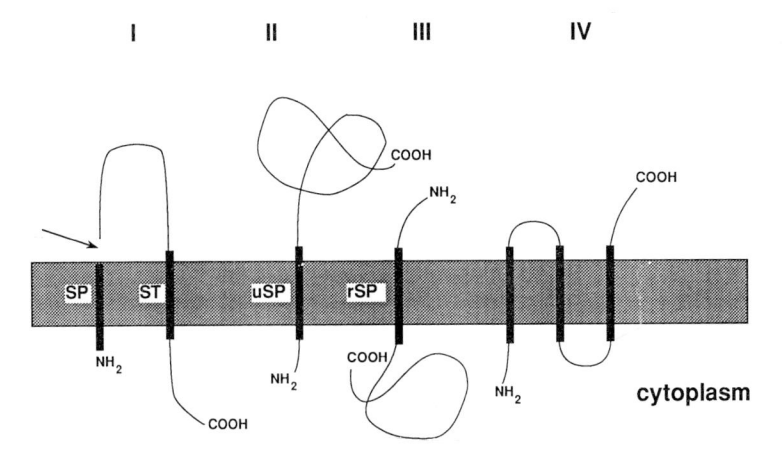

Fig. 2. Classification of membrane proteins according to their assembly signals. SP, signal peptide; ST, stop-transfer sequence; uSP, uncleaved signal peptide; rSP, reverse signal peptide.

terminal apolar transmembrane anchor (or reverse signal peptide [4]) rather close to the N-terminus. In contrast to the class II proteins, they nevertheless have an N_{out}-C_{in} orientation. Class IV proteins, finally, span the membrane multiple times and can, at least in a heuristic sense, be thought of as being constructed from a signal peptide or reverse signal peptide followed by a succession of alternating uncleaved signal peptides and stop-transfer sequences.

2.2. The positive inside-rule

From an early statistical analysis of bacterial inner membrane proteins of known sequence and topology [13], it was concluded that the positively charged amino acids (Lys and Arg) tend to be highly enriched in those regions of the protein that flank the transmembrane segments on their cytoplasmic side, and are found with a much lower frequency in the periplasmic flanking regions. Over a fairly large sample of proteins, the average difference in the frequency of Lys + Arg is about four-fold between the cytoplasmic and periplasmic flanking regions. This positive inside-rule has since been found to hold also for eukaryotic plasma membrane proteins [14], for thylakoid membrane proteins [15], and for proteins of the inner mitochondrial membrane (our unpublished data), although in these cases the difference in Lys + Arg frequency between the cytoplasmic and extra-cytoplasmic flanking regions is more like 2–3-fold. These statistical results do not by themselves prove that positively charged residues can act as topological determinants, but they at least provide a strong clue as to the orientation of any membrane protein and do in fact predict the correct topology for almost all known bacterial inner membrane proteins where experimental data on the topology are available (our unpublished work).

2.3. Positively charged residues control membrane protein topology

Direct demonstrations that positively charged amino acids can act as potent topological determinants are now available from a number of systems, both prokaryotic and eukaryotic. The *E. coli* inner membrane protein leader peptidase (Lep) has been the molecule of choice for many of these studies (Fig. 3a). With its two transmembrane segments, it provides a certain topological complexity that is lacking in the single-spanning proteins. A second useful property of Lep is that it requires the *sec*-machinery for translocation of its large periplasmic P2-domain [16,17], while the first transmembrane segment H1 apparently can insert into the membrane in its correct orientation even in the absence of a functional *sec*-machinery (R. Dalbey, pers. commun.). Lep thus provides some opportunities for studying differences between *sec*-dependent and *sec*-independent translocation events.

One of the first examples where a membrane protein was shown to become reoriented in response to changes in the distribution of positively charged residues around a transmembrane segment was based on a derivative of Lep lacking the

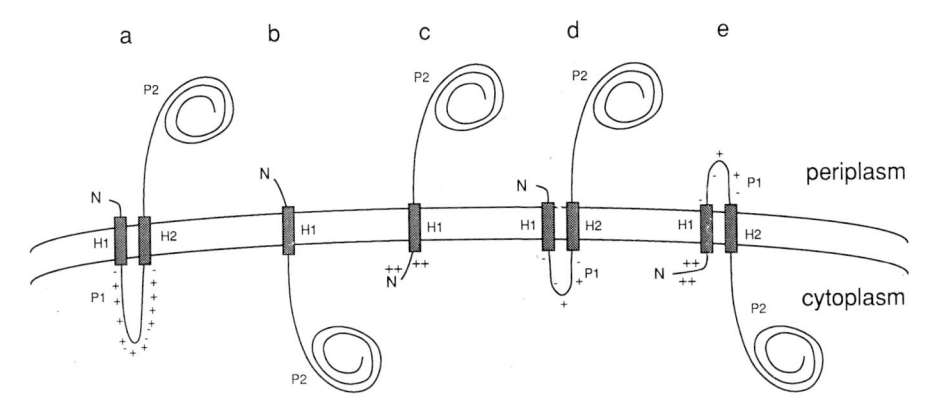

Fig. 3. Leader peptidase constructs with different topologies. (a) Wildtype; (b) class III construct; (c) class II construct; (d) class IV N_{out}–C_{out} construct; (e) class IV N_{in}–C_{in} construct.

second transmembrane segment H2 (Fig. 3b,c). By a series of deletions and insertions, this single spanning class III protein was turned into a class II protein; in essence, positively charged residues had to be added on the N-terminal side of H1 and simultaneously removed from its C-terminal, downstream region to effect the switch [18,19] We have later shown that the class II construct can be further modified to convert its uncleaved signal peptide into a cleavable signal peptide, resulting in the release of the periplasmic domain from the membrane [20]. Membrane insertion of the class II (but probably not the class III) construct is *sec*-dependent, as expected from the pattern of *sec*-dependence for membrane insertion of the H1- and H2-segments of wildtype Lep (c.f. above).

An even more satisfying result was obtained with Lep-derivatives retaining both membrane-spanning regions H1 and H2 [21]. By deleting a region in the P1-domain, eight out of ten of its positively charged residues could be removed while still maintaining the original topology; when subsequently four lysines were added to the N-terminus of H1, a complete reversal of the topology from N_{out}–C_{out} to N_{in}–C_{in} took place (Fig. 3d,e). The new construct further turned out not to require the *sec*-machinery for membrane insertion.

These Lep-derivatives have made it possible to carry out rather detailed studies of the role of charged amino acids in the membrane insertion process. In essence, by addition or removal of a single charged residue at a time, one can perform topological titrations and study quantitatively how the molecule changes from one orientation to the other as a function of the number of charged residues in P1 and on the N-terminal side of H1. For example, addition of one N-terminal lysine to the original construct with two lysines in the P1-domain results in a mixed topology: about half of the molecules insert in the wildtype orientation, while the other half have the inverted topology [22]; with the addition of a second N-terminal lysine, all molecules insert N_{in}–C_{in} (i.e. in the inverted topology). When the same titration is carried out on a construct with three lysines in the P1-domain, three N-terminal lysines are required to force all molecules to insert N_{in}–C_{in}.

In a similar series of experiments, we have recently shown that arginines and lysines are nearly equally potent topological determinants [23], and that negatively charged residues (Asp and Glu) are much weaker than Lys and Arg in their topological effects. Nevertheless, it is interesting to note that both Asp and Glu can promote an inversion of the topology, albeit only when they are present in high numbers (at least four) on the N-terminus of H1. This observation speaks against a simple electrophoretic model, where the electrical transmembrane potential (negative inside) would be responsible for orienting the protein.

Much less work has been done on the topology of eukaryotic membrane proteins, but it has been reported that eukaryotic class II proteins can at least partly be turned into class III proteins (and vice versa) by alterations in the number of positively charged residues in the regions flanking the membrane anchor segment [24–28].

2.4. A membrane protein with pH-dependent topology

In our studies of Lep-derivatives, we have also tested the effect of histidines. At a normal cytoplasmic pH (~ 7.5), His residues carry only a very small positive net charge ($+0.03$) and consequently have little if any effect on the topology when placed on the N-terminal side of H1 in a construct with the wildtype N_{out}–C_{out} topology [23]. However, when the intracellular pH is dropped by about one pH unit by incubation in an appropriate low pH buffer (thus charging up each His to about $+0.2$), a mutant with four N-terminal histidines suddenly becomes largely re-oriented ($\sim 60\%$ N_{in}–C_{in}). This not only underlines the importance of positive charge for controlling topology, but also shows that it is possible to design membrane proteins with a topology that can be manipulated rather easily by changes in, for example, the growth medium.

2.5. Position-specific charge-pairing can affect the topology

Although negatively charged residues by themselves have very little effect on the topology of our model Lep constructs, we have found that they can in certain cases neutralize the topological effects of a nearby positively charged residue. Thus, when an Asp residue was placed at different distances relative to a lysine already present at the N-terminus of H1, we found that the percentage of wildtype oriented (as opposed to inverted) molecules increased significantly when the Asp was either next to the Lys, or even more markedly when it was four residues upstream, but not when it was two or three residues distant in the sequence (our unpublished data). A possible interpretation of these results is that the entire H1-region, including the N-terminal tail, inserts into the membrane as an α-helix, thus making it possible for the Asp and Lys residues to form a salt-bridge when they are present on the same side of the helix but not otherwise. Apparently, such a charge-pair is easier to move through the membrane than an isolated, positively charged residue.

2.6. Sec-dependent versus sec-independent assembly

Our studies on the topology of Lep-derived constructs, as well as other similar studies on other proteins [4,5,10], provide strong support for the general validity of the positive inside-rule. However, the mechanism(s) by which positively charged residues control the topology remains elusive. Obvious possibilities range from electrophoretic effects (the membrane potential), energetic costs of neutralizing a charged residue by protonation/de-protonation before transfer through a membrane, interactions between positively charged residues with acidic phospholipids, and differential effects of the membrane dipole field on positively and negatively charged amino acids, to sensor proteins (SecA?) that might interact more or less strongly with a potential apolar translocation signal depending on the number of positively charged residues in its N- and C-terminal flanking regions.

Although still a far cry from a detailed understanding of these factors, we have at least been able to begin separating the *sec*-dependent and *sec*-independent insertion events that result in the membrane insertion of various Lep-derived model proteins. As mentioned above, in wildtype Lep, H1 seems not to require the *sec*-machinery for insertion, whereas H2 can only act as an uncleaved signal peptide for the translocation of the P2-domain in the presence of a functional *sec*-pathway. On the other hand, H1 is not an intrinsically *sec*-independent signal, since when re-designed into a (cleaved or uncleaved) signal peptide in mutants lacking H2, translocation of P2 is still *sec*-dependent (see above). Finally, in the inverted N_{in}–C_{in} constructs, the periplasmically exposed loop derived from the P1-domain can be translocated independently of a functional *sec*-pathway. These results clearly indicate that it is not the signal peptide itself but rather the properties of the translocated region that determines whether the *sec*-machinery will be required; one attractive possibility is that the length of this region plays an important role in this regard [13,14,29].

Statistical analysis of translocated regions in bacterial inner membrane proteins has shown that the under-representation of Lys and Arg residues noted above is only observed for loops or tails shorter than about 60–80 residues [14]. Longer periplasmic loops have more normal frequencies of these residues, similar to what is found for soluble periplasmic proteins. This might indicate that longer segments can accommodate a higher number of positively charged residues because they can use the *sec*-machinery for translocation, whereas shorter segments have to penetrate the membrane unaided. In agreement with this idea, a lengthening of the translocated segment of the phage M13 coat protein from some 20 up to around 100 residues converted the protein from being *sec*-independent to becoming *sec*-dependent [29]. We have taken a more step-wise approach, and have successively lengthened the translocated loop of a *sec*-independent inverted Lep-construct in steps of about 10 residues; so far, we have found it to become increasingly *sec*-dependent up to a loop length of about 50 residues (our unpublished data).

Another aspect of *sec*-dependent/independent processes is whether the topological effects of charged residues are the same or different in the two contexts. Again, we have used our panel of Lep-derived constructs to begin exploring this question. When H2 is deleted from a construct lacking positively charged residues on its N-terminus (thus with essentially 100% N_{out}-C_{out} topology), we find that the P2-domain now becomes translocated in an appreciable fraction ($\sim 30\%$) of the resulting molecules; when one lysine is added to the N-terminus of H1, the P2-domain is quantitatively translocated (i.e. H1 is oriented N_{in}) in a *sec*-dependent manner, although only about 50% of the molecules have this orientation when H2 is present and H1 integration is *sec*-independent (our unpublished data).

3. Discussion

It is by now clear that positively charged residues (Lys and Arg) can act very efficiently as topological determinants, at least in *E. coli*. Negatively charged residues have little effect on the topology, except possibly when placed in a position such that they can form salt-bridges with a nearby positively charged residue. The positive inside-rule thus seems to be a useful generalization for describing the topology of the helical bundle integral membrane proteins.

Organelles such as chloroplasts and mitochondria are thought to be descendants of once free-living prokaryotes; thus one would expect organellar membrane proteins to behave much as bacterial proteins. Indeed, the same positive inside-rule that was first discovered in bacterial inner membrane proteins is also statistically valid for thylakoid membrane proteins and mitochondrially encoded proteins of the inner mitochondrial membrane. In these cases, the more highly charged regions are oriented towards the stroma (thylakoids) or matrix (mitochondria), i.e. they are not translocated across the respective membrane.

Eukaryotic plasma membrane proteins are also characterized by a skewed distribution of positively charged residues, and some experimental support for their importance as topological determinants exist; yet, given that protein translocation across the ER membrane is strictly co-translational, in contrast to the situation in *E. coli*, it would seem that the most N-terminal membrane integration signals would be more important than more C-terminal ones [30], and that membrane insertion would progress in an N-terminal to C-terminal direction. In *E. coli*, on the other hand, the available data seem to suggest that membrane insertion is a local process, where individual apolar helices or helical hairpins [31] composed of two adjacent apolar helices insert into the membrane more or less independently of their neighbors. Thus, when either the first, second, or third of the eight transmembrane segments in the MalF protein is deleted, most of the remaining transmembrane segments apparently insert in the same orientation as in the wildtype protein, except possibly the apolar segment next to the deleted one, which, in the absence of its helical hairpin partner, may not insert at all [32,33].

Future studies should be able to shed more light on how such complicated, multi-spanning proteins insert in both prokaryotic and eukaryotic cells, what role(s) are played by the bacterial *sec*-machinery and the components of the eukaryotic SRP-pathway [34,35], and whether organellar membrane protein biogenesis is a carbon-copy of the bacterial system. Beyond the primary insertion event and its attendant determination of the transmembrane topology, one would also like to learn more about how the transmembrane α-helices pack together in the apolar milieu of the membrane, and ultimately to arrive at a full understanding of how the amino acid sequences of membrane proteins dictate their final three-dimensional structures.

Acknowledgement

This work was supported by grants from the Swedish Natural Sciences Research Council and the Swedish Board for Technical Development.

References

1. Deisenhofer, J., Epp, O., Miki, K., Huber, R. and Michel, H. (1985) Nature 318, 618–624.
2. Rees, D.C., Komiya, H., Yeates, T.O., Allen, J.P. and Feher, G. (1989) Annu. Rev. Biochem. 58, 607–633.
3. Weiss, M.S., Kreusch, A., Schiltz, E., Nestel, U., Welte, W., Weckesser, J. and Schulz, G.E. (1991) FEBS Lett. 280, 379–382.
4. Dalbey, R.E. (1990) Trends Biochem. Sci. 15, 253–257.
5. Boyd, D. and Beckwith, J. (1990) Cell 62, 1031–1033.
6. von Heijne, G. (1990) J. Membr. Biol. 115, 195–201.
7. Hardy, S.J.S. and Randall, L.L. (1991) Science 251, 439–443.
8. Andersson, H. and von Heijne, G. (1991) Proc. Natl. Acad. Sci. USA, in press.
9. von Heijne, G. (1988) Biochim. Biophys. Acta 947, 307–333.
10. von Heijne, G. and Manoil, C. (1990) Protein Eng. 4, 109–112.
11. Kuroiwa, T., Sakaguchi, M., Mihara, K. and Omura, T. (1991) J. Biol. Chem. 266, 9251–9255.
12. Shaw, A.S., Rottier, P.J. and Rose, J.K. (1988) Proc. Natl. Acad. Sci. USA 85, 7592–7596.
13. von Heijne, G. (1986) EMBO J. 5, 3021–3027.
14. von Heijne, G. and Gavel, Y. (1988) Eur. J. Biochem. 174, 671–678.
15. Gavel, Y., Steppuhn, J., Herrmann, R. and von Heijne, G. (1991) FEBS Lett. 282, 41–46.
16. Wolfe, P.B. and Wickner, W. (1984) Cell 36, 1067–1072.
17. Wolfe, P.B., Rice, M. and Wickner, W. (1985) J. Biol. Chem. 260, 1836–1841.
18. Laws, J.K. and Dalbey, R.E. (1989) EMBO J. 8, 2095–2099.
19. von Heijne, G., Wickner, W. and Dalbey, R.E. (1988) Proc. Natl. Acad. Sci. USA 85, 3363–3366.
20. Nilsson, I.M. and von Heijne, G. (1991) J. Biol. Chem. 266, 3408–3410.
21. von Heijne, G. (1989) Nature 341, 456–458.
22. Nilsson, I.M. and von Heijne, G. (1990) Cell 62, 1135–1141.

23. Andersson, H., Bakker, E. and von Heijne, G. (1992) J. Biol. Chem. 267, 1491–1495.
24. Haeuptle, M.T., Flint, N., Gough, N.M. and Dobberstein, B. (1989) J. Cell Biol. 108, 1227–1236.
25. Parks, G.D., Hull, J.D. and Lamb, R.A. (1989) J. Cell Biol. 109, 2023–2032.
26. Parks, G.D. and Lamb, R.A. (1991) Cell 64, 777–787.
27. Sato, T., Sakaguchi, M., Mihara, K. and Omura, T. (1990) EMBO J. 9, 2391–2397.
28. Szczesna-Skorupa, E. and Kemper, B. (1989) J. Cell Biol. 108, 1237–1243.
29. Kuhn, A. (1988) Eur. J. Biochem. 177, 267–271.
30. Hartmann, E., Rapoport, T.A. and Lodish, H.F. (1989) Proc. Natl. Acad. Sci. USA 86, 5786–5790.
31. Engelman, D.M. and Steitz, T.A. (1981) Cell 23, 411–422.
32. Ehrmann, M. and Beckwith, J. (1991) J. Biol. Chem. 266, 16530–16533.
33. McGovern, K., Ehrmann, M. and Beckwith, J. (1991) EMBO J. 10, 2773–2782.
34. High, S., Flint, N. and Dobberstein, B. (1991) J. Cell Biol. 113, 25–34.
35. High, S., Gorlich, D., Wiedmann, M., Rapoport, T.A. and Dobberstein, B. (1991) J. Cell Biol. 113, 35–44.

CHAPTER 8

Lipid involvement in protein translocation

B. DE KRUIJFF, E. BREUKINK, R.A. DEMEL, R. VAN 'T HOF,
H.H.J. DE JONGH, W. JORDI, R.C.A. KELLER, J.A. KILLIAN,
A.I.P.M. DE KROON, R. KUSTERS and M. PILON

*Department of Biomembranes, Centre for Biomembranes and Lipid Enzymology,
University of Utrecht, Padualaan 8, 3584 CH Utrecht, The Netherlands*

1. Introduction

In membrane biogenesis and protein targeting, polypeptide chains very often insert into and move across biological membranes. Correct membrane assembly and functioning critically depends on the spatial and temporal interactions between the two main membrane components, i.e. proteins and lipids. It can therefore a priori be reasoned that protein–lipid interactions must be involved in membrane insertion and translocation of newly synthesized proteins. Such an involvement will be of a general nature such as to correctly assemble a putative proteinaceous insertion/translocation device in a membrane (e.g. a proteinaceous tunnel is depicted in Fig. 1) and to provide and maintain the essential barrier function of the membrane during the translocation process. In addition, membrane lipids could play more direct roles to provide alone or in combination with proteins an insertion and trans-

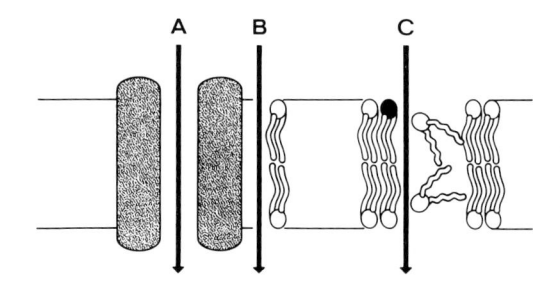

Fig. 1. Some ways in which lipids might be involved in protein translocation (A). A proteinaceous tunnel is integrated in a lipid bilayer. (B) The interface between proteins and lipids provides a translocation pathway. (C) Specific lipid classes (symbolized by the dark headgroup) or structures (non-bilayer) mediate membrane insertion and translocation. Combinations between the different pathways provide additional possibilities.

location pathway (for example see the types shown in Fig. 1). The specific, dynamical and complex lipid composition of biomembranes together with the unique structural and motional properties of individual membrane lipid classes offer fascinating possibilities for protein–lipid interactions to be involved in binding, insertion, translocation and release of proteins in transit across membranes. The approach we have chosen to analyze these possibilities is to study selected protein transport pathways in the main cellular protein trafficking routes via a combination of biochemical and biophysical techniques using both model and biological membranes. This review summarizes our current insights into the involvement of lipids in protein translocation, in protein secretion of prokaryotes and in protein import in mitochondria and chloroplasts.

2. Results and discussion

2.1. Prokaryotic protein secretion

Our studies in this area have concentrated on translocation of the outer membrane pore protein PhoE across the *Escherichia coli* inner membrane (for a recent review see [1]). This protein follows a route shared by many periplasmic and outer membrane proteins. The protein is synthesized in the cytosol as a precursor (prePhoE) carrying a signal sequence. During or shortly after synthesis the protein is translocated across the inner membrane. The signal sequence and the Sec proteins A, B, D, E, F and Y [2] are essential for membrane translocation, which is driven by ATP and facilitated by the proton motive force [3]. During or shortly after translocation, the enzyme leader peptidase removes the signal sequence and the mature protein is finally assembled as a functional trimeric pore in the outer membrane. The first direct proof for the involvement of a particular class of membrane lipids in protein translocation came from studies [4] on PhoE using an *E. coli* mutant (strain HD3122) with defects in the biogenesis of the main anionic lipid phosphatidylglycerol (PG) and the anionic cardiolipin (CL) due to disruption of the *pgsA* gene (Fig. 2). Such cells have a very low PG and CL content, but are still viable in the proper genetic background. Inner membrane translocation of prePhoE was decreased both in vivo and, even more pronounced, in vitro using inverted inner membrane vesicles as acceptor for the in vitro synthesized protein. From these and other experiments [5], it could be concluded that PG is essential for efficient inner membrane translocation of different *E. coli* outer membrane proteins.

Recently, it became possible to regulate the PG content of *E. coli* in a more systematic way by the use of cells carrying one copy of the *pgsA* gene, which is under the regulation of the Lac operon, such that synthesis of the enzyme and thereby synthesis of PG can be induced by growing the cells (strain HDL11) in the presence of isopropylthiogalactoside (IPTG) [6]. In the absence of IPTG in the growth medium,

```
              G-3-P   CMP        Pi    PG   GLYCEROL
                ↘  ↗         ↘  ↗
CDP-DIGLYCERIDE ────→ PGP ─→ PG ──────→ CL
                pgsA                cls
```

	PE	PG	CL	PA	Rest
WT (SD12)	85.4	10.9	3.4	–	0.3
PG⁻ (HD3122)	95.3	1.8	0.5	1.9	0.5

Fig. 2. The last steps in the biosynthesis of the main acidic phospholipids in *E. coli* and the lipid composition (mol%) of wild-type (WT) and PG-depleted (PG⁻) cells [65].

the PG content is very low as well as prePhoE translocation efficiency assayed using purified precursor and isolated inner membrane vesicles (Fig. 3). Increasing the PG level by increasing the IPTG concentration results in an increase in translocation efficiency that is proportional to the PG content. The direct involvement of PG in the translocation process could be unambiguously demonstrated by reintroducing PG into PG-depleted inner membrane vesicles using a lipid transfer protein based method [6]. Introduction of PG but not of the overall neutral zwitterionic PC restored prePhoE translocation (Fig. 4). The negative charge on the PG molecule appears to be the essential factor as a large variety of both natural and artificial anionic lipids restored prePhoE translocation in this system [6].

What is the molecular basis for the involvement of PG in the translocation process? Wickner's group came up with the original suggestion [5] that the SecA protein requires PG for correct functioning in the translocation process. This cytosolic and

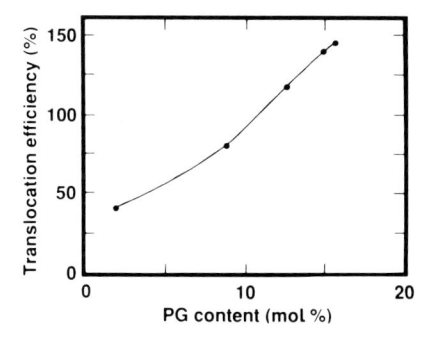

Fig. 3. prePhoE translocation depends on the PG content of the inner membrane. Translocation efficiency, expressed as the percentage of protein translocated across wild-type SD12 inner membrane vesicles is presented as a function of the PG content of inner membranes of strain HDL11 grown at different concentrations of IPTG (see [6] for details).

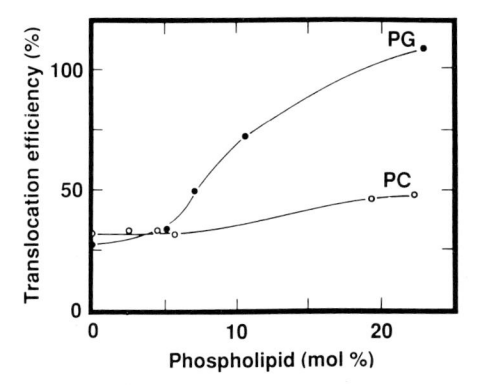

Fig. 4. Lipid transfer protein mediated incorporation of PG, but not of PC stimulates translocation of prePhoE across inner membranes of HDL11 cells grown in the absence of IPTG. Inner membrane vesicles with a composition similar to strain HD3122 (Fig. 2) were enriched in PG or PC via incubation with lipid donor vesicles and a lipid transfer protein. The amount of phospholipid introduced is given as a mol% of the total phospholipid. Translocation efficiency is presented relative to the amount of translocated protein across wild-type (SD12) vesicles (see [6] for details).

peripherally membrane associated ATPase plays an important role in translocation because it interacts with both the chaperonin SecB [7] and the integral protein SecY [8] as well as with the precursor. It probably uses ATP to initiate membrane translocation of precursor proteins [9]. Negatively charged lipids specifically stimulate the SecA ATPase activity [5] and trigger a conformational change resulting in an enhanced susceptibility towards a protease [10]. We recently demonstrated that SecA can efficiently penetrate into a lipid monolayer but only when negatively charged lipids are present [11], the surface pressure increase being much smaller in the case of

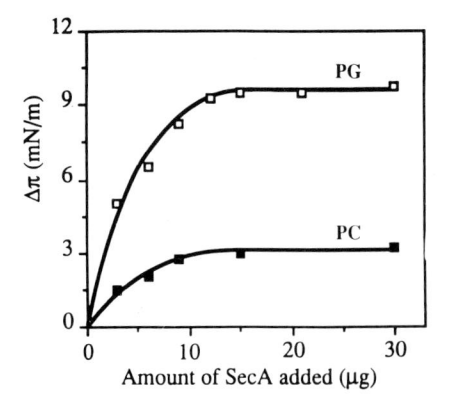

Fig. 5. Increase in surface pressure ($\Delta\pi$) of monolayers of dioleoyl phosphatidylcholine (PC) and dioleoylphosphatidyl glycerol (PG) mediated by SecA injection in the subphase (5 ml 100 mM NaCl, 50 mM Tris HCl, pH 7.6, temperature 31°C). Initial surface pressure 20 mN/m (see [11] for details).

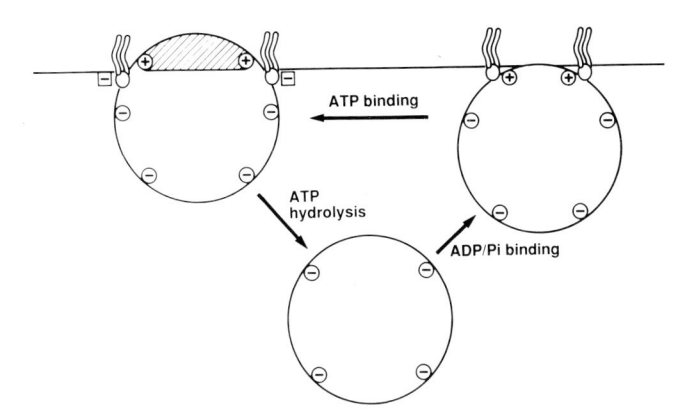

Fig. 6. Modulation of SecA-lipid interaction by nucleotides. The model is based on data presented by Breukink et al. [11], in which binding and penetration of SecA was measured in the presence of different nucleotides. Upon ATP binding, the overall anionic secA is proposed to expose a positively charged domain which preferentially interacts with anionic phospholipids resulting in membrane penetration. Upon ATP hydrolysis, the protein undergoes a conformational change resulting in a decreased affinity for the interface. In the presence of ADP and P_i the SecA has an efficient binding with the lipids but penetration is restricted. Upon ATP binding, the SecA again inserts deeper in the lipid layer.

zwitterionic lipids such as PC (Fig. 5). Efficient penetration only occurs for inner membrane lipid extracts isolated from wild-type cells and not for lipid extracts isolated from HDL11 cells, consistent with the difference in translocation efficiency of these membranes [6]. From the modulation of penetration of SecA in between anionic lipids by various nucleotides [11], a cyclic model of SecA membrane penetration can be proposed (Fig. 6).

In addition to the SecA protein, there are many more candidates for functional interaction with PG in the translocation process [12–14]. In early translocation models, an interaction of the signal sequence with the membrane lipids was proposed as an obligatory step in the translocation process. This view originated from considerations on the overall chemical structure of the signal sequence with its positive N-terminus and hydrophobic central region and was substantiated by studies on synthetic peptides corresponding to the signal sequence of precursor proteins (signal peptides). These studies revealed that such peptides have a high affinity for lipids, and anionic lipids in particular (for review see [15]).

The consequences of this interaction for the structure of the peptide–lipid complex have been extensively studied using different biophysical techniques. The results immediately pointed to possibilities on how these interactions could play a role in the translocation process. One example of such studies is illustrated in Fig. 7. This figure shows that injection of the signal peptide of the prePhoE protein underneath a mixed monolayer of PC and PG at the air–water interface results in a large increase in surface pressure above a certain threshold amount of PG [16]. This result can be interpreted in terms of a penetration of the signal peptide between the lipids mediated

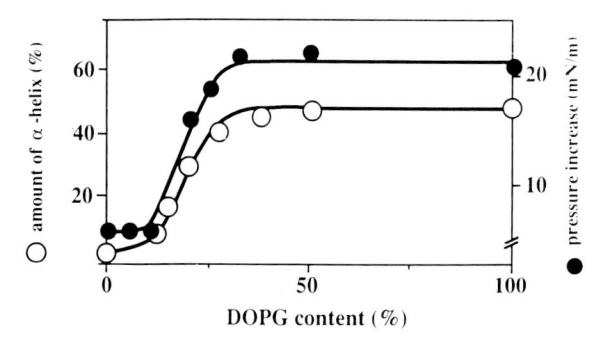

Fig. 7. Penetration and α-helix formation of a signal peptide in a lipid–water interface requires a threshold amount of anionic phospholipids. The synthetic signal peptide of prePhoE (MKKSTLALVVMGIVA-SASVQA) was injected at room temperature underneath a lipid monolayer (initial surface pressure 20 mN/m) or was added to a solution of small unilamellar vesicles (9 mM lipid), whereafter the increase in surface pressure or the amount of α-helix formation was determined via the Wilhelmy plate method or circular dichroism, respectively (for additional details see [16,18]).

by electrostatic interactions between the lipid head group and the N-terminus of the signal peptide. This conclusion is supported by similar studies on other signal peptides and on fluorescence studies on a Trp-containing PhoE signal peptide analog interacting with lipid vesicles [17]. This penetration, which is strongly decreased under conditions of tight lipid packing, is apparently accompanied by induction of α-helix formation as analyzed by CD in a lipid vesicle system (Fig. 7). Again α-helix formation is observed only above a certain threshold value of PG [18].

The ability of a signal peptide to adopt an α-helical configuration in the presence of lipids has been correlated with the functionality of the signal peptide in the translocation process [19]. The idea that membrane penetration and α-helix formation in the lipid phase of the membrane is functionally important is further suggested from the observations that (i) efficient penetration is only observed for lipid extracts of inner membranes from wild-type *E. coli* and not from PG$^-$ mutant strains [16,17]; (ii) that penetration is impaired for a signal peptide which corresponds to a less functional signal sequence of prePhoE in which the two lysines are replaced by aspartic acid residues [16]; and (iii) that signal peptide insertion [20] and protein translocation [21–24] is greatly decreased when the membrane lipids are in the gel state. The implications of these results are that the signal sequence adopts a transmembrane α-helical orientation in the target membrane such that the N-terminus is anchored at the cytoplasmic side by interactions with PG and the C-terminus and thereby the first part of the mature protein is translocated across the membrane, features proposed in several translocation models. Theoretical and monolayer studies [25] on the prePhoE signal peptide lead to the proposal that the signal peptide initially inserts in a looped conformation. Unlooping of the signal sequence would facilitate the movement of the C-terminus of the signal sequence across the membrane [1,25].

Recently a novel property of signal peptides was described [26]. These peptides when added to bilayers made of the total lipid extract of *E. coli* were found to be profound inducers of type II non-bilayer lipid structures. The potency to disrupt bilayer structure appeared to be related to the functionality of the signal sequence [26]. This observation resulted in the logical suggestion of a signal sequence induced local loss of bilayer structure thereby facilitating passage of (part of) the mature region, consistent with some early hypotheses on the mechanism of protein translocation [27,28].

The challenge in the near future will be to test these ideas in the functional process and to analyze at which steps and at which sites the membrane lipids are involved in the translocation process. A central question to be answered is whether the signal sequence does interact with the membrane lipids during or prior to translocation. In view of the similarities in the pro- and eukaryotic protein secretion pathways, it can be anticipated that at the level of the endoplasmic reticulum membrane, protein–lipid interactions will also be important for protein translocation.

2.2. Mitochondrial protein import

Most mitochondrial proteins are encoded by nuclear genes and are synthesized as precursor proteins in the cytosol. Their import into mitochondria constitutes one of the main cellular protein trafficking routes and involves specific organellar targeting, membrane translocation and intramitochondrial sorting (Chapter 21). In the general import pathway, N-terminal temporal sequences play an essential role. These presequences share a basic and amphipathic character with signal sequences but differ from them by the more uniform distribution of the positive charges and their relative high content in hydroxylated amino acids (for review see [29]. The periodicity in the location of the basic amino acids and their ability to adopt α-helical structures in a membrane mimicking environment [30–33] suggested that an amphipathic helical structure of the presequence is important in the import process. Together with other studies on peptides corresponding to such presequences, this suggested that interactions between presequences and anionic lipids are involved in membrane binding and insertion of the precursor [29]. Studies using precursor proteins pointed to a specific role of the unique mitochondrial lipid cardiolipin in the import process [34–36].

The membrane potential ($\Delta\Psi$) across the mitochondrial inner membrane is essential to initiate membrane translocation and moves the presequence into the matrix where it is removed by a specific protease [37–39]. One attractive possibility given by the direction of the potential and the basic character of the presequence is that $\Delta\Psi$ can exert a direct effect on the presequence and facilitates its insertion into the lipid phase of the membrane. We recently investigated this possibility using a synthetic peptide corresponding to the presequence of yeast cytochrome *c* oxidase subunit IV in which leu-18 was replaced by tryptophan in order to make use of the fluorescence properties of this amino acid [40]. The primary structure of preCOXIV 1-25 W^{18} is

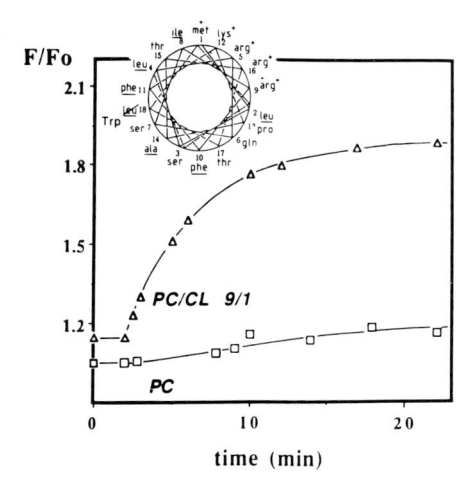

Fig. 8. Cardiolipin renders a mitochondrial presequence susceptible to a membrane potential across a lipid bilayer. PreCOX IV 1-25 W^{18} (2 μM) was added to large unilamellar vesicles (300 μM lipid) experiencing a K^+_{in}/Na^+_{out} gradient. The tryptophan fluorescence change at 340 nm was followed in time. At $t = 2$ min valinomycin was added to generate a membrane potential (negative inside). The fluorescence measured (F) is related to F_0, the fluorescence intensity in the absence of vesicles. The 9:1 ratio of PC/CL reflects the ratio of lipid phosphorus (see [40] for details).

given in Fig. 8 in the form of an α-helical wheel projection emphasizing the amphipathic character of the helix. It is known that this peptide adopts an α-helical structure upon interaction with lipids [29]. When a membrane potential (negative inside) is generated across the bilayer of large unilamellar phosphatidylcholine vesicles only a minor increase in tryptophan fluorescence intensity of the peptide is observed. Incorporation of a low amount of CL in the bilayer results in an increase in spontaneous peptide–vesicle interaction as evidenced by the increase in initial tryptophan fluorescence intensity, thus emphasizing the electrostatic contribution to the presequence–lipid interaction. Most interestingly, application of a membrane potential in this case causes a very large increase in fluorescence, demonstrating that the peptide is now entering a more hydrophobic environment, i.e. it is pulled into the bilayer. ΔpH does not contribute to this effect [40] suggesting that the membrane potential exerts its effect directly on the presequence, fully consistent with very recent observations on the role of ΔΨ in mitochondrial protein import [41]. These experiments suggest that interactions between the presequence and negatively charged lipids render the peptide susceptible to a ΔΨ driven membrane insertion. These observations, together with expectations [42] concerning the lipid structure modulating activity of presequences suggest a direct involvement of presequence–lipid interactions in the import process.

Much insight into the way lipids can actively participate in precursor protein translocation has come from studies on the import of cytochrome c into mitochondria. This intermembrane space localized protein follows a unique pathway. The

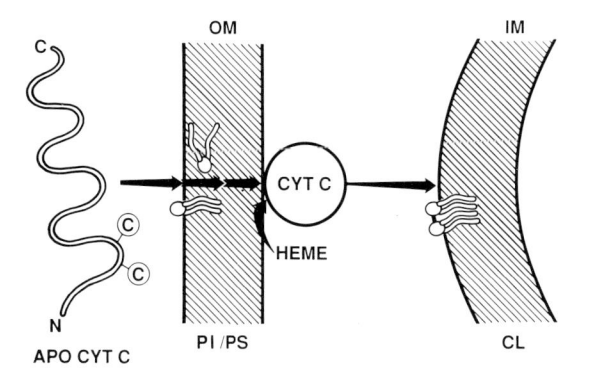

Fig. 9. The current model for lipid involvement in apocytochrome c import into mitochondria. The unfolded highly basic precursor binds preferentially to the anionic lipids [44] in the outer membrane. The protein inserts [48] into the membrane, thereby causing a change in lipid packing/organization [52]. The deep insertion of the N-terminus exposes the cysteines 14 and 17 to the heme lyase which has its active center in the intermembrane space. Enzymatic coupling of the heme results in formation of the holoprotein which is released in a folded state from the inside of the outer membrane to the intermembrane space. Interactions with cardiolipin in the inner membrane might finally target the protein to its functional location.

precursor (apocytochrome c) does not contain a cleavable presequence and its import does not require a proton motive force or ATP (for review see [43,44]. Covalent coupling of the heme group by the enzyme heme lyase is an essential step in import because it couples translocation to holoprotein formation. The import process which is schematically represented in Fig. 9 can be divided into several steps. In vitro studies showed that the precursor binds with high affinity ($K_d \approx \mu M$) to mitochondria and model systems containing anionic phospholipids [43,44]. No proteinaceous receptors on the outside of the mitochondria involved in apocytochrome c uptake could be detected [45] using methods which identified other import receptors [46]. This suggests that binding of the highly basic precursor to anionic lipids in the mitochondrial outer membrane is a first step in the import process [47]. Lipid vesicles made of the negatively charged phosphatidylserine (PS) indeed compete efficiently for uptake of in vitro synthesized precursor into Neurospora crassa mitochondria (Fig. 10). No such competition is observed for PC vesicles. Biophysical studies demonstrated that the precursor is able to recruit anionic lipids in bilayers in order to preferentially associate with them [44]. The next step in the process is membrane insertion. Both biochemical experiments on the apocytochrome c import process and biophysical studies on the precursor–lipid systems demonstrated that the protein moves into a more hydrophobic environment and penetrates the membrane [43,44,48].

In bilayer vesicles containing negatively charged lipids, the precursor is able to spontaneously reach the opposite membrane–water interface, a remarkable property of this polypeptide [49,50]. This can for instance be derived from experiments in which digestion of the in vitro synthesized protein at the inside of the lipid vesicles is

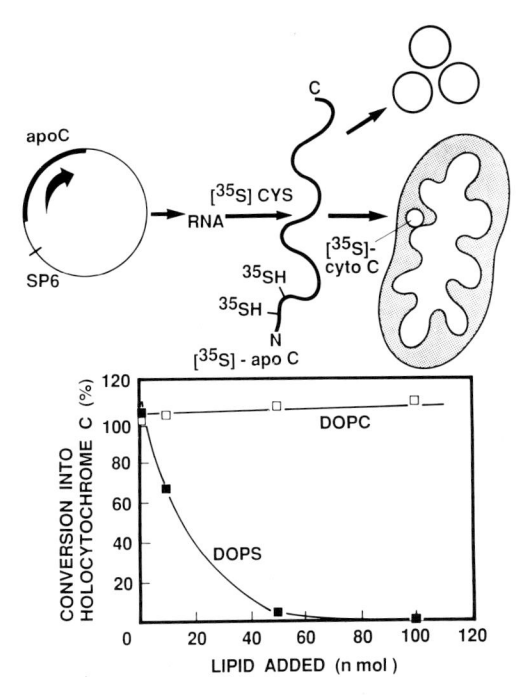

Fig. 10. Effect of increasing amounts of dioleoylphosphatidylserine (DOPS, ■) and dioleoylphosphatidylcholine (DOPC, □) liposomes on the conversion of in vitro synthesized apocytochrome c into holocytochrome. Transcription of *Neurospora crassa* apocytochrome c cDNA cloned into pGEM3 was carried out as previously described [66]. Apocytochrome c was synthesized in a reticulocyte lysate in the presence of [^{35}S]cysteine. Import of the protein into mitochondria in the absence and presence of liposomes was determined by measuring holoprotein formation [66] (see [67] for details).

monitored (Fig. 11). A precursor following the general pathway (pCOX IV-DHFR, a fusion protein containing the presequence of yeast cytochrome c oxidase subunit IV and mouse dihydrofolate reductase) [51] does not show this behavior, despite the high affinity of the presequence for anionic lipids. The insertion of apocytochrome c into the bilayer results in a dramatic change in lipid organization as illustrated by ^{2}H NMR in Fig. 12. Addition of the precursor (and not the holoprotein) causes a large reduction in quadrupolar splitting of [11,11-^{2}H$_2$]-labelled dioleoylphosphatidylserine bilayer [52]. Such an effect can be anticipated in view of the overall polar character of the entire protein and its bilayer penetrating capacity.

Covalent attachment of the heme blocks deep insertion into the membrane lipid phase [52] and thus makes the translocation process irreversible. Studies with apocytochrome c fragments demonstrated that the ability to move across a lipid bilayer is confined to the N-terminal part of the protein [53]. Translated to the import process, this implies that the N-terminus can spontaneously move across the outer membrane and exposes its two cysteines to the heme lyase, which has its active center in the intermembrane space. A deletion analysis of apocytochrome c suggested that

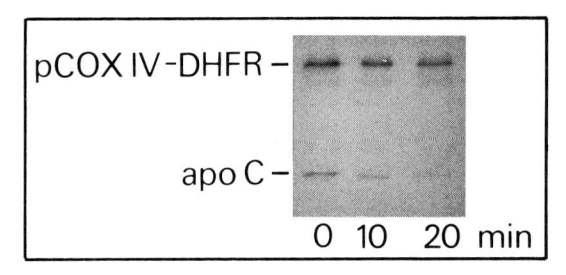

Fig. 11. Digestion of in vitro synthesized pCOX IV-DHFR and apocytochrome c by trypsin enclosed in vesicles composed of an outer mitochondrial membrane lipid mixture. A mixture of in vitro synthesized pCOX IV-DHFR and *Neurospora crassa* apocytochrome c was added to large unilamellar vesicles composed of an outer membrane lipid mixture with entrapped trypsin and incubated for various times. Outside protease activity was fully inhibited. The samples were analyzed on a polyacrylamide gel and visualized by autoradiography (see [67] for details).

the entire polypeptide is required for import [54]. Because also all parts of the polypeptide also strongly and differentially interact with lipids, this suggests that the specific structure of the apocytochrome c–lipid complex is essential for membrane translocation.

It is commonly believed that in the main import route, proteins move across the membrane in an **unfolded** state. This state is mediated by chaperonins and probably requires ATP hydrolysis. The observation that cardiolipin partially unfolds a presequence-containing precursor [36] suggests that lipid–protein interactions are also involved. Surprisingly apocytochrome c **folds**, i.e. undergoes a transition from a random coil to a partially α-helical conformation upon interaction with anionic lipids [55]. Most interestingly, the induced α-helices are localized in the same region

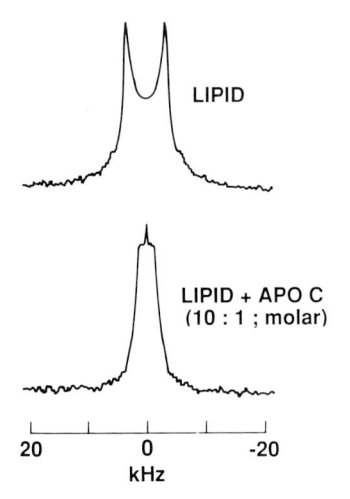

Fig. 12. ^{2}H NMR spectra (46.1 MHz) at 30°C of 1,2-[11,11-^{2}H$_2$]-dioleoylphosphatidylserine dispersions in the absence or presence of apocytochrome c (see [52] for details).

of the polypeptide as in the folded holoprotein [53,56]. This immediately suggests that the interaction with lipids prefolds the polypeptide towards the holoprotein.

Photo-CIDNP NMR experiments localized the aromatic amino acids, which are evenly distributed over the precursor, in the water–micelle interface [57]. This implies that the α-helices including the N-terminal one which encompasses cysteine 14 and 17 are present in the interface. Recent ^{2}H–^{1}H exchange experiments unambiguously demonstrated the extreme dynamics in the helices of the lipid associated precursor [58]. The rate of amide ^{2}H→^{1}H exchange of the unfolded amide deuterated precursor is extremely rapid as expected for a non-structured protein (Fig. 13). However, also in the micelle-associated form, 90% of the ^{2}H is exchanged within 1 min at 4°C, despite the fact that 42% of the polypeptide is estimated (from circular dichroism measurements) to be in an α-helical conformation [58]. In contrast, the ^{2}H→^{1}H exchange of the amide deuterons present in the interior of the tightly folded cytochrome c is extremely slow as expected (Fig. 13). Surprisingly, in the presence of the detergent micelles, rapid ^{2}H→^{1}H exchange is also observed for the holoprotein.

These and other data [59,60] demonstrate that anionic detergents and phospholipids greatly increase the dynamics of the secondary structure of cytochrome c [58]. It can then be postulated that the lipid–protein interactions and heme attachment mediate a sequential folding of the protein in the membrane translocation process, finally resulting in the stably structured functional cytochrome c molecule. Release from the inside of the outer membrane is the final step in the translocation process. Because cytochrome c in contrast to apocytochrome has only a very limited affinity for the lipid mixture present in the outer membrane, as evidenced in Fig. 14 by the inability to increase the surface pressure of a lipid monolayer, we proposed that the

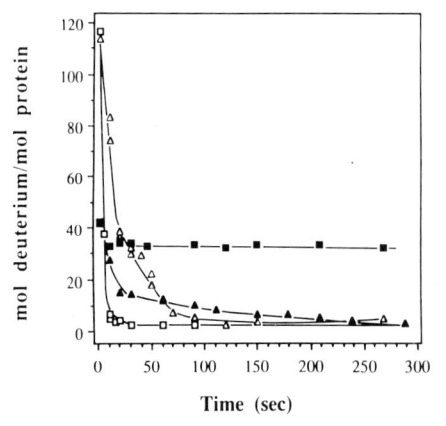

Fig. 13. ^{2}H→^{1}H amide exchange of apocytochrome c (□, △) and cytochrome c (■, ▲) in 10 mM acetate (pH 4.75) at 0°C in the absence (□, ■) and presence (△, ▲) of dodecylphosphoglycol/dodecylphosphocholine (9:1, molar) micelles (lipid to protein molar ratio of 120). The deuterated proteins were diluted in H$_2$O. At the indicated time points the samples were quenched in liquid N$_2$, whereafter the residual deuterium content of the samples was analyzed after freeze drying and unfolding (see [58] for details).

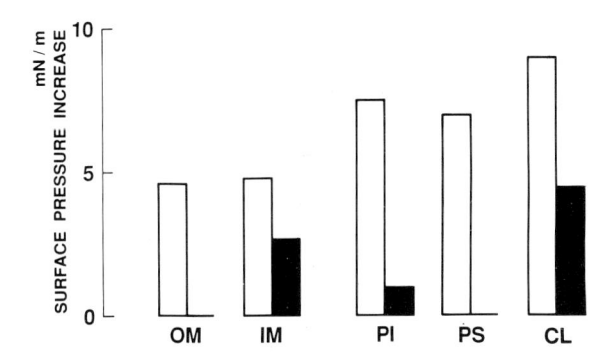

Fig. 14. Surface pressure increases induced by apocytochrome c (open bars) and cytochrome c (closed bars) in monolayers prepared from the total lipid extract of rat liver mitochondrial outer (OM) and inner (IM) membranes, phosphatidylinositol (PI), phosphatidylserine (PS) or cardiolipin (CL). Initial surface pressure 30 mM/m, subphase 10 mM Pipes (pH 7.0), 50 mM NaCl, protein concentration 50 nM (see [47] for details).

holoprotein is spontaneously released into the intermembrane space once it is formed [47]. The high affinity and specific interactions between cytochrome c and cardiolipin (Fig. 14), which is preferentially localized in the inner membrane, might play a role in targeting the protein towards the inner membrane [47].

2.3. Chloroplast protein import

As in the case of mitochondria, most chloroplast proteins are synthesized as precursors in the cytosol and are subsequently imported into the organelle (Chapter 22). Similarly, temporal N-terminal extensions, called transit sequences, are essential and sufficient for import. Transit sequences lack clear structure consensus features but are extremely rich in the hydroxylated amino acids Ser and Thr and have a significantly lower amount of acidic residues [61]. From the membrane lipid point of view, chloroplast membranes are particularly intriguing because they are rich in lipids found exclusively in this organelle. The glycolipids mono- and digalactosyl diglyceride (MGDG and DGDG) and the sulfolipid sulfoquinovosyl diglyceride (SQDG) are examples of such lipids present in large amounts both in the inner and outer membrane of the chloroplast envelope (Fig. 15). They will provide the surfaces of the chloroplast membranes with unique properties not found elsewhere in the plant cell. The obvious speculation is that these lipids could play a role in specific targeting and import of chloroplast precursor proteins. Specific interactions between transit sequences and chloroplast lipids are a likely candidate to consider in analyzing such possibilities.

We tested this hypothesis by analyzing the interactions between synthetic peptides corresponding to the different parts of the transit sequence of a small subunit of the stromal enzyme ribulose-1,5-biphosphate carboxylase/oxygenase and the individual

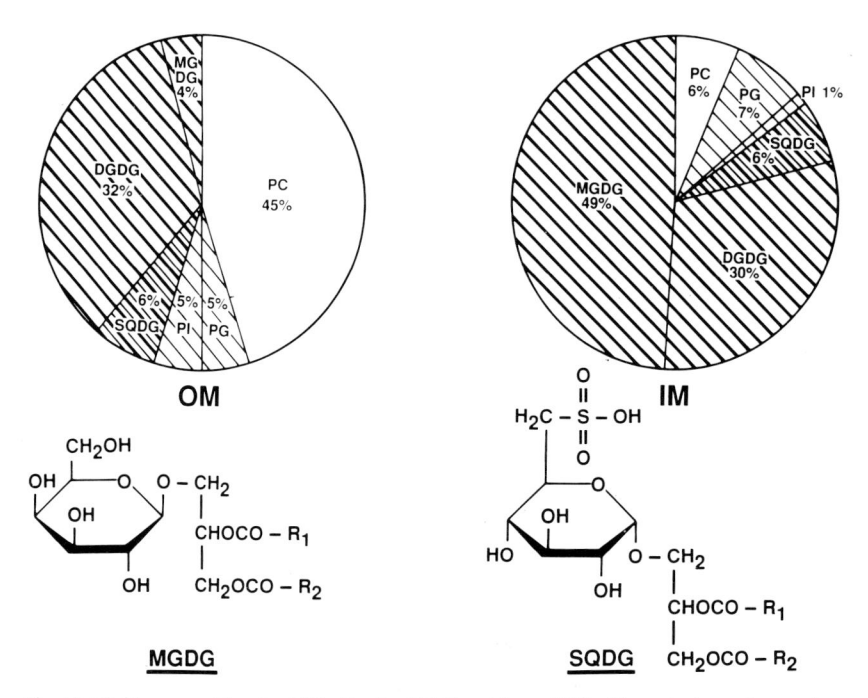

Fig. 15. Lipid composition (mol%) of outer (OM) and inner (IM) chloroplast membranes from pea leaves emphasizing the specific chloroplast lipid classes (shaded dark) MGDG, DGDG and SQDG and the anionic lipids (shaded light). Data from [62].

lipids of the chloroplast envelope in monolayers at the air–water interface [62]. The results were intriguing in that all fragments studied showed a substantial increase in surface pressure of negatively charged SQDG and DOPG monolayers, whereas in particular fragments corresponding to residues 21–40 and 41–60 of the transit sequence showed considerable interaction with MGDG and DGDG. The significance of these results is very apparent from the virtual lack of interaction with phosphatidylcholine monolayers, which like MGDG and DGDG are also overall neutral. These results indeed suggest that the transit sequence can efficiently interact with the specific chloroplast lipids. In agreement with this idea is the observation that when the purified translocation-competent precursor of the stromal ferredoxin (preFd) [63,64] is injected underneath a monolayer of envelope outer membrane chloroplast lipids, a large increase in surface pressure is observed which is absent for the mature holo ferredoxin and does not occur for a mitochondrial outer membrane extract (Fig. 16). The picture emerging from these data is that specific interactions between the transit sequence and lipid facilitate recognition and insertion of the precursor into the outer membrane of the chloroplast.

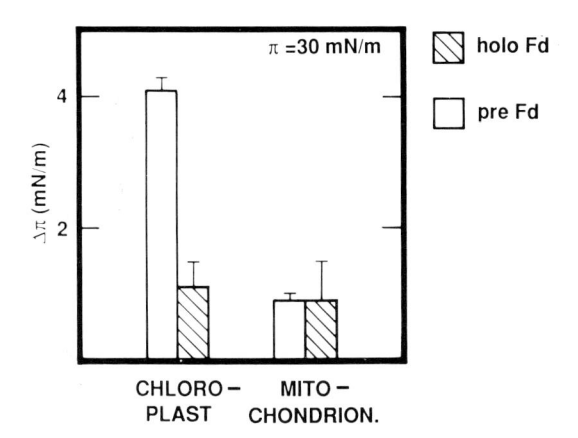

Fig. 16. Surface pressure increases at 25°C induced by purified preFd and holoFd underneath monolayers prepared of outer membrane lipid extracts of pea chloroplasts and rat liver mitochondria. Initial surface pressure 30 mN/m. Subphase as in legend of Fig. 14.

3. Concluding remarks

The results and insights obtained so far strongly suggest that membrane lipids play direct roles in different cellular protein translocation pathways. The data complement the insights emerging from studies on polypeptide toxin insertion and translocation across membranes and allow now common themes for lipid involvement of protein insertion and translocation to be formulated. Protein–lipid interactions are at the heart of these processes and are involved in different steps in the pathway. These include membrane binding, insertion and translocation and involve specific lipid classes of which the acidic ones appear to be of general importance. Changes in lipid organization and packing as well as changes in the folding pattern of the proteins to be translocated as a result of their interaction with lipids, appear to be other common features in these processes. However, it should be realized that the number of systems studied is still limited and that direct measurements of lipid–protein interactions in the functional process are still lacking. The interplay between lipids and proteins within the translocation machinery is still largely a mystery but a very challenging and fruitful area to explore.

References

1. De Vrije, G.J., Batenburg, A.M., Killian, J.A. and De Kruijff, B. (1990) Mol. Microbiol. 4, 143–150.
2. De Cock, H. (1991) Ph.D. Thesis, University of Utrecht, NL.
3. De Vrije, T., Tommassen, J. and De Kruijff, B. (1987) Biochim. Biophys. Acta 900, 63–72.
4. De Vrije, T., De Swart, R.L., Dowhan, W., Tommassen, J. and De Kruijff, B. (1988) Nature 334, 173–175.

5. Lill, R., Dowhan, W. and Wickner, W. (1990) Cell 60, 271–280.
6. Kusters, R., Dowhan, W. and De Kruijff, B. (1991) J. Biol. Chem. 266, 8659–8662.
7. Hartl, F.-U., Lecker, S., Schiebel, E., Hendrick, J.P. and Wickner, W. (1990) Cell 63, 269–279.
8. Lill, R., Cunningham, K., Brundage, L.A., Ito, K., Oliver, D. and Wickner, W. (1989) EMBO J. 8, 961–966.
9. Schliebel, E., Driessen, A.J.M., Hartl, F.-U. and Wickner, W. (1991) Cell 64, 927–939.
10. Shinkai, A., Mei, L.H., Tokuda, H. and Mizushima, S. (1991) J. Biol. Chem. 266, 5827–5833.
11. Breukink, E., Demel, R.A., De Korte-Kool, G. and De Kruijff, B. (1992) Biochemistry 31, 1119–1124.
12. Wickner, W. (1979) Annu. Rev. Biochem. 48, 23–45.
13. Engelman, D.M. and Steiz, T.A. (1981) Cell 23, 411–422.
14. Inouye, H. and Halegoua, S. (1980) CRC Crit. Rev. Biochem. 7, 339–371.
15. Gierasch, L.M. (1989) Biochemistry 28, 923–930.
16. Demel, R.A., Goormaghtigh, E. and De Kruijff, B. (1990) Biochim. Biophys. Acta 1027, 155–162.
17. Killian, J.A., Keller, R.C.A., Struyvé, M., De Kroon, A.I.P.M., Tommassen, J. and De Kruijff, B. (1990) Biochemistry 29, 8131–8137.
18. Keller, R.C.A., Killian, J.A. and De Kruijff, B. (1992) Biochemistry 31, 1672–1677.
19. Briggs, M.S. and Gierasch, L.M. (1986) Adv. Protein Chem. 38, 109–180.
20. Batenburg, A.M., Demel, R.A., Verkleij, A.J. and De Kruijff, B. (1988a) Biochemistry 27, 5678–5685.
21. Kimura, K. and Izui, K. (1976) Biochem. Biophys. Res. Commun. 70, 900–906.
22. Ito, K., Sato, T. and Yura, T. (1977) Cell 11, 551–559.
23. Pagès, J.M., Piovant, M., Varenne, S. and Lazdunski, C. (1978) Eur. J. Biochem. 86, 589–602.
24. DiRienzo, J.M. and Inouye, M. (1979) Cell 17, 155–161.
25. Batenburg, A.M., Brasseur, R., Ruysschaert, J.M., Van Scharrenburg, G.J.M., Slotboom, A.J., Demel, R.A. and De Kruijff, B. (1988b) J. Biol. Chem. 263, 4202–4207.
26. Killian, J.A., De Jong, A.M.Th., Bijvelt, J., Verkleij, A.J. and De Kruijff, B. (1990) EMBO J., 9, 815–819.
27. Nesmeyanova, M.A. (1982) FEBS Lett. 142, 189–193.
28. De Kruijff, B., Verkleij, A.J., Van Echteld, C.J.A., Gerritsen, W.J., Noordam, P.C., Mombers, C., Rietveld, A., De Gier, J., Cullis, P.R., Hope, M.J. and Nayar, R. (1981) in: H.G. Schweiger (Ed.), International Cell Biology 1980–1981, Springer-Verlag, Berlin.
29. Roise, D. and Schatz, G. (1988) J. Biol. Chem. 263, 4509–4511.
30. Roise, D., Horvath, S.J., Tomich, J.M., Richards, J.H. and Schatz, G. (1986) EMBO J. 5, 1327–1334.
31. Epand, R.M., Hui, S.-W., Argon, C., Gillespie, L.L. and Shore, G. (1986) J. Biol. Chem. 261, 10017–10020.
32. Goormaghtigh, E., Martin, I., Van den Branden, M., Brasseur, R. and Ruysschaert, J.M. (1989) Biochem. Biophys. Res. Commun. 158, 610–616.
33. Endo, T., Shimada, I., Roise, D. and Inagaki, F. (1989) J. Biochem. 106, 396–400.
34. Ou, W.-J., Ito, A., Umeda, M., Inoue, K. and Omura, T. (1987) J. Biochem. 103, 589–595.
35. Endo, T. and Schatz, G. (1988) EMBO J. 7, 1153–1158.
36. Endo, T. and Oya, M. (1989) FEBS Lett. 249, 173–177.
37. Schleyer, M. and Neupert, W. (1985) Cell 43, 339–350.
38. Chen, W.J. and Douglas, M.G. (1987) Cell 49, 651–658.
39. Eilers, M. and Schatz, G. (1988) Cell 52, 481–483.
40. De Kroon, A.I.P.M., De Gier, J. and De Kruijff, B. (1991) Biochim. Biophys. Acta 1068, 111–124.
41. Martin, J., Mahlke, K. and Pfanner, N. (1991) J. Biol. Chem. 266, 18051–18057.
42. Batenburg, A.M. and De Kruijff, B. (1988) Biosci. Rep. 8, 299–307.
43. Stuart, R.A. and Neupert, W. (1990) Biochimie 72, 115–121.
44. De Kruijff, B., Rietveld, A., Jordi, W., Berkhout, T.A., Demel, R.A., Görissen, H. and Marsh, D. (1988) NATO ASI Series H16, 257–269.
45. Nicholson, D.W., Hergersberg, C. and Neupert, W. (1988) J. Biol. Chem. 263, 19034–19042.

46. Söllner, T., Griffith, G., Pfaller, R., Pfanner, N. and Neupert, W. (1989) Cell 59, 1061–1070.
47. Demel, R.A., Jordi, W., Lambrechts, H., Van Damme, H., Hovius, R. and De Kruijff, B. (1989) J. Biol. Chem. 264, 3988–3997.
48. Görrissen, H., Marsh, D., Rietveld, A. and De Kruijff, B. (1986) Biochemistry 25, 2904–2910.
49. Dumont, M.E. and Richards, F.M. (1984) J. Biol. Chem. 259, 4147–4156.
50. Rietveld, A. and De Kruijff, B. (1984) J. Biol. Chem. 259, 6704–6707.
51. Vestweber, D. and Schatz, G. (1988) J. Cell. Biol. 107, 2045–2049.
52. Jordi, W., De Kroon, A.I.P.M., Killian, J.A. and De Kruijff, B. (1990) Biochemistry 29, 2312–2321.
53. Jordi, W., Li-Xin, Z., Pilon, M., Demel, R.A. and De Kruijff, B. (1989) J. Biol. Chem. 264, 2292–2301.
54. Sprinkle, J.R., Hakvoort, T.B.M., Koshy, T.I., Miller, D.D. and Margoliash, E. (1990) Proc. Natl. Acad. Sci. 87, 5729–5733.
55. Rietveld, A., Ponjee, G.A.E., Schiffers, P., Jordi, W., Van de Coolwijk, P.J.F.M., Demel, R.A., Marsh, D. and De Kruijff, B. (1985) Biochim. Biophys. Acta 818, 398–409.
56. De Jongh, H.H.J. and De Kruijff, B. (1990) Biochim. Biophys. Acta 1029, 105–112.
57. Snel, M.M.E., Kaptein, R. and De Kruijff, B. (1991) Biochemistry 30, 3387–3395.
58. De Jongh, H.H.J., Killian, J.A. and De Kruijff, B. (1992) Biochemistry 31, 1636–1643.
59. Spooner, P.J.R. and Watts, A. (1991) Biochemistry 30, 3871–3879.
60. Muga, A., Mantsch, H.H. and Surewicz, W.K. (1991) Biochemistry 30, 2629–2635.
61. Von Heyne, G. and Nishikawa, K. (1991) FEBS Lett. 278, 1–3.
62. Van 't Hof, R., Demel, R.A., Keegstra, K. and De Kruijff, B. (1991) FEBS Lett. 291, 350–354.
63. Pilon, M., De Boer, A.D., Knols, S.L., Koppelman, M.H.G.M., Van der Graaf, R.M., De Kruijff, B. and Weisbeek, P.J. (1990) J. Biol. Chem. 265, 3358–3361.
64. Pilon, M., De Kruijff, B. and Weisbeek, P.J. (1992) J. Biol. Chem. 267, 2548–2556.
65. De Vrije, G.J. (1989) Ph.D. Thesis, University of Utrecht, NL.
66. Nicholson, D.W., Köhler, H. and Neupert, W. (1987) Eur. J. Biochem. 164, 147–157.
67. Jordi, W., Hergersberg, C. and De Kruijff, B. (1992) Eur. J. Biochem. 204, 841–846.

Part B

Endoplasmic reticulum

CHAPTER 9

Membrane protein insertion into the endoplasmic reticulum: signals, machinery and mechanisms

STEPHEN HIGH and BERNHARD DOBBERSTEIN

European Laboratory for Molecular Biology, Postfach 102209, 6900 Heidelberg, Germany

Abstract

The targeting, membrane insertion and translocation of ER-proteins is determined by a limited number of signal sequence types characterized by a central core of hydrophobic amino acid residues. The residues flanking the hydrophobic stretch determine whether the signal sequence is removed by signal peptidase. In the absence of signal sequence cleavage they determine the ultimate orientation of the protein in the membrane. Protein factors in both the cytosol and the membrane mediate the targeting of nascent chains to the endoplasmic reticulum and their insertion into the membrane.

1. Introduction

The insertion of proteins into the membrane of the rough endoplasmic reticulum (ER) plays a pivotal role in the biosynthesis of secreted proteins, plasma membrane proteins, all organelles of the secretory pathway, lysosomes, and endosomes. The orientation that a protein assumes in the ER membrane is normally irreversible and it will therefore determine a protein's final topology. In this review, we focus on the signals that determine the targeting, insertion and orientation of a protein in the ER membrane and on the approaches used to identify and characterize the components involved in these cellular processes.

2. Types of membrane proteins and their topological signals

A protein sequence motif is responsible for the specific targeting of nascent chains to

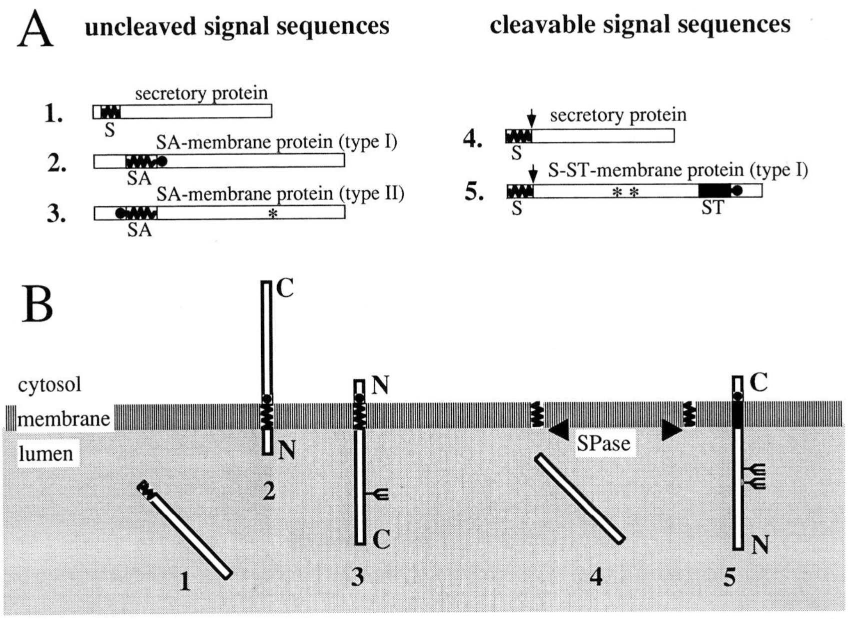

Fig. 1. Signal sequences and protein topologies across the membrane of the endoplasmic reticulum. (A) Outline of signals on secretory and membrane proteins. The hydrophobic cores of signal and signal-anchor sequences are indicated by a zig-zag line and the hydrophobic core of the stop transfer sequence by a filled oblong. Clusters of charged amino acid residues are indicated by filled circles. Potential N-glycosylation sites are indicated by a star and arrows indicate signal peptidase cleavage sites. S, signal sequence; SA, signal-anchor sequence; ST, stop-transfer sequence. (B) Topologies of secretory and membrane proteins. Topologies of proteins as outlined in (A) are shown. The cluster of charged amino acids, indicated by the filled circle, remains on the cytosolic side of the membrane. Oligosaccharide side chains are indicated by a fork. N, NH_2-terminus; C, COOH-terminus; SPase, signal peptidase.

the membrane of the ER [1]. The distinctive feature of the signal sequences responsible for targeting to the ER has proved not to be a strict consensus sequence, but rather a continuous stretch of 6–20 apolar amino acids with little sequence homology. Such signal sequences are found on secreted and membrane proteins. Upon membrane insertion or translocation, the signal sequence can either remain a part of the protein or be cleaved off (Fig. 1). These two possibilities are discussed in detail.

2.1. Proteins with uncleaved signal sequences

When the signal sequence is uncleaved, it can either be completely translocated across the ER membrane with the rest of the nascent chain, or become inserted into the membrane and stably anchor the protein in the lipid bilayer. There are few examples known of completely translocated (i.e. secreted) proteins with an uncleaved signal sequence. The best studied is ovalbumin where an uncleaved signal sequence is located close to the NH_2-terminus [2]. Uncleaved signal sequences that anchor the protein into the membrane are referred to as signal-anchor (SA) sequences (indicated by the zig-zag line in Fig. 1). SA sequences can mediate membrane insertion in two possible orientations, either with the NH_2-terminus exposed on the lumenal side of the membrane and the COOH-terminus on the cytosolic side (type I SA proteins) or in the reverse orientation with the NH_2-terminus on the cytosolic side and the COOH-terminus lumenal (type II SA proteins) (Fig. 1, examples 1–3). In most proteins so far analysed, the SA sequence is found close to the NH_2-terminus. There are also, however, examples of a type II SA sequence being located at the COOH-terminus of a protein.

The orientation that a SA protein assumes is determined by the properties of the 10–20 amino acid residues that flank each side of the hydrophobic core of the SA sequence [3,4]. Using site directed mutagenesis, it has been shown that an increase in the number of charged amino acids, particularly positively charged ones, within a flanking region will reduce the efficiency of translocation of this region causing it to remain on the cytosolic side of the membrane. This is true for both the NH_2- and COOH-terminal flanking regions [5]. The flanking region causing retention on the cytosolic side of the membrane is indicated by a filled circle in the SA proteins outlined in Fig. 1. Statistical analysis using a number of proteins of known topology has shown that the more positively charged of the two domains flanking the hydrophobic core of a SA sequence is most likely to remain on the cytosolic side of the membrane [4]. Negatively charged residues, when present in a domain in sufficient number, can also inhibit its translocation although they are less effective than positively charged ones. Thus, the nature of the charged residues flanking the hydrophobic core of a SA sequence determines the final orientation of a protein in the membrane.

2.2. Proteins with cleavable signal sequences

A signal sequence present at the NH_2-terminus of a protein is usually cleaved off by signal peptidase before the polypeptide chain is completed [6]. After signal sequence cleavage, the protein has two possible fates, it can either be completely translocated (secretory proteins) or retained in the lipid bilayer (membrane proteins) (Fig. 1, examples 4 and 5). In the latter case, complete translocation of the protein is prevented by the presence of a stop-transfer sequence which is usually toward the COOH-terminus of the protein [7]. The stop-transfer sequence, like a SA sequence, usually consists of approximately 20 apolar amino acid residues and it is flanked on the COOH-terminal side by positively charged amino acids [8]. Since a cleavable signal sequence exclusively mediates NH_2-terminal translocation, any membrane protein bearing such a signal will always expose its NH_2-terminus on the lumenal side of the membrane with the COOH-terminus remaining cytosolic. Cleavable signal sequences and SA sequences are very closely related; they can be converted from one to the other by altering the length of the hydrophobic core [9,10] or the hydrophilic flanking sequences [11,12].

The site of signal peptidase cleavage is largely determined by two parameters: the amino acid residues preceding the cleavage site [13] and the properties of the hydrophilic segment NH_2-terminal of the hydrophobic core of the signal sequence [11]. Thus the function of both SA sequences and cleavable signal sequences is defined by three structural elements: the hydrophobic core and both of the hydrophilic flanking sequences.

2.3. The loop model for protein insertion into the membrane

The proposal that the insertion of proteins into the ER membrane requires the nascent chain to adopt a loop configuration was first made for secretory proteins [14,15] and is supported by convincing experimental evidence [12]. The signal sequence pairs with the COOH-terminal hydrophilic flanking sequence and the resulting loop inserts into the membrane (see Fig. 2a,b) [14,15]. If a suitable signal

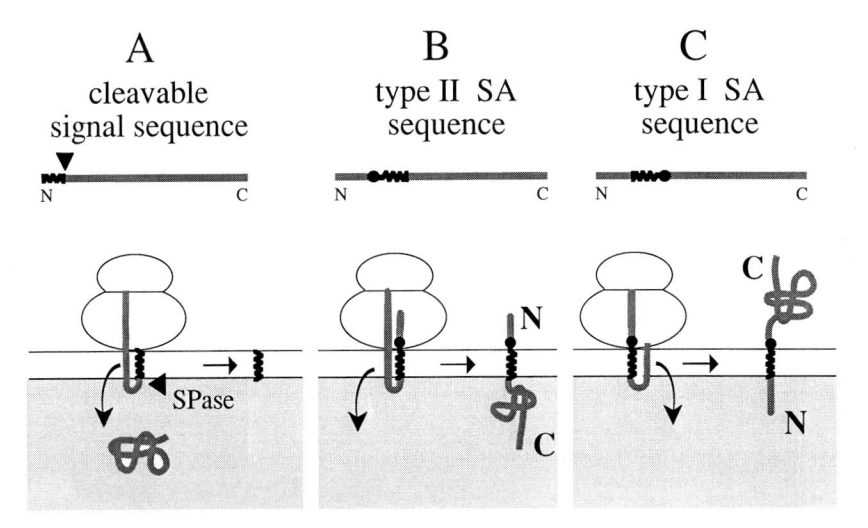

Fig. 2. Loop model for the membrane insertion of secretory and SA type II and type I membrane proteins. Nascent chains bearing a cleavable signal sequence (A), a type II SA sequence (B), or a type I SA sequence (C) are postulated to form different membrane insertion loops: in secretory and type II SA membrane proteins the loop is formed between the hydrophobic core of the SA sequence and the COOH-terminal flanking portion (A and B) and in type I SA proteins between the hydrophobic core of the SA sequence and the NH_2-terminal flanking region (C). Retention of the charged flanking region (filled circle) of the SA sequence on the cytosolic side of the membrane may be mediated by interaction with a component of the postulated translocation complex. The zig-zag line indicates the hydrophobic core of the cleaved signal sequence or SA sequence, the adjacent filled circle indicates a cluster of charged amino acid residues. The region of the nascent polypeptide that is translocated is indicated by an arrow. N, NH_2-terminus; C, COOH-terminus; SPase, signal peptidase.

peptidase cleavage site is exposed, cleavage within the lumenally exposed portion of the loop will occur, removing the signal sequence and resulting in a secreted protein (Fig. 2a). If a stop-transfer sequence is also present, a type I membrane protein will result. In the absence of a suitable cleavage site, the signal sequence serves as a membrane anchor. The COOH-terminus of the protein is translocated and a type II SA protein results (see Fig. 2b) [16].

A simple loop model does not readily explain the insertion of type I SA protein (Fig. 2c). This can, however, be explained by assuming that the formation of different types of loops can occur. A type I SA protein would be generated by the insertion of a loop formed between the hydrophobic core of the SA sequence and the preceding NH_2-terminal flanking region (Fig. 2c). This is in contrast to the type II SA sequence where loop formation occurs between the hydrophobic core of the SA sequence and the following COOH-terminal flanking region (Fig. 2b). The deciding factor as to which type of loop could be formed seems likely to be the number of charged residues in the flanking regions. The flanking region with the most charged residues (indicated by the filled circle at the appropriate side of the hydrophobic core of the SA sequence in Fig. 2b,c) would be retained on the cytosolic side of the membrane. This would allow the other domain of the protein to be translocated across the membrane (cf. Fig. 2b,c). The charge distribution of a type I SA sequence would retain the COOH-terminus of the protein on the cytosolic side of the membrane allowing the NH_2-terminus to be translocated (Fig. 2c). The observation that some artificially constructed SA proteins can insert in both a type I and a type II orientation [5] shows that the two possible orientations of a SA protein are not mutually exclusive. The statistical analysis of proteins of known topology showed that the charge difference between the two flanking regions of naturally occurring SA sequences is sufficiently large to ensure that only one orientation is usually found in the membrane [4].

2.4. Biosynthesis of multiple spanning membrane proteins

The sequence of a protein that spans the membrane multiple times (multi-spanning protein) is usually characterized by the presence of several regions comprised of 12–20 apolar amino acid residues. Two approaches have been used to study the function of these hydrophobic segments in the biogenesis of these proteins: (1) the hydrophobic segments from multi-spanning proteins have been fused to reporter proteins and their function as cleavable signals or SA sequences tested. It has been found that in most cases the segments can perform one or the other of these functions [17]. (2) On the basis of such results, it has been proposed that the hydrophobic segments of multi-spanning proteins function as alternating SA and stop-transfer sequences [18]. To further test this hypothesis, several SA and stop-transfer sequences or several identical type II SA sequences were linked together and the resulting proteins were tested for their ability to insert into the ER membrane. The most essential finding was that multi-spanning proteins can indeed be constructed by

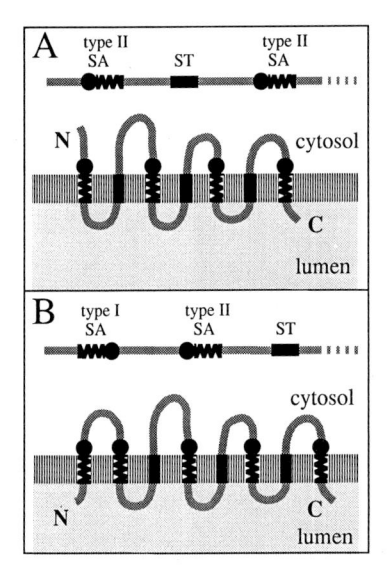

Fig. 3. Signals in multiple spanning membrane proteins. The membrane insertion of multiple spanning membrane proteins is determined by alternating type II SA and stop-transfer (ST) sequences (A). If the NH$_2$-terminus is exposed on the cytosolic side of the membrane the first signal sequence can be either a type I SA (B) or a cleaved signal sequence (not shown). The zig-zag line indicates the hydrophobic core of the SA sequence and the adjacent filled circle indicates a cluster of charged amino acid residues. The filled oblong indicates the stop-transfer (ST) sequence.

joining together either alternate type II SA and stop-transfer sequences or type II SA sequences alone [17,18]. Thus a SA sequence can serve as a stop-transfer sequence but its functions is dependent upon how it is located within the nascent chain relative to other topological signals.

Based upon the structural and functional similarity between the topological signals in single spanning and multi-spanning membrane proteins, their mode of insertion is also likely to be similar. The integration of multi-spanning membrane proteins into the ER is thought to occur via consecutive rounds of type II SA sequence mediated loop insertion followed by stop-transfer sequence mediated termination of translocation [18]. If the NH$_2$-terminus of a multi-spanning protein is cytosolic, the first signal will be a type II SA sequence (Fig. 3A). If, however, the NH$_2$-terminus is lumenal, the first signal could be either a type I SA sequence (Fig. 3B) or a cleavable signal sequence. While this is an appealing model, rigorous experimental evidence is lacking. By applying the predictions based on the loop model of insertion, it will be possible to construct artificial multi-spanning proteins of predicted topologies and test it further.

3. Components involved in the insertion of proteins into the ER membrane

The process of protein insertion into the membrane of the ER consists of two phases: (a) targeting and (b) insertion. Components that mediate these processes have been identified by genetic approaches in yeast and by biochemical techniques in mammalian cells. Here, we concentrate on the components identified in mammalian cells.

3.1. Targeting

Shortly after synthesis, both cleavable signal sequences and SA sequences interact with a cytoplasmic component, the signal recognition particle (SRP) [21–23]. SRP is a ribonucleoprotein particle consisting of a 7S RNA molecule and six polypeptide subunits of 9, 14, 19, 54, 68 and 72 kDa. The protein of SRP that interacts with signal sequences has been identified using a photocrosslinking technique. A chemically modified lysine residue is incorporated into the nascent chain and can be activated into a highly reactive radical upon UV irradiation. This radical will react covalently with any protein next to the nascent chain at the time of irradiation [24–27] and the resulting crosslinking products can be analysed. Using this approach, the 54 kDa subunit of SRP (SRP54) was shown to bind to cleavable signal sequences [24,25] and SA sequences [28]. The methionine rich COOH-terminal domain of SRP54 has now been shown to interact with both cleavable signal sequences [29,30] and SA sequences [47]. After binding of SRP to a signal sequence, the entire nascent chain/ribosome/SRP complex is specifically targeted to the rough ER membrane via an interaction with a membrane bound receptor, the docking protein or SRP receptor (Fig. 3). The function of SRP and docking protein is to ensure that the nascent chain is correctly targeted to the ER membrane in a translocation competent state.

3.2. Membrane insertion

Protein insertion into the mammalian ER usually occurs co-translationally. In order to study this process independently of ongoing protein synthesis, a post-translational assay system has been developed that makes use of preformed nascent chain/ribosome complexes [31]. The continued presence of the ribosome allows a nascent chain to be trapped at any stage of membrane targeting or insertion. Two procedures have been used to form these complexes: in the first, nascent chain synthesis was arrested by SRP and further elongation inhibited by the addition of cycloheximide [31]. Alternatively, nascent chain/ribosome complexes were obtained by using mRNA that was truncated within the coding region [26,32,33].

In both cases, stable nascent chain/ribosome complexes were obtained which could be isolated and subsequently incubated with microsomal membranes. Successful

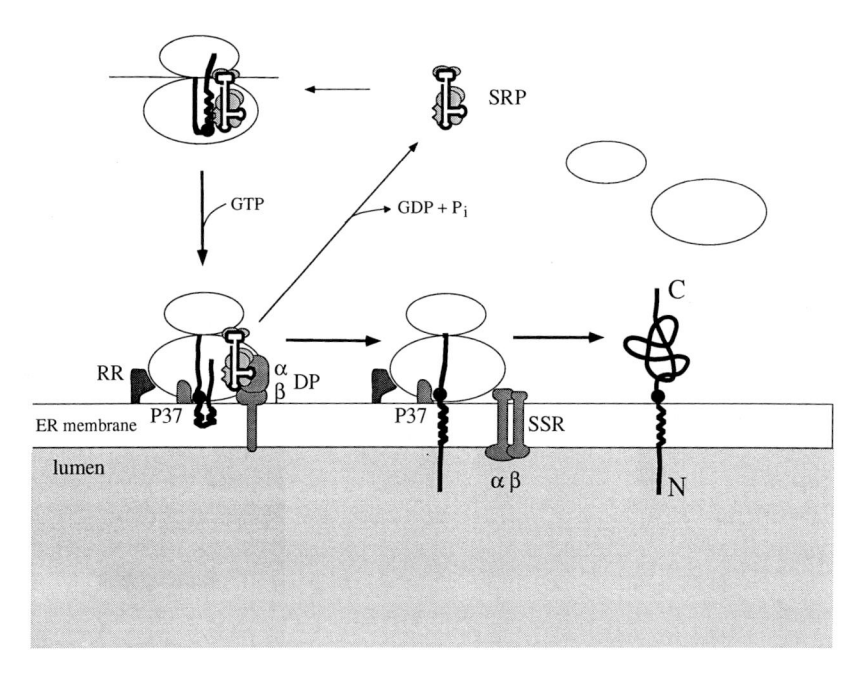

Fig. 4. Summary of events occurring during membrane insertion of a type I SA protein. Signal recognition particle (SRP) interacts with the SA sequence (zig-zag line) after it has emerged from the ribosome. The nascent chain/ribosome/SRP complex then interacts with the docking protein α subunit (DPα) and the postulated ribosome receptor (RR) in the ER membrane. The SA sequence is released from SRP in a GTP dependent fashion and subsequently inserts into the membrane. The major component in close proximity to the nascent chain during membrane insertion is a non-glycosylated protein of 37 kDa (P37). Upon the completion of synthesis of the SA protein, P37 detaches from the nascent protein. Other components of the translocation site are the signal peptidase (SPase) and the signal sequence receptor (SSR or mp39).

membrane insertion was then tested by resistance of the nascent chain to exogenously added protease, by the presence of the nascent chain in the membrane pellet after extraction with alkaline sodium carbonate solution and centrifugation [26,27] or by the acquisition of asparagine linked oligosaccharide side chains [34].

It was found that ongoing protein synthesis is not required for the membrane insertion of secretory or membrane proteins [32,33]. Thus, in each case, part of the nascent secretory or membrane protein was protected against exogenously added proteases. Nascent chains containing SA sequences were found in the membrane pellet fraction after extraction at pH 11.5 and an asparagine linked oligosaccharide side chain was added to the translocated NH_2-terminal domain of a type I SA protein.

The same experimental approach has also been used to study the environment of proteins during their insertion into the membrane of the ER. A nascent chain/ribosome/SRP complex was allowed to insert into the ER membrane prior to activation of a biosynthetically incorporated photocrosslinking reagent [26–28,35] or the addition of soluble bifunctional crosslinking reagents [36,37]. Crosslinking products to soluble components were then separated from those crosslinked to membrane integrated components by extraction with alkaline sodium carbonate and centrifugation. Presence of the crosslinking products in the resulting membrane pellet was then taken as successful membrane insertion. The nascent chains of a number of secretory and membrane proteins have by now been used as crosslinking probes in such studies. The general finding is that nascent chains of both types are adjacent to ER membrane proteins throughout their insertion and translocation. As the length of the nascent chain was increased the number of crosslinked ER proteins was also found to increase [26]. However, for most of the nascent chains tested, there were only one or two major crosslinked proteins found. These major crosslinked polypeptides can be placed into two groups: glycoproteins of about 35 kDa [26,27] and non-glycosylated polypeptides with molecular weights ranging from 34 to 39 kDa [28,35,37].

By using photocrosslinking reagents incorporated biosynthetically into nascent preprolactin, a 35 kDa glycosylated integral membrane protein of the ER was identified as a neighbouring protein during the initial stage of preprolactin membrane insertion. Because of its proximity to the preprolactin signal sequence, this protein was named the signal sequence receptor α (SSRα) [32] also known as mp39 [25]. SSRα was subsequently shown to be in close proximity to the translocating nascent chain of preprolactin throughout its length and not just to the signal sequence [25,37]. Very recently Görlich et al. [48] have shown that the major glycoprotein which can be crosslinked to nascent preprolactin is the 37 kDa TRAM (translocating chain-associating membrane protein).

Using a membrane impermeable, bifunctional, crosslinking reagent, the next neighbours of the translocation intermediates of two proteins with cleavable signal sequences, preprolactin (secreted) and VSV G protein (type I membrane protein) have been studied [37]. A 34 kDa non-glycosylated ER membrane protein (imp34) was found in close proximity to both nascent chains while no crosslinking of the nascent chain to SSRα was found. Imp34 could clearly be distinguished from SSRα by its lack of asparagine linked oligosaccharide moieties and its greater resistance to proteinase K. Using the photocrosslinking approach, a protein of similar properties was also found as a major cross-linking partner of nascent preprolactin chains comprising 90 or more amino acid residues in length [26] .

Using a translocation intermediate of the type I SA protein (IMC-CAT) as a photocrosslinking probe, one major crosslink to a non-glycosylated ER membrane protein of 37 kDa (P37) was observed [28]. A crosslink to a protein of the same size and properties was also obtained with a type II SA membrane protein (Ii) (S. High and B. Dobberstein, unpublished results). With both of these proteins, only minor crosslinks to SSRα were seen. The IMC-CAT nascent chain was shown to be photocrosslinked to P37 on the cytosolic side of the ER membrane. The cytosolically exposed domain of P37 responsible for this interaction was protected from protease digestion by the ribosome and became accessible upon ribosome dissociation.

It must be kept in mind that components that are found in close proximity of a nascent chain during membrane insertion need not play a direct role in this process. By reconstituting functional microsomes from isolated ER membrane proteins, it has already been possible to test the function of glycoproteins and of the SSR in the translocation process. Vesicles reconstituted from isolated ER membrane proteins and lipids have been shown to translocate proteins [39]. Removal of glycosylated proteins prior to reconstitution resulted in vesicles greatly reduced in their capacity for translocating nascent polypeptides [39]. To date SSRα [26,27] and two of the subunits of the signal peptidase complex [6] are the only identified glycoproteins known to be in close proximity of nascent chains during their membrane insertion and translocation. Removal of 95% of SSRα prior to the reconstitution of membrane vesicles did not reduce the targeting or translocation efficiency of these vesicles compared to control vesicles [49]. This indicates that SSRα, although in close proximity to the ribosome [40] and the ER translocation site, is not an essential component for translocation. In contrast, reconstituted rough microsomal vesicles lacking the TRAM protein are unable to translocate several secretory proteins [48]. Thus TRAM may be one of the components of the translocation site or postulated translocation tunnel [1,46].

The strongest remaining candidates for proteins of the postulated translocation site of the ER membrane are the non-glycosylated proteins with molecular weights of around 35 kDa (imp34 and P37). To date these proteins largely fulfil the prediction that a general component of the translocation site would be in close proximity to all, or at least most, of the translocating nascent chains tested.

Using a genetic selection system, it has been shown that in the yeast *Saccharomyces cerevisiae*, the proteins encoded by the SEC61, SEC62 and SEC63 genes have essential functions during protein translocation. Together with two other proteins, they form a multimeric complex in the yeast ER membrane [41]. Sequence analysis of SSRα [42] has revealed no sequence homology to any of the SEC proteins. Whether any of the other proteins crosslinked to signal sequences are homologous to SEC gene products remains to be elucidated. The mobilities of both SEC61p and SEC62p on gels make them both possible candidates as potential homologues of imp34 or P37.

The results of the crosslinking experiments described above together with the observation that translocation intermediates can be extracted from the membrane with protein denaturing agents [31] all support the presence of membrane proteins in the close proximity to nascent chains throughout their translocation and membrane insertion. This is consistent with the idea that translocation and insertion are protein-mediated events and that in the ER, they do not occur by unassisted entry into the lipid bilayer.

3.3. GTP requirement

The membrane insertion of a secretory protein and type I and type II SA proteins was found to require GTP. Since the non-hydrolysable analogue of GTP, GMPPNP, could substitute for GTP, this suggested that GTP hydrolysis was not required for membrane insertion as assayed in an in vitro system (see [43,34]. Using photocrosslinking, it was shown that in the absence of GTP, both nascent secreted and SA type proteins remained bound to SRP54. Only in the presence of GTP or its analogue was the nascent chain released from the SRP54 and transferred to the components of the ER membrane described above [34]. Connolly and Gilmore [43] showed that GTP was necessary for the docking protein dependent release of SRP from a nascent chain bearing a cleavable signal sequence. It still remains unclear which of the GTP binding proteins involved in this process (SRP54, and the α and β subunits of docking protein) mediates the GTP dependent release of the signal sequence from SRP54 [34]. The subsequent release of SRP from the docking protein complex, allowing it to re-enter the cytoplasmic pool, has been shown to require GTP hydrolysis [44].

4. Discussion

Both cytosolic and membrane components are involved in protein insertion into the membrane of mammalian ER. The initial steps involved in signal recognition and subsequent ER targeting up to and including the GTP dependent release of SRP from the nascent chain, appear to be identical for all secretory and membrane proteins tested so far (Fig. 4). Exceptions may be small secretory proteins that can by-pass the SRP mediated step when tested in in vitro systems [45] and certain membrane proteins with a COOH-terminal membrane anchor sequence [50]. SRP may keep the signal sequence exposed until contact has been made with components of the ER membrane. In addition to the membrane bound receptor for SRP, the docking protein, other proteins are involved in the translocation and membrane insertion of proteins. A complex of heterologous proteins is present in close proximity to the nascent chain/ribosome complex. Some components of the complex may play a direct role in the process of membrane insertion. Others may mediate events that occur concurrently with membrane insertion, for example glycosylation, signal sequence cleavage and catalysed folding.

Proteins with both SA sequences and cleavable signal sequences are in close proximity to a small subset of ER proteins throughout membrane insertion (Fig. 4). Components interacting with nascent type I and type II SA proteins were shown to be of similar size suggesting that the same machinery may mediate membrane insertion in two orientations (cf. Fig. 2). What could be the molecular mechanism by which different SA proteins assume their respective topologies in the ER membrane. One possibility is that either the NH_2- or the COOH-terminal portion of a SA sequence interacts with the cytosolically exposed domain of a protein in the membrane insertion complex. In this way, either the NH_2- or the COOH-terminal end of the protein would be retained on the cytosolic side of the membrane and the other end of the SA sequence would be translocated (see also Fig. 2B,C). If the interaction was of an electrostatic nature, this would explain both the orientation and its charge dependence for type I and type II SA proteins. Since the region of P37 that is in close proximity to the nascent chain of a type I SA protein has been mapped to the cytosolic side of the membrane [28], it is a candidate for such a protein. An alternative is that different types of SA sequences could interact with different domains of the same translocator protein. In this way, subtle differences in the translocation machinery, although composed of the same subunits, could allow the insertion of SA proteins with different orientations.

It has been proposed that the nascent chain/ribosome complex itself may play a role in the formation of the membrane insertion machinery [18]. In the absence of the nascent chain, the components of this machinery would be in only a loose association. The arrival of the nascent chain/ribosome complex at the ER membrane would lead to the recruitment of a functional membrane insertion complex from its subunit components (Fig. 4). If this is correct, then the type of signal sequence present on the nascent chain could influence the exact nature of the membrane insertion machinery. Coupling the formation of a functional membrane insertion complex to the presence of the nascent chain would also ensure that the integrity of the ER compartment was maintained and that established gradients of ions and cofactors were not dissipated. Consistent with this view is the finding that the release of the nascent chain by chain termination [35] or by ribosome dissociation [28] leads to a displacement of the nascent membrane protein from the vicinity of the identified crosslinking partners. Electrophysiological experiments have shown that nascent chain release alone may not be sufficient to disassemble a proposed protein conducting channel and that release of the ribosome from the membrane may also be required [46].

The stage at which SA sequences become stably integrated into the lipid bilayer is unknown. Since the main feature of a signal sequence is a core of hydrophobic amino acid residues, it is conceivable that already during the formation of the insertion loop, the SA sequence is in contact with lipids of the bilayer and that components of the insertion complex may only interact with the hydrophilic parts of the inserted polypeptide. This view is supported by the fact that the translocation intermediates formed by truncated SA proteins cannot be released from the membrane by reagents that disrupt protein–protein interactions such as extraction with alkaline sodium carbonate or 4 M urea [28].

Many open questions about the insertion of proteins into the ER membrane remain: (1) what is the role of GTP in the targeting process and what are the functions of the GTP binding proteins; (2) which ER proteins are essential for membrane protein insertion in mammalian cells; (3) what is the molecular architecture of the insertion site? When more answers to these questions are available, we may be in a position to understand not only how proteins insert into the membrane of the ER, but also how membrane protein topology across the ER is generated.

Acknowledgements

We would like to thank Ari Helenius and Jean Pieters for their careful reading of the manuscript and their useful suggestions for improvements. This work was supported by the Deutsche Forschungsgemeinschaft (Do 19915-3 and SFB 352).

References

1. Blobel, G. and Dobberstein, B. (1975) J. Cell Biol. 67, 835–851, 852–862.
2. Meek, R.L., Walsh, K.A. and Palmiter, R.D. (1982) J. Biol. Chem. 257, 12245–12251.
3. Haeuptle, M.-T., Flint, N., Gough, N.M. and Dobberstein, B. (1989) J. Cell Biol. 108, 1227–1236.
4. Hartmann, E., Rapoport, T.A. and Lodish, H.F. (1989) Proc. Natl. Acad. Sci. USA 86, 5786–5790.
5. Parks, G.D. and Lamb, R.A. (1991) Cell 64, 777–787.
6. Evans, E.A., Gilmore, R. and Blobel, G. (1986) Proc. Natl. Acad. Sci. USA 83, 581–585.
7. Rapoport, T.A. and Wiedmann, M. (1985) Application of the Signal Hypothesis to the Incorporation of Integral Membrane Protein, Vol. 24, Academic Press, New York, pp. 1–63.
8. Kuroiwa, T., Sakaguchi, M., Katsuyoshi, M. and Omura, T. (1991) J. Biol. Chem. 266, 9251–9255.
9. Lipp, J. and Dobberstein, B. (1988) J. Cell Biol. 106, 1813–1820.
10. Sato, T., Sakaguchi, M., Mihara, K. and Omura, T. (1990) EMBO J. 9, 2391–2397.
11. Lipp, J. and Dobberstein, B. (1986) Cell 46, 1103–1112.
12. Shaw, A.E., Rottier, P.J.M. and Rose, J.K. (1988) Proc. Natl. Acad. Sci. USA 85, 7592–7596.
13. von Heijne, G. (1983) Eur. J. Biochem. 133, 17–21.
14. Engelmann, D.M. and Steitz, T.A. (1981) Cell 23, 411–422
15. Inouye, M. and Halegoua, S. (1980) Crit. Rev. Biochem. 7, 339–371.
16. Wickner, W.T. and Lodish, H.F. (1985) Science 230, 400–407.
17. Friedlander, M. and Blobel, G. (1985) Nature 318, 338–343.
18. Blobel, G. (1980) Proc. Natl. Acad. Sci. USA 77, 1496–1500.
19. Wessels, H.P. and Spiess, M. (1988) Cell 55, 61–70.
20. Lipp, J., Flint, N., Haeuptle, M.-T. and Dobberstein, B. (1989) J. Cell Biol. 109, 2013–2022.
21. Sakaguchi, M., Mihara, K. and Sato, R. (1984) Proc. Natl. Acad. Sci. USA 81, 3361–3364.
22. Lipp, J. and Dobberstein, B. (1986) J. Cell Biol. 102, 2169–2175.
23. Walter, P. and Lingappa, V. R. (1986) Annu. Rev. Cell Biol. 2, 499–516.
24. Krieg, U.C., Walter, P. and Johnson, A.E. (1986) Proc. Natl. Acad. Sci. USA 83, 8604–8608.
25. Kurzchalia, T.V., Wiedmann, M., Girshovich, A.S., Bochkareva, E.S., Bielka, H. and Rapoport, T.A. (1986) Nature 320, 634–636.
26. Krieg, U.C., Johnson, A.E. and Walter, P. (1989) J. Cell Biol. 109, 2033–2043.
27. Wiedmann, M., Kurzchalia, T.V., Hartmann, E. and Rapoport, T.A. (1987) Nature 328, 830–833.
28. High, S., Görlich, D., Wiedmann, M., Rapoport, T. A. and Dobberstein, B. (1991) J. Cell Biol. 113, 35–44.
29. High, S. and Dobberstein, B. (1991) J. Cell Biol. 113, 229–233.
30. Zopf, D., Bernstein, H.D., Johnson, A.E. and Walter, P. (1990) EMBO J. 9, 4511–4517.
31. Gilmore, R. and Blobel, G. (1985) Cell 42, 497–505.
32. Connolly, T. and Gilmore, R. (1986) J. Cell Biol. 103, 2253–2261.
33. Perara, E., Rothman, R.E. and Lingappa, V.R. (1986) Science 232, 348–352.
34. High, S., Flint, N. and Dobberstein, B. (1991) J. Cell Biol. 113, 25–34.
35. Thrift, R.N., Andrews, D.W., Walter, P. and Johnson, A.E. (1991) J. Cell Biol. 112, 809–821.
36. Görlich, D., Prehn, S., Hartmann, E., Herz, J., Otto, A., Kraft, R., Wiedmann, M., Knespel, S., Dobberstein, B. and Rapoport, T.A. (1990) J. Cell Biol. 111, 2283–2294.

37. Kellaris, K.V., Bowen, S. and Gilmore, R. (1991) J. Cell Biol. 114, 21–33.
38. Wiedmann, M., Görlich, D., Hartmann, E., Kurzchalia, T.V. and Rapoport, T.A. (1989) FEBS Lett. 257, 263–268.
39. Nicchitta, C.V. and Blobel, G. (1990) Cell 60, 259–269.
40. Collins, P.G. and Gilmore, G. (1991) J. Cell Biol. 114, 639–649.
41. Deshaies, R.J., Sanders, S.L., Feldheim, D.A. and Schekman, R. (1991) Nature 349, 806–808.
42. Prehn, S., Herz, J., Hartmann, E., Kurzchalia, T. V., Frank, R., Römisch, K., Dobberstein, B. and Rapoport, T.A. (1990) Eur. J. Biochem. 188, 439–445.
43. Connolly, T. and Gilmore, R. (1989) 57, 599–610.
44. Connolly, T., Rapiejko, P.J. and Gilmore, R. (1991) Science 252, 1171–1173.
45. Schlenstedt, G. and Zimmermann, R. (1987) EMBO J. 6, 699–703.
46. Simon, S.M. and Blobel, G. (1991) Cell 65, 371–380.
47. Lütcke, H., High, S., Römisch, K., Ashford, A.J. and Dobberstein, B. (1992) EMBO J. 11, 1543–1551.
48. Görlich, D., Hartmann, E., Prehn, S. and Rapoport, T.A. (1992) Nature 357, 47-52.
49. Migliaccio, G., Nicchitta, C.V. and Blobel, G. (1992) J. Cell Biol. 117, 15–25.
50. Anderson, D.J., Mostov, K.E. and Blobel, G. (1983) Proc. Natl. Acad. Sci. USA 80, 7249–7253.

CHAPTER 10

Translocation of proteins through the endoplasmic reticulum membrane: investigation of their molecular environment by cross-linking

ENNO HARTMANN and TOM A. RAPOPORT

Institut für Molekularbiologie, Robert-Rössle-Straße 10, 1115 Berlin-Buch, Germany

Abstract

We discuss the application of cross-linking methods for the investigation of the molecular environment of nascent polypeptide chains as they pass through the endoplasmic reticulum (ER) membrane. Several membrane proteins have been identified, purified and characterized which seem to be located at the translocation site. The results suggest that the latter is more complex and more dynamic than previously assumed.

1. Introduction

Transport across the endoplasmic reticulum (ER) membrane is a crucial step in the fate of many proteins destined for secretion or localization in the plasma membrane, the lysosome, the Golgi apparatus as well as the ER itself. The polypeptides generally carry hydrophobic signal sequences which are often located at the N-terminus of the precursor molecules and cleaved off after membrane transfer. The recognition of signal sequences and the targeting of nascent polypeptides to the membrane have been elucidated to some extent (for review see [1,2]) but the subsequent process of protein transfer across the ER membrane is not clear at all. Different hypotheses have been proposed concerning the molecular environment that polypeptides meet during their membrane passage. One suggestion is that they may cross the hydrophobic core of the phospholipid bilayer without involvement of membrane proteins [3,4]. According to another idea, translocating polypeptides may face the hydrophilic head groups of phospholipids arranged in an inverted micelle structure [5,6].

According to still other models, integral ER membrane proteins transiently form an aqueous channel through which polypeptides are transported across the membrane [7,8].

Indirect evidence indicates that the environment of translocating polypeptide chains is made up at least in part of membrane proteins. It is accessible to aqueous perturbants [9]. Short nascent translocating polypeptides are protected against proteases even after solubilization of the ER membrane unless protein denaturants are added [10], suggesting that they are surrounded by membrane proteins. The existence of a protein-conducting channel has received support from the observation that upon fusion of microsomes into planar lipids, large ion channels occur which increase in number after addition of puromycin, suggesting that they had been plugged by nascent chains [11].

A more direct way to study the molecular environment of translocating polypeptides has been pursued by cross-linking techniques. Membrane proteins in the proximity of translocating chains have been identified by two different cross-linking methods. One is a classical method employing bifunctional reagents which has often been used to identify next neighbors in different complexes (for review see [12]). The other is a photocross-linking method in which photoreactive chemical groups are introduced into newly synthesized polypeptides and are cross-linked to neighboring proteins upon irradiation (for review see [13]). In the present review, results obtained with these approaches are summarized.

2. Results

2.1. Experimental strategies

The application of cross-linking methods is based on previously established in vitro translation/translocation systems [14]. Nascent polypeptide chains were labeled with a radioactive amino acid and trapped at specific points of their passage through the ER membrane. Such translocation intermediates were obtained with nascent chains of defined lengths by the addition of signal recognition particles (SRP), which arrest elongation when the chain is about 70 residues long [15], or by the use of truncated mRNAs that lack a termination codon [16]. Cross-linking was then carried out either in a classical manner with bifunctional chemical reagents [17,18] or with photoactivatable cross-linkers [19–23]. For the latter, photoreactive chemical groups were introduced into the nascent polypeptide chain using lysyl-tRNA that was modified in the ϵ-amino group of the amino acid. The photoreactive lysine derivatives could be positioned either into the signal sequence or into the mature portion of the polypeptide chain.

The two cross-linking methods employed have both their advantages and disadvantages. Cross-linking with bifunctional chemical reagents is easy to perform and generally results in high yields. However, it depends on the juxtaposition of suitable chemical groups, on their spatial accessibility and is prone to artefacts owing to the extensive modifications introduced into the system which may occur prior to cross-linking of the molecules of interest. On the other hand, photocross-linking with the probes incorporated into the nascent chain results in only small perturbations of the system. It should be noted, however, that point mutations are introduced into the nascent chain that change a positively charged residue into a hydrophobic lysine derivative. The photoreactive probes can be precisely positioned and their activation produces nitrenes or carbenes which potentially react with almost any chemical bond although they prefer electrophiles [24,25]. The yields of cross-links are generally low because of the competing reaction with water.

2.2. The SSR-complex

The first integral membrane protein demonstrated to be in the proximity of translocating chains was a glycoprotein of about 34 kDa [19]. Since in initial experiments, cross-linking to this protein was obtained with short preprolactin chains bearing the photoreactive groups in the signal sequence, it was provisionally termed the signal sequence receptor (SSR) (as SSR is now known to be a complex of four polypeptides, the 34 kDa component is called SSRα) [19]. In later experiments, it was found that cross-linking also occurred with preprolactin chains containing the photoreactive groups exclusively in the mature portion [20,21]. This indicates, that SSRα is not, or not only, a signal sequence receptor. Cross-linking to SSRα could also be demonstrated for β-lactamase (Görlich, unpublished results) and for different signal-anchor type membrane proteins [22], although the yields were low in some cases.

Several results support the assumption that SSRα is a component of the translocation site of the ER membrane. Particularly striking is its relationship to membrane-bound ribosomes that are in the process of delivering nascent chains through the phospholipid bilayer. SSRα is in the vicinity of membrane-bound ribosomes as indicated by cross-linking with bifunctional reagents [26]. Furthermore, it is segregated to the rough portion of the ER membrane [27]. SSRα is a major membrane protein in the ER, being present in a slight molar excess over membrane-bound ribosomes [28]. It is in a more than tenfold excess over SRP and SRP-receptor which are only involved in the initiation of the translocation process. SSRα is found in all cells of mammals, birds and fish investigated [28]. F_{ab}-fragments produced from antibodies against SSRα inhibit the translocation of several secretory proteins in vitro [28], again indicating its spatial proximity to the translocation site. Finally, the fact that SSRα could be cross-linked to translocating chains suggests that it is close to nascent chains during their passage through the membrane [20,21].

Purification of SSRα under non-denaturing conditions yielded a protein complex that contains three other polypeptides in stoichiometric relation (SSRβ, SSRγ, SSRδ) with the SSRβ more tightly associated with SSRα than the other two subunits [17,

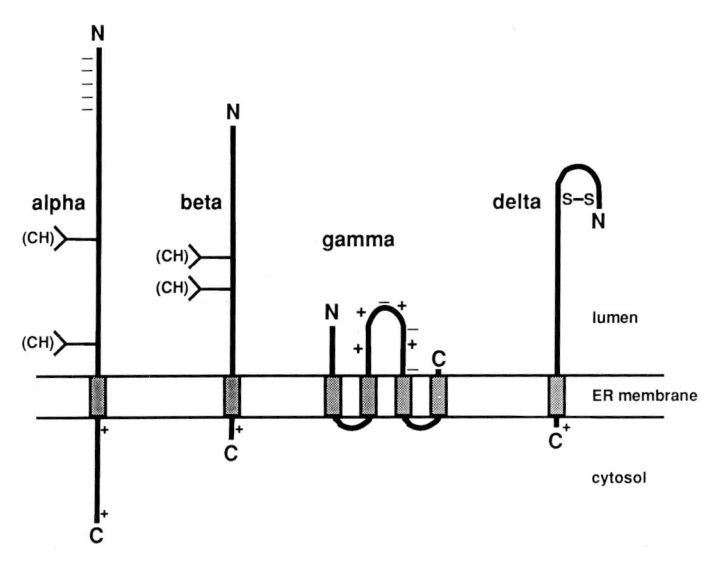

Fig. 1. Membrane topology of the subunits of the SSR-complex.

Görlich, unpublished results]. Cross-linking with a cleavable bifunctional reagent demonstrated that the four proteins are genuine neighbors in intact microsomal membranes. The primary structure of all four subunits was deduced from the nucleotide sequences of the corresponding cDNAs (Prehn et al., unpublished results). The α-, β- and δ-subunits are single-spanning membrane proteins with a cleavable signal sequence (Fig. 1). Their N-termini are in the ER-lumen and the C-termini in the cytosol [17,29, Prehn et al., unpublished results]. The cytoplasmic tail of SSRα is 56 amino acids long whereas those of SSRβ and SSRδ are much shorter. SSRα can be phosphorylated both in vivo and in vitro [29], but the significance of this modification is unclear as yet. Both, SSRα and SSRβ contain two N-linked sugars [17,29]; the δ-subunit has a disulfide bridge in its lumenal domain (Prehn et al., unpublished results). The most N-terminal part of the SSRα is highly negatively charged and, from comparison of the sequences from dog, man and trout, it seems that only the charge and not the exact amino acid sequence is important (Hartmann et al., unpublished results). SSRγ contains an uncleaved signal-anchor sequence and is predicted to span the membrane four times (Prehn et al., unpublished results). The orientation of the protein remains to be confirmed. In total, the SSR-complex has seven membrane anchors distributed to different subunits which may associate in the membrane in a dynamic manner.

2.3. The TRAM protein

Although SSRα gave cross-links to all nascent chains investigated, it does not seem to be the major glycoprotein cross-linked to short chains since the amount of product immunoprecipitated with SSRα-antibodies was low compared with that bound to concanavalin A. For somewhat longer chains, however, the yields of immunoprecipitated and lectin-bound material were almost equal, indicating that SSRα is a major glycoprotein cross-linking partner at later stages of the translocation process.

To investigate the possibility that short nascent chains are cross-linked to another glycoprotein of about the same size, an assay was established that is based on the reconstitution system by Nicchitta and Blobel [30]. Canine microsomes were solubilized in cholate and the resulting extract was depleted from all glycoproteins using a lectin column. It was then replenished with fractions of glycoproteins obtained from various purification steps. After dialysis, reconstituted proteoliposomes were produced and tested with a short fragment of preprolactin for the appearance of cross-linked product. It turned out that indeed another glycoprotein (termed translocating chain-associating membrane (TRAM) protein) was responsible for the major cross-links. The TRAM protein was purified to homogeneity and its sequence determined by molecular cloning of the corresponding cDNA. It is a multi-spanning membrane protein and has one N-linked carbohydrate chain [48]. Its similarity with SSRα in several respects (mobility in SDS-gels, presence of carbohydrate, size of the cytoplasmic tail) explains why only one protein was thought to be responsible for the cross-links.

2.4. Other glycoproteins

Other cross-linked products containing integral membrane glycoproteins of about 70 kDa and 140 kDa [23] have been detected, but not characterized as yet.

Recently, cross-linked products of prepro-α-factor and microsomal glycoprotein(s) of the yeasts *Saccharomyces cerevisiae* and *Candida maltosa* were detected (Müsch et al., unpublished). The estimated size of these proteins of about 32–34 kDa raises the possibility that homologs to SSRα and/or the TRAM protein may be among them.

2.5. Unglycosylated proteins

By cross-linking, several unglycosylated proteins were found to be located in the vicinity of translocating chains. None has been isolated as yet.

An imp34 was found to be cross-linked to nascent chains by the bifunctional reagent DSS [18]. This protein has a cytosolic portion accessible to high concentrations of trypsin [18] and its interaction with nascent chains is inhibited by treatment of the membranes with 3 mM NEM [18], indicating that either a previously described NEM-sensitive factor [31] is involved or that imp34 itself is sensitive to NEM.

A protein of about 40 kDa was detected by photocross-linking of prepro-α-factor chains to canine membranes (Müsch, unpublished results).

A number of other unglycosylated proteins (p19, p37, p42, p60 [22]; p15, p20, p24, p27, p39 [23]) were also found to give photocross-linked products with nascent proteins that use their signal sequence as membrane anchor (signal-anchor type). Although it is not yet clear if all these proteins are integral membrane proteins, p37, p39 and p42, which are perhaps identical to the 40 kDa protein mentioned above, seem to be of particular interest because of the similarity of their apparent sizes with Sec61p (see below).

2.6. The Sec proteins of yeast microsomes

Photocross-linking was also employed to investigate the molecular environment of translocating nascent chains in the ER-membrane of the yeast *S. cerevisiae* [49]. It was demonstrated that the two unglycosylated proteins Sec61p and Sec62p, found previously by genetic means to be essential for protein translocation in vivo [32,33], are in proximity to translocating prepro-α-factor. At late stages of the translocation process Sec61p was found to be the major cross-linking partner. No cross-links to Sec62p were observed under these conditions. The relationship of nascent chains to the Sec proteins at earlier stages of the translocation process was studied in two ways. First, shorter prepro-α-factor chains were investigated. They gave cross-links to both Sec61p and Sec62p. Second, long prepro-α-factor chains were cross-linked to microsomal proteins in the absence of ATP under conditions known to allow only their initial membrane insertion [34]. Cross-links to Sec62p but not to Sec61p were observed. These data led to a model according to which Sec61p is in continuous contact with translocating nascent chains whereas Sec62p is only transiently involved.

Sec61p is an abundant multi-spanning membrane protein [35, Schekman, pers. commun.], which may have up to nine membrane-spanning segments. It has some homology to the SecY protein family which is essential for protein translocation in prokaryotes [36, Hartmann, unpublished results].

Sec62p spans the membrane twice with both the N- and the C-termini being located in the cytosol [37]. Its amount in ER-membranes is only one-tenth that of Sec61p [35]. It forms a complex with Sec63p [38] and two other proteins (gp31.5 and p23) [35]. Sec61p seems to be close to the complex as demonstrated by cross-linking with a bifunctional reagent [35].

3. Discussion

The identification of genuine neighbors depends on the occurrence of cross-linked products in only limited numbers in contrast to the expectation if random collisions prevailed. It is therefore of particular importance that, regardless of the cross-linking method employed, only a small set of membrane proteins could be cross-linked to translocating nascent chains. A convincing argument for the validity of the conclusions drawn from cross-linking is provided by the identification in proximity of translocating nascent chains of two yeast Sec-proteins which were previously found to be essential for ER-translocation on genetic grounds. It is thus likely that the membrane proteins identified by cross-linking (SSRα, TRAM protein, Sec61p, Sec62p, various other unglycosylated proteins) are genuine neighbors of nascent chains as they pass through the ER membrane. A caveat, however, is that the translocation intermediates used for cross-linking provide an artificial situation with a greatly prolonged arrest in the membrane of the nascent chain during which its environment may have changed.

In addition to the membrane proteins identified by cross-linking, several other ER components have been implicated in the process of protein transfer across the membrane. These include the β-subunit of the SRP-receptor [39], the signal peptidase [40–42], the ribophorins I and II [43] and a mp30 with affinity to SRP [39] (see Table I). Assuming that all components can be found in a single organism (Sec61p homologs seem to exist in mammals, Müsch et al., unpublished results), the translocation site contains at least 17 different polypeptides. This number is certainly a lower estimate since only purified, reasonably well characterized proteins have been included in Table I whereas others, like the ribosome receptor, the NEM-sensitive factor, the imp34 etc. have been omitted. Thus, the translocation apparatus seems to be rather complex. Of course, the question arises as to whether all components are mechanistically involved. Some proteins may indeed only be needed for modification

TABLE I

Purified proteins which may be involved in the actual translocation process (for references, see text)

Protein	Size (kDa)	Proximity to nascent chains detected	Source
SRP-receptor β-subunit	30		Mammals
Ribophorin I	73		Mammals
Ribophorin II	70		
mp 30	30		Mammals
Signalpeptidase	12, 18, 21, 22/23, 25		Mammals
Sec 61	53	Yes	Yeast
Sec 62	32	Yes	Yeast
Sec 63	76		Yeast
SSR-complex	18, 20, 22, 36	Yes	Mammals, birds, fish
TRAM protein	45	Yes	Mammals

reactions of newly synthesized proteins. For example, the glycosyltransferase, recently shown to consist of the ribophorins and a 40 kDa protein [50] is most likely located in or near the translocation site as demonstrated by the inhibition of translocation by antibodies to the ribophorins [44]. Other proteins could function as receptors for stop-transfer sequences allowing membrane proteins to integrate into the phospholipid bilayer [45]. Still others, which may be called transmembrane chaperons, could stabilize weak membrane anchors of multi-spanning or oligomeric membrane proteins containing hydrophilic or charged amino acid residues. Such chaperons would bind to the trans-membrane domain during translocation and remain associated with it until the final binding partner is encountered [46].

The cross-linking experiments have provided evidence that the molecular environment of a translocating nascent chain is dynamic. This conclusion could be derived both for the mammalian and for the yeast system by the demonstration that the yields of cross-links to different membrane proteins (TRAM protein and SSRα or Sec62p and Sec61p, respectively) change with the stage of the translocation process. The molar ratio between Sec62p and Sec61p as well as their spatial proximity also support the assumption of their dynamic interaction. These data seem to preclude the existence of a rigid channel structure. On the other hand, a dynamic structure would be consistent with the multiple functions postulated for a protein-translocating machinery, which include signal sequence recognition, determination of the orientation of signal-anchor type membrane proteins, lowering of the energetic barrier for membrane transfer of polypeptide chains, their pushing or pulling across the membrane, signal sequence cleavage and insertion of membrane-spanning segments into the phospholipid bilayer. Some of these functions may be needed only transiently, others during the entire process and, correspondingly, membrane proteins would be expected to join and leave the translocation site at different stages. Such a concept is an extension of a previous idea according to which the translocation complex would only assemble and disassemble at the beginning and at the end of the translocation process, respectively [14]. A dynamic model would also leave room for the role of lipids in the translocation process, be it in a passive way to provide a fluid environment for movement of proteins in the plane of the membrane, or in a more active way as interacting partners of signal sequences [5,6,47].

References

1. Walter, P. and Lingappa, V.R. (1986) Annu. Rev. Cell Biol. 2, 499–516.
2. Bernstein, H.D., Rapoport, T.A. and Walter, P. (1989) Cell 58, 1017–1019.
3. von Heijne, G. and Blomberg, C. (1979) Eur. J. Biochem. 97, 175–181.
4. Engelman, D.M. and Steitz, T.A. (1981) Cell 23, 411–422.
5. Nesmeyanova, M.A. (1982) FEBS Lett. 142, 189–193.
6. de Vrije, G.J., Batenburg, A.M., Killian, J.A. and de Kruiff, B. (1990) NATO ASI Series H40, 247–258.
7. Blobel, G. and Dobberstein, B. (1975a) J. Cell Biol. 67, 835–851.

8. Rapoport, T.A. (1985) FEBS Lett. 187, 1–10.
9. Gilmore, R. and Blobel, G. (1985) Cell 42, 497–505.
10. Connolly, T., Collins, P. and Gilmore, R. (1989) J. Cell Biol. 108, 299–307.
11. Simon, S.M. and Blobel, G. (1991) Cell 65, 371–380.
12. Tae, H.J. (1983) Methods Enzymol. 91, 580–609.
13. Görlich, D., Kurzchalia, T.V., Wiedmann, M. and Rapoport, T.A. (1991) Methods Cell Biol. 34, 241–262.
14. Blobel, G. and Dobberstein, B. (1975b) J. Cell Biol. 67, 852–862.
15. Walter, P. and Blobel, G. (1981) J. Cell Biol. 91, 557–561.
16. Connolly, T. and Gilmore, R. (1986) J. Cell Biol. 103, 2253–2261.
17. Görlich, D., Prehn, S., Hartmann, E., Herz, J., Otto, A., Kraft, R., Wiedmann, M., Knespel, S., Dobberstein, B. and Rapoport, T.A. (1990) J. Cell Biol. 111, 2283–2294.
18. Kellaris, K.V., Bowen, S. and Gilmore, R. (1991) J. Cell Biol. 114, 21–33.
19. Wiedmann, M., Kurzchalia, T.V., Hartmann, E. and Rapoport, T.A. (1987) Nature 328, 830–833.
20. Krieg, U.C., Johnson, A.E. and Walter, P. (1989) J. Cell Biol. 109, 2033–2043.
21. Wiedmann, M., Görlich, D., Hartmann, E., Kurzchalia, T.V. and Rapoport, T.A. (1989) FEBS Lett. 257, 263–268.
22. High, S., Wiedmann, M., Görlich, D., Rapoport, T.A. and Dobberstein, B. (1991) J. Cell Biol. 113, 229–233.
23. Thrift, R.N., Andrews, D.W., Walter, P. and Johnson, A.E. (1991) J. Cell Biol. 112, 809–821.
24. Brunner, J., Senn, H. and Richards, F.M. (1980) J. Biol. Chem. 255, 3313–3318.
25. Brunner, J. and Richards, F.M. (1980) J. Biol. Chem. 255, 3319–3329.
26. Collins, P.G. and Gilmore, R. (1991) J. Cell Biol. 114, 639–650.
27. Vogel, F., Hartmann, E., Görlich, D. and Rapoport, T.A. (1990) Eur. J. Biochem. 188, 439–455.
28. Hartmann, E., Wiedmann, M. and Rapoport, T.A. (1989) EMBO J. 8, 2225–2229.
29. Prehn, S., Herz, J., Hartmann, E., Kurzchalia, T.V., Frank, R., Römisch, K., Dobberstein, B. and Rapoport, T.A. (1990) Eur. J. Biochem. 188, 439–455.
30. Nicchitta, C. and Blobel, G. (1990) Cell 60, 259–269.
31. Nicchitta, C. and Blobel, G. (1989) J. Cell Biol. 108, 789–795.
32. Deshaies, R.J. and Schekman, R. (1987) J. Cell Biol. 105, 633–645.
33. Deshaies, R.J. and Schekman, R. (1989) J. Cell Biol. 109, 2653–2664.
34. Sanz, P. and Meyer, D.I. (1989) J. Cell Biol. 108, 2101–2106.
35. Deshaies, R.J., Sanders, S.L., Feldheim, D.A. and Schekman, R. (1991) Nature 346, 806–808.
36. Akiyama, Y. and Ito, K. (1987) EMBO J. 6, 3465–3470.
37. Deshaies, R.J. and Schekman, R. (1990) Mol. Cell. Biol. 10, 6024–6035.
38. Sadler, I., Chiang, A., Kurihara, T., Rothblatt, J., Way, J. and Silver, P. (1989) J. Cell Biol. 109, 2665–2675.
39. Tajima, S., Lauffer, L., Rath, V.L. and Walter, P. (1986) J. Cell Biol. 103, 1167–1178.
40. Lively, M.O. and Walsh, K.A. (1983) J. Biol. Chem. 258, 9488–9495.
41. Evans, E.A., Gilmore, R. and Blobel, G. (1986) Proc. Natl. Acad. Sci. USA 83, 581–585.
42. Böhni, P.C., Deshaies, R.J. and Schekman, R. (1988) J. Cell Biol. 106, 1035–1042.
43. Kreibich, G., Czako-Graham, M., Grebenau, R., Mok, W., Rodriguez-Boulan, E. and Sabatini, D.D. (1978) J. Supramol. Struct. 8, 279–302.
44. Yu, Y.H., Sabatini, D.D. and Kreibich, G. (1990) J. Cell Biol. 111, 1335–1342.
45. Lingappa, V.R. (1991) Cell 65, 527–530.
46. Cosson, P., Lankford, S.P., Bonifacio, J.S. and Klausner, R.D. (1991) Nature 351, 414–416.
47. Briggs, M.S. and Gierasch, L.M. (1986) Adv. Protein Chem. 38, 109–180.
48. Görlich, D., Hartmann, E., Prehn, S. and Rapoport, T.A. (1992) Nature 357, 47–52.
49. Müsch, A., Wiedmann, M. and Rapoport, T.A. (1992) Cell 69, 343–352.
50. Kelleher, D.J., Kreibich, G. and Gilmore, R. (1992) Cell 69, 55–65.

CHAPTER 11

The role of GTP in protein targeting to the endoplasmic reticulum

STEPHEN C. OGG, JODI M. NUNNARI, JOSHUA D. MILLER
and PETER WALTER

Department of Biochemistry and Biophysics, University of California Medical School, San Francisco, CA 94143-0448, USA

Abstract

GTP has been shown to be required for the translocation of nascent secretory and membrane proteins into the lumen of the endoplasmic reticulum (ER). To date, three components known to be involved in protein translocation, the 54 kDa subunit of SRP (SRP54) and both the alpha and the beta subunits of the SRP receptor (SRα and SRβ) have been shown to be GTPases. Here, we briefly review what is known about the role of GTP in protein translocation, and we present a speculative model that incorporates these findings.

In recent years, biochemical, molecular biological and genetic techniques have revealed the global importance of GTPases in regulating biological processes. GTP binding and hydrolysis are used in cells to convey extracellular signals that affect growth and differentiation, to promote and regulate the translation of proteins, to direct protein transport through the eukaryotic secretory pathway, and to aid in the transport of nascent secretory and membrane proteins into the ER [1].

Through GTP binding and hydrolysis, GTPases can exist in two discrete conformations, a GTP bound conformation and a GDP bound conformation. Interconversion between these two states causes the GTPase to interact in temporal succession with its effectors, thereby regulating the biological process. In most cases, the GTPase switch is triggered from the outside; the exchange of bound GDP for GTP requires the action of specific guanine nucleotide release factors (GNRFs), and the hydrolysis of bound GTP requires the action of specific GTPase activating proteins. Here we focus on the role that three recently identified GTPases play during the translocation of nascent secretory and membrane proteins into the lumen of the rough endoplasmic reticulum.

Proteins that are destined to be secreted outside of the cell or inserted into the plasma membrane; and those that reside in the lumen or the membrane of the ER, the Golgi apparatus, lysosomes, endosomes and various transport vesicles are synthesized on ribosomes bound to the ER membrane [2]. Ribosomes synthesizing these proteins are selectively targeted to the ER by the action of two well defined components, the signal recognition particle (SRP), a soluble ribonucleoprotein composed of six polypeptides and SRP RNA, and the SRP receptor (SR), an ER membrane protein composed of two nonidentical polypeptides [3–7]. Together these components function to bring the ribosome to the ER membrane and to engage it with ER membrane proteins, collectively termed the translocon, which then catalyze the translocation of the nascent polypeptide across the lipid bilayer.

Targeting of a ribosome to the ER is specified by a signal sequence that is part of the growing polypeptide chain [8,9]. Once the signal sequence emerges from the translating ribosome in the cytoplasm, SRP binds to the nascent chain–ribosome complex [10]. UV crosslinking studies have demonstrated that the 54 kDa protein subunit of SRP (SRP54) binds directly to the signal sequence [11,12]. The binding of the SRP54 to the signal sequence results in a pause or arrest of elongation of the nascent chain and the formation of a translocation-competent complex [13,14]. Once the SRP–ribosome–nascent chain complex interacts specifically with the SRP receptor in the ER membrane, the nascent chain dissociates from SRP and elongation arrest is released [3,15]. Subsequently, the ribosome with its nascent chain protruding becomes tightly bound to the ER membrane via interactions with components of the translocon, and translocation of the nascent chain proceeds [16].

The amino acid sequences derived from cDNA clones of SRP54 and both SR subunits, SRα and SRβ, reveal that these proteins are members of the GTPase superfamily as defined by Bourne et al. [1] (see also [17–19]; J. Miller, unpublished results). All three proteins contain short segments of amino acids (regions G-1 to G-4 in Fig. 1A) that are highly conserved among GTPases [20,21]. These regions are known from the crystal structure of H-ras and EF-Tu to form loops on the surface of the protein which directly contact the bound GTP [22–24]. Based on the conservation of these regions, it is reasonable to assume that the basic folding unit of the GTPase domains of SRP54 and SRα resembles that of H-ras. This notion is plausible because H-ras and EF-Tu, although highly divergent in amino acid sequence, have a virtually superimposable tertiary structure. Indeed, SRP54, SRα and SRβ can be shown experimentally to bind GTP [18; J. Miller, unpublished results], although at this time it remains a conjecture that all three proteins hydrolyze the bound GTP during a functional cycle.

Interestingly, SRP54 and SRα form a unique subfamily of GTPases. Elements of the loops involved in GTP binding (e.g., the G-2 motif (DTFRAGA) and the G-4 motif (TKxD)) are conserved between SRP54 and SRα, but differ from the corresponding motifs of GTPases required for other cellular functions. The sequence similarity between SRP54 and SRα, however, extends well beyond the loops involved

A

	G-1	G-2	G-3	G-4
Consensus	OOOOGXXGXGKS/T	D-(X)$_n$-T	OJOODXAGJX	OOOONKXD
SRP54	imfvGlqGsGKT	DTFRAGA	iiivDtsGrh	svivTKlD
SRα	vtfcGvnGvGKS	DTFRAGA	vvlvDtaGrm	givlTKfD

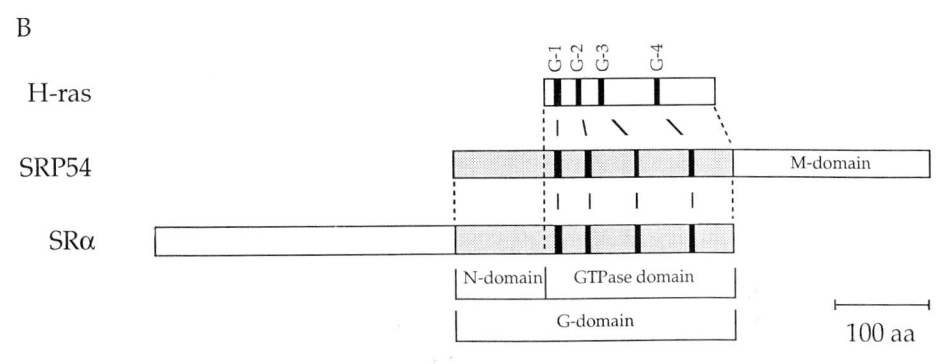

B

Fig. 1. Conserved sequence motifs of SRP54 and SRα. Panel A: the amino acid sequences of regions G-1 to G-4 of a consensus sequence, SRP54 and SRα are shown. Bold face type indicate amino acids conserved in nearly all GTPases; other upper-case letters indicate residues conserved between SRP54 and SRα. Amino acids are indicated according to the single letter amino acid code. In the consensus sequences, X represents any amino acid, while O and J represent hydrophobic and hydrophilic residues, respectively. Panel B: Schematic sequence lineup of H-ras, SRP54, and SRα. Bars represent the regions G-1 to G-4, shaded domains represent the G-domain which is conserved between SRP54 and SRα.

in GTP binding such that both proteins are significantly similar to one another. The regions of similarity (shaded in Fig. 1B) include a ras-like GTPase domain, with an additional 100 amino acids (N-domain) of weaker homology N-terminal to the GTPase domain [19,25]. Here, we refer to the N- and GTPase domains collectively as the G-domain. The similarities in the G-domains of SRP54 and SRα suggest that both proteins were evolutionarily derived from a common ancestor.

In addition to the G-domains, both SRP54 and SRα contain other domains that are unique to either protein. For SRP54, a methionine-rich domain, the M-domain, is attached to the C-terminus of the G-domain. This domain functions to bind SRP54 to SRP RNA and contains the signal sequence binding site [26,27]. For SRα, a unique domain is found at the N-terminus of the G-domain and functions to attach SRα specifically to the ER membrane [28]. We speculate that the homologous G-domains of SRP54 and SRα contribute a similar function to the two proteins, while the unique domains provide for their specific function and localization.

The GTPase domain of SRβ contains the conserved loops (G-1 to G-4) characteristic of GTPases, but beyond these few amino acids is not homologous to that of SRP54 and SRα, or any other GTPase subfamily (J. Miller, unpublished results). The only other characteristic feature is an N-terminal transmembrane segment, which presumably anchors the SR in the ER membrane.

Experimental data support a functional role of GTP in the translocation of nascent polypeptide chains across the ER. Initially, it was demonstrated that GTP is required for the insertion of the nascent polypeptide chain into the ER membrane [29,30]. A more detailed study revealed that GTP is required for the release of SRP from the nascent chain–ribosome complex [18]. A non-hydrolyzable analogue, GppNHp, allows targeting of SRP to the SRP receptor and the insertion of the nascent chain into the translocon, however, GppNHp prevents SRP from subsequently leaving the membrane [18,31]. Thus, following nascent chain targeting, GTP hydrolysis is required to release SRP from the SRP receptor to allow both components to be recycled.

Neither SRP nor its receptor have significant GTPase activity when assayed separately, however, when they are combined, GTP is hydrolyzed [32; Connolly and Gilmore, unpublished results]. A minimal SRP, consisting of only SRP54 and a small RNA, is sufficient to stimulate GTPase activity when combined with SR [32]. The possibility that other SRP proteins are responsible for the observed GTPase activity is thereby ruled out. Thus, there are three GTPases, SRα, SRβ, and SRP54, which may be responsible for the observed GTP hydrolysis. Preliminary experiments indicate that the GTP bound to SRP54 is hydrolyzed in this reaction (J.Miller, unpublished results). Consistent with this observation, synthetic signal peptides, which bind to SRP54, inhibit the receptor dependent GTPase activity (H. Wilhelm, and P. Walter, unpublished results). This observation is teleologically appealing, when considered in the context of the overall protein targeting reaction. Since GTP binding to SRP54 causes a release of SRP from the signal sequence and ribosome, it would not be advantageous for this to occur prior to targeting of the nascent chain to the ER membrane and interaction of the nascent chain–ribosome with components of the translocon.

In Fig. 2 we show a cartoon that depicts our current working model of how two of the GTPases, SRP54 and SRα may function to promote protein targeting. This model is consistent with all available data; however, it is almost certainly incomplete and many of its details are still very speculative.

As the nascent chain emerges from the ribosome, its signal sequence binds to SRP54. We propose that at this stage in the reaction SRP54 is in its GDP bound state. The binding of the signal sequence to SRP54 stabilizes this state and prevents the spontaneous exchange of GDP for GTP (Fig. 2, step 1). As previously shown, subsequent targeting of the SRP–ribosome–nascent chain complex to the ER membrane occurs via an interaction of SRP with SRα (Fig. 2, step 2). SRα, in association with other components of the translocon (see below) acts as a GNRF for SRP54 (Fig. 2, step 3). The release of GDP from SRP54 results in immediate binding of GTP and,

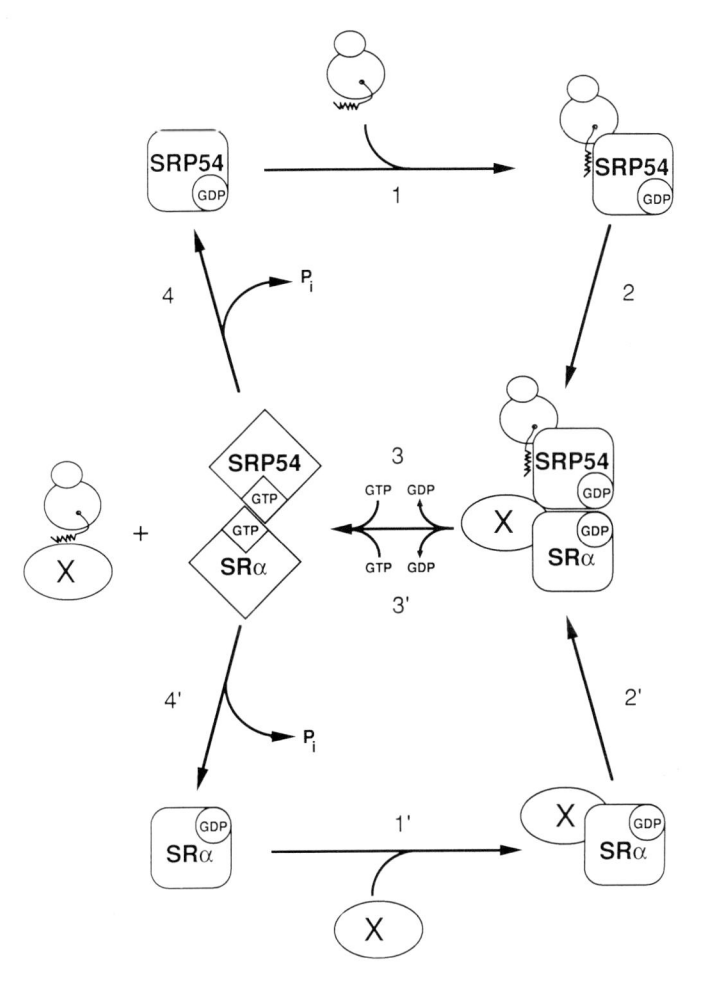

Fig. 2. Speculative model depicting the putative role of GTP in protein targeting and translocon assembly. The model is drawn to emphasize the roles of SRP54 and SRα, yet both components are part of their respective multicomponent complexes, SRP and SR. X represents a putative component of the translocon and may consist of one or more proteins.

consequently, a change in SRP54 conformation. This conformational change promotes the release of the signal sequence from SRP54, allowing the signal sequence to engage with other components of the translocon (in Fig. 2 depicted as component(s) X). After release of the SRP/SRP receptor complex from the translation and translocation machinery, hydrolysis of the bound GTP allows dissociation of SRP from its receptor (Fig. 2, step 4). Thus, SRP can return to a soluble pool, ready to interact with a nascent secretory protein emerging from the ribosome.

Similar to SRP54, SRα in its GDP bound form encounters components of the translocon (labeled X) in the ER membrane (Fig. 2, step 1') with which it interacts transiently to form a complex. As in the case of SRP54, binding of these components acts to stabilize SRα in its GDP bound form. Upon interaction of the SRα–X complex with the SRP–ribosome–nascent chain complex (Fig. 2, step 2'), GTP is exchanged for GDP on SRα causing X to be released from SRα (Fig. 2, step 3'). This release allows X to interact with the signal sequence, which has been recently released from SRP54. Hydrolysis of the bound GTP releases SRα from SRP, such that the SRP receptor can promote another round of translocation by interacting with a new X.

In the model, the functions of SRP54 and SRα are envisioned as symmetric in that both components function to preassemble components that are building blocks for the ribosome–membrane junction, produced as the result of the targeting reaction. It is an attractive notion in this regard that the required components in the cytoplasm (Fig. 2, step 1) and in the membrane (Fig. 2, step 1') obligatorily have to be assembled prior to their interaction (Fig. 2, steps 2 and 2'). Furthermore, we envision that the concomitant release of the signal sequence from SRP54 and X from SRα results in the binding of the signal sequence to a constituent of X in a reaction that may be controlled by the rate of GTP hydrolysis of SRP54 and SRα; the inherent rate of GTP hydrolysis by SRP54 and SRα must be slow enough to allow the signal sequence and X to functionally interact before either SRP54 or SRα return to their GDP bound states and thus regain their ability to rebind the previously released components.

In our model, SRP54 and SRα are proposed to effect GTP binding to one another. A similar scenario where two GTPases interact is observed in eukaryotic translation initiation [33]. The interaction of SRα with SRP54 may have evolved from an originally homotypic interaction of two identical G-domains which then diverged and became part of SRP54 and SRα, respectively. In the context of a homotypic interaction, the relatively high degree of sequence conservation in the N-domain (see Fig. 1A) suggests that these sequences may provide an additional function of the G-domain. One possibility is that the N-domains of SRα and SRP54 could be reciprocally required to release the bound guanine nucleotide.

The proposed model considerably extends our views regarding the role of SRP and SR in protein targeting. Previously, the role of SRP and SR in translocation was limited to targeting the ribosomes that synthesize a secretory protein to the translocon in the ER membrane. In contrast, we emphasize here the importance of SRP and SR in preassembling subcomplexes which then become joined in a precisely coordinated reaction. Thus, both SRP54 and SRα act as classical GTPases (such as eIF-2, for example) to effect the specific unidirectional assembly of macromolecular complexes required for translocation. We now postulate that SRα also regulates the assembly of the membrane components required for translocation by virtue of its GTP binding capability.

As presented, the model does not account for the GTPase domain of SRβ. More experimentation is required before we can speculate which step may need to be further subdivided to add yet another level of complexity. To date, the interaction of the components of the protein targeting machinery of the ER reflects the most complex system of interacting GTPases described. It will be fascinating to decipher the individual steps in their molecular detail.

Acknowledgements

We thank Drs. Henry Bourne and Reid Gilmore for helpful discussions. S.C.O. is a Howard Hughes Medical Institute Predoctoral fellow. J.M.N. is supported by a Gordon Tomkins Fellowship from the Department of Biochemistry and Biophysics of the University of California, San Francisco. This work was supported by grants from NIH and the Alfred P. Sloan Foundation.

References

1. Bourne, H.R., Sanders, D.A. and McCormick, F. (1990) Nature 348, 125–132.
2. Palade, G. (1975) Science 189, 347–358.
3. Gilmore, R., Blobel, G. and Walter, P. (1982) J. Cell Biol. 95, 463–469.
4. Meyer, D.I., Krause, E. and Dobberstein, B. (1982) Nature 297, 647–650.
5. Tajima, S., Lauffer, L., Rath, V.L. and Walter, P. (1986) J. Cell Biol. 103, 1167–1178.
6. Walter, P. and Blobel, G. (1980) Proc. Natl. Acad. Sci. USA 77, 7112–7116.
7. Walter, P. and Blobel, G. (1982) Nature 99, 691–698.
8. Blobel, G. and Dobberstein, B. (1975) J. Cell Biol. 67, 835–851.
9. Milstein, C., Brownlee, G.G., Harrison, T.M. and Mathews, M.B. (1972) Nature 239, 117–120.
10. Walter, P. and Blobel, G. (1981a) J. Cell Biol. 91, 551–556.
11. Krieg, U.C., Walter, P. and Johnson, A.E. (1986) Proc. Natl. Acad. Sci. USA 83, 8604–8608.
12. Kurzchalia, T.V., Wiedmann, M., Girshovich, A.S., Bochkareva, E.S., Bielka, H. and Rapoport, T.A. (1986) Nature 320, 634–636.
13. Walter, P. and Blobel, G. (1981b) J. Cell Biol. 91, 557–561.
14. Wolin, S.L. and Walter, P. (1989) J. Cell Biol. 109, 2617–2622.
15. Gilmore, R. and Blobel, G. (1983) Cell 35, 677–685.
16. Gilmore, R. and Blobel, G. (1985) Cell 42, 497–505.
17. Bernstein, H.D., Poritz, M.A., Strub, K., Hoben, P.J., Brenner, S. and Walter, P. (1989) Nature 340, 482–486.
18. Connolly, T. and Gilmore, R. (1989) Cell 57, 599–610.
19. Römisch, K., Webb, J., Herz, J., Prehn, S., Frank, R., Vingron, M. and Dobberstein, B. (1989) Nature 340, 478–482.
20. Dever, T.E., Glynias, M.J. and Merrick, W.C. (1987) Proc. Natl. Acad. Sci. USA 84, 1814–1818.
21. Halliday, K.R. (1984) J. Cyclic Nucleotide Res. 9, 435–448.
22. de Vos, A.M. et al. (1988) Science 239, 888–893.
23. LaCour, T.F.M., Nyborg, J., Thirup, S. and Clark, B.F.C. (1985) EMBO J. 4, 2385–2388.
24. Pai, E.F., Kabsch, W., Krengel, U., Holmes, K.C., John, J. an d Wittinghofer, A. (1989) Nature 341, 209–214.
25. High, S. and Dobberstein, B. (1991) J. Cell Biol. 113, 229–233.
26. Römisch, K., Webb, J., Lingelbach, K., Gausepohl, H. and Dobberstein, B. (1990) J. Cell Biol. 111, 1793–1802.
27. Zopf, D., Bernstein, H.D., Johnson, A.E. and Walter, P. (1990) EMBO J. 9, 4511–4517.

28. Andrews, D.W., Lauffer, L., Walter, P. and Lingappa, V.R. (1989) J. Cell Biol. 108, 797–810.
29. Connolly, T. and Gilmore, R. (1986) J. Cell Biol. 103, 2253–2261.
30. Hoffman, K.E. and Gilmore, R. (1988) J. Biol. Chem. 263, 4381–4385.
31. Connolly, T., Rapiejko, P.J. and Gilmore, R. (1991) Science 252, 1171–1173.
32. Poritz, M.A., Bernstein, H.D., Strub, K., Zopf, D., Wilhelm, H. and Walter, P. (1990) Science 250, 1111–1117.
33. Dholakia, J.N. and Wahba, A.J. (1989) J. Biol. Chem. 264, 546–550.

CHAPTER 12

Consecutive steps of nucleoside triphosphate hydrolysis are driving transport of precursor proteins into the endoplasmic reticulum

PETER KLAPPA, GÜNTER MÜLLER[#], GABRIEL SCHLENSTEDT[*],
HANS WIECH and RICHARD ZIMMERMANN

*Zentrum Biochemie/Abteilung Biochemie II der Universität, Gosslerstraße 12d,
W-3400 Göttingen, Germany*

Abstract

Transport of secretory proteins into the mammalian endoplasmic reticulum can be visualized as a sequence of various steps which include membrane association, membrane insertion and completion of translocation. It turns out that this transport depends on the hydrolysis of nucleoside triphosphates at various stages: (i) There is a GTP requirement in ribonucleoparticle-dependent transport. This GTP effect is related to the GTP binding proteins signal recognition particle (SRP) and docking protein. (ii) There is an ATP requirement in ribonucleoparticle-independent transport. This ATP effect is related to the cytosolic (termed cis-acting) molecular chaperone hsp70. (iii) Recently we addressed the question of whether there are additional nucleoside triphosphate requirements in protein transport into mammalian microsomes. We observed that a microsomal protein which depends on ATP hydrolysis is involved in membrane insertion of both, ribonucleoparticle-dependent and -independent precursor proteins. The azido-ATP sensitive protein was shown to be distinct from the lumenal (termed trans-acting) molecular chaperone BiP.

1. Introduction

Every polypeptide has a unique intra- or extracellular location where it fulfills its function. The following facts complicate our attempts to understand this situation:

[#] *Present address:* Hoechst AG, W-6230 Frankfurt am Main 80, Germany.
[*] *Present address:* Department of Molecular Biology, Princeton University, Princeton, New Jersey, USA.

(i) most proteins are synthesized in the cytosol, however, non-cytosolic proteins must subsequently be directed to a variety of different subcellular locations, and (ii) in the case of non-cytosolic proteins the sites of synthesis and of functional location are separated by at least one biological membrane. Consequently, mechanisms exist which ensure the specific transport of proteins across membranes. Here we discuss the mechanisms involved in export of newly synthesized secretory proteins.

There appear to be different ATP-dependent transport mechanisms for protein export [1]. One can distinguish between transport mechanisms involving signal peptides and those that do not. The signal peptide-independent mechanism takes place at the plasma membrane. It involves transport components which are related to the multiple drug resistance proteins, i.e. a family of ATP-dependent membrane proteins. The signal peptide-dependent mechanism, however, operates at the level of the membrane of the endoplasmic reticulum. From there, secretory proteins reach the extracellular space by vesicular transport. There are at least two different mechanisms for the transport of secretory proteins into the mammalian endoplasmic reticulum. Both mechanisms depend on the presence of a signal peptide on the respective precursor protein and involve a signal peptide receptor on the cytosolic surface of the membrane and a membrane component that is sensitive towards photoaffinity modification by azido-ATP. The decisive feature of the precursor protein with respect to which of the two mechanisms is used is the chain length of the polypeptide. The critical size seems to be around 70 amino acid residues (including the signal peptide). One mechanism is used by precursor proteins larger than about 70 amino acid residues and relies on the hydrolysis of GTP and two cytosolic ribonucleoparticles (ribosome and signal recognition particle) and their receptors on the microsomal surface (ribosome receptor and docking protein). The other mechanism is used by small precursor proteins and involves the hydrolysis of ATP and cytosolic molecular chaperones such as hsp70.

2. Results

We focus on the following presecretory proteins as tools for gaining insight into the molecular details of how proteins are transported into the mammalian endoplasmic reticulum: preprocecropin A [2–4], prepromelittin [5–7], and prepropeptide GLa [8]. All three precursor proteins contain a cleavable signal peptide and about 70 amino acid residues (including the signal peptide). We employ in vitro systems which are derived from mammalian organisms such as rabbit reticulocyte lysates and dog pancreas microsomes.

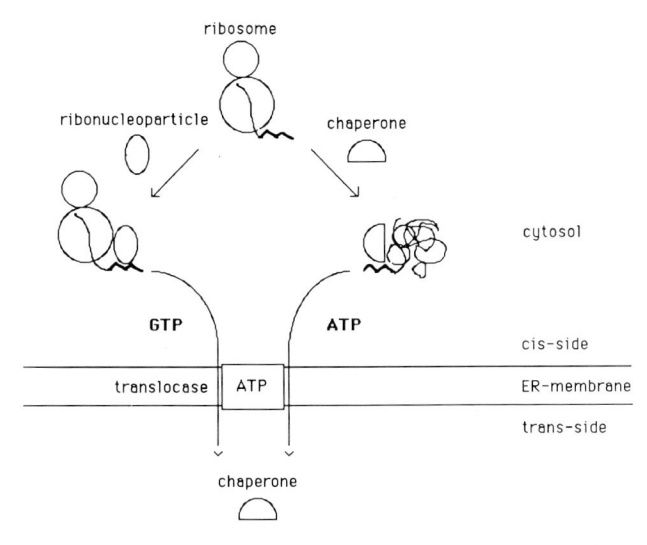

Fig. 1. Signal peptide-dependent transport of secretory proteins into the mammalian endoplasmic reticulum involves nucleoside triphosphate hydrolysis. Refer to Results for details.

2.1. Ribonucleoparticles versus molecular chaperones

It is clear that precursor proteins are not transported in their native (i.e. folded) state and that signal peptides are involved in preserving the unfolded state as well as in facilitating membrane recognition. Furthermore, it appears that there are two mechanisms preserving transport competence in the cytosol (Fig. 1, Table I). The mechanisms differ in how transport competence is preserved. In the first case protein synthesis is slowed down, in the second case protein folding and/or aggregation is slowed down. The first mechanism involves the hydrolysis of GTP and ribonucleoparticles and their receptors on the microsomal surface, the second mechanism does not involve ribonucleoparticles and their receptors but depends on the hydrolysis of ATP and on molecular chaperones. Small presecretory proteins (i.e. precursor proteins which contain less than 75 amino acid residues) such as preprocecropin A are the best substrates for the latter mechanism.

The ribonucleoparticle-dependent pathway seems to be used by the majority of presecretory proteins and has been analyzed in great detail (refer to Chapters 9 and 10 for references). It involves SRP and its receptor in the microsomal membrane, docking protein (SRP receptor) and the ribosome and its receptor. In addition, ribophorins I and II seem to be involved in this mechanism [9]. There is a GTP requirement in the transport of ribonucleoparticle-dependent precursor proteins [10–12]. This GTP effect is related to the GTP binding proteins, SRP and docking protein [11,13,14].

The first observations with respect to ribonucleoparticle-independent transport

TABLE I

Components involved in protein transport into the mammalian endoplasmic reticulum

Signal recognition particle	7S RNA
	SRP 72 kDa subunit
	SRP 68 kDa subunit
	SRP 54 kDa subunit
	SRP 19 kDa subunit
	SRP 14 kDa subunit
	SRP 9 kDa subunit
SRP receptor	DPα subunit
	DPβ subunit
Ribosome	
Ribosome receptor	
cis-Acting chaperone	hsp70
Translocase	Signal peptide receptor
	NEM-sensitive component
	Azido-ATP-sensitive component
	SSRα subunit
	SSRβ subunit
trans-Acting chaperone	BiP

were that the loosely folded (unfolded, denatured) precursor is the best substrate for transport and that the hydrolysis of ATP by cytosolic factors is involved in preserving this state [2,7,8,15]. In collaboration with M. Lewis and H. Pelham, we were able to demonstrate that hsp70 is part of what we had termed a cytosolic ATPase and that a second cytosolic protein (which in contrast to hsp70 is NEM-sensitive) is involved [16]. Our current working model proposes that hsp90 may be the protein of interest, the main reason being that it is enriched in a fraction that contains the desired activity (Fig. 2). We find this to be an attractive hypothesis for two reasons: (i) hsp70 and hsp90 were shown to cooperate with respect to hormone receptors and (ii) BiP (grp78, a member of the hsp70 family) and grp94 (a member of the hsp90 family) are present in the microsomal lumen.

The decisive feature of the precursor protein with respect to which of the two mechanisms is used is the chain length of the polypeptide. This conclusion was based on the observation that carboxy-terminal extension of a small precursor protein in size, typically leads to the phenotype of a large precursor protein [6,8]. If one takes into account that approximately 40 amino acid residues of a nascent polypeptide chain are buried in the ribosome [17–19] and that a signal peptide contains 20–30 amino acid residues [20–22] and, furthermore, that SRP can bind to signal peptides

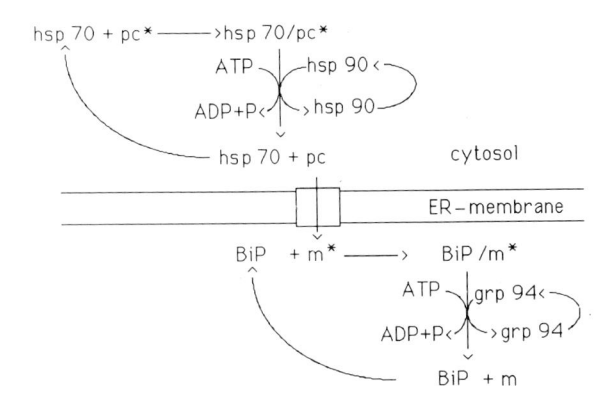

```
hsp 70 + pc* ────────>hsp 70/pc*
         ATP ─┐ ┌─hsp 90 <─┐
         ADP+P<┘ └─>hsp 90 ─┘
         └── hsp 70 + pc            cytosol
─────────────────────┬──┬──────────────────────
                     │  │           ER-membrane
─────────────────────┴──┴──────────────────────
         BiP  + m* ────────>   BiP /m*
         ATP ─┐ ┌─grp 94 <─┐
         ADP+P<┘ └─>grp 94 ─┘
         └── BiP + m
```

Fig. 2. Molecular chaperones are involved in ribonucleoparticle-independent transport. pc, precursor after release from cis-acting molecular chaperone (molten globule state); pc*, precursor during or after release from ribosome; m, mature protein after release from trans-acting molecular chaperone (native state); m*, mature protein during or after release from translocase. Refer to Results for details.

only as long as they are presented by a ribosome [23,24], one can imagine that precursor proteins with less than 60–70 amino acids cannot make use of the two ribonucleoparticles; they are released before SRP can bind to the signal peptide. However, the ribonucleoparticle-independent mechanism can also be used by a large precursor protein [2]. A synthetic hybrid between preprocecropin A and dihydrofolate reductase, translocates post-translationally (without the involvement of signal recognition particle and ribosome). This was directly demonstrated by adding methotrexate to the translocation reaction. Methotrexate and related drugs bind to ppcecDHFR after it is completed and released from the ribosome, stabilize the native conformation of the DHFR domain and allow membrane insertion but block completion of translocation.

2.2. Translocase

We assume that the two pathways converge at the level of a putative signal peptide receptor which may be identical to the 45 kDa protein that was characterized as a signal sequence binding protein in microsomal membranes [25]. Besides this protein, biochemical evidence points to additional membrane proteins as parts of a general translocase (Table I).

There is an ATP-requiring step at the microsomal level which is involved in both mechanisms and which is not related to the lumenal molecular chaperone BiP [4].

After solubilization in DMSO and subsequent dilution into an aqueous buffer, the transport of the chemically synthesized and purified precursor protein preprocecropin A* occurs in the absence of molecular chaperones but depends on the hydrolysis of ATP. The concentration of ATP that leads to half-maximal stimulation is in the order of 10 μM. At this concentration other nucleotides cannot substitute for ATP. In other words, the effect appears to be specific for ATP. Furthermore, non-hydrolyzable ATP analogs, such as AMP-PCP or AMP-PNP, cannot substitute for ATP. Since these analogs compete with ATP, one can conclude that the hydrolysis of ATP is required. Photoaffinity modification of dog pancreas microsomes with 8-azido-ATP leads to inactivation of the microsomes with respect to membrane insertion of preprocecropin A* as well as of prepro-α-factor and preprolactin. Therefore, we concluded that a hitherto unknown microsomal protein that depends on ATP hydrolysis is involved in membrane insertion of both ribonucleoparticle-dependent and -independent precursor proteins (Fig. 1, Table I). We are currently employing a combination of two approaches in order to identify the ATP-dependent component of interest: photoaffinity modification of microsomal proteins with ^{32}P-8-azido-ATP and affinity purification of ATP-binding proteins from microsomal extracts.

Although BiP is an ATP-binding protein and is modified by azido-ATP, it appears to be distinct from the azido-ATP sensitive component that is involved in protein transport. Treatment of dog pancreas microsomes with octyl glucoside and subsequent removal of the detergent leads to depletion of the lumenal content. Under these conditions more than 90% of BiP is removed. Protein transport, however, is unaffected. Since it is very unlikely that photoaffinity modification leads to more than 90% derivatization of its targets, BiP cannot be the target of the observed inhibition of protein transport after photoaffinity modification of microsomes. However, this result does not rule out the possibility that BiP is involved in protein transport under these conditions.

In addition, ribonucleoparticle-independent transport of presecretory proteins involves a membrane component which is sensitive to chemical alkylation with N-ethylmaleimide, i.e. which has an essential sulfhydryl [3]. The sulfhydryl is cytoplasmically exposed and is involved in membrane insertion but not in membrane binding of the precursor proteins (M. Zimmermann, unpublished observation). This component may be identical to an N-ethylmaleimide-sensitive component which acts past docking protein and ribosome receptor in ribonucleoparticle-dependent transport [26,27].

The so-called SSR subunits appear to be part of the translocase and can be expected to be generally involved [28–32]. We addressed the question of what stage of ribonucleoparticle-dependent transport is affected after photoinactivation of microsomes by azido-ATP [33]. Thus, a nascent presecretory protein was employed. We observed that the nascent precursor protein does not become associated with the SSR complex after photoaffinity labeling of microsomes with azido-ATP. We concluded that the microsomal protein, which is sensitive to photoaffinity labeling with azido-ATP, acts prior to the SSR complex.

3. Discussion

3.1. Components involved in protein transport into yeast endoplasmic reticulum

With respect to yeast microsomes, genetic and biochemical evidence demonstrate a role for the cis-acting chaperone hsp70 and a second, NEM-sensitive, protein [34,35]. However, there also is ribonucleoparticle-dependent protein transport in yeast [36–38]. We assume that the two pathways converge at the level of a putative signal peptide receptor [39]. Genetic evidence suggests that the membrane proteins sec61, sec62 and sec63 (also termed ptl1 or npl1) are generally involved in protein transport [40–43]. Biochemical evidence suggests that the sec61, sec62 and sec63 proteins transiently form complexes with a 31.5 kDa glycoprotein and a 23 kDa protein, i.e. two proteins that are reminiscent of two mammalian ER proteins which have been termed SSR α- and β-subunit [44]. Furthermore, the trans-acting chaperone BiP (KAR2 gene product) has been shown to have a role in transport [45].

3.2. Model for ribonucleoparticle-independent transport

It is clear that precursor proteins have to be unfolded to be translocated and that unfolding has to occur on the cis-side of the respective membrane (Fig. 3). It ap-

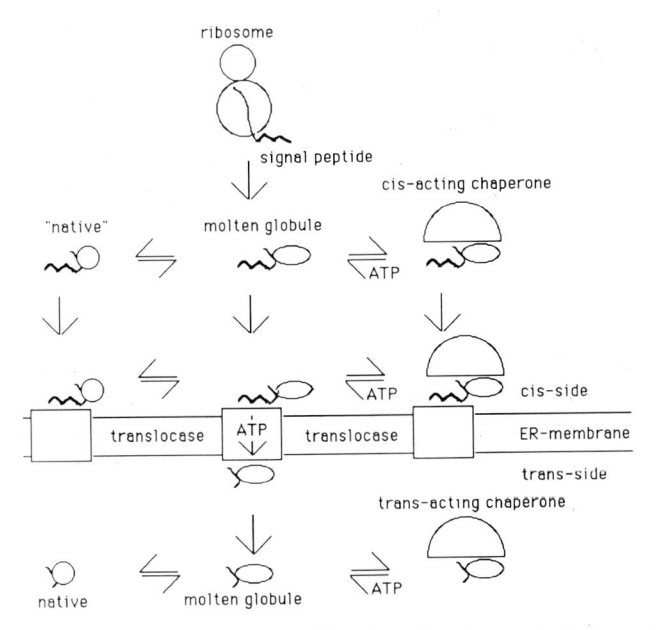

Fig. 3. Model for ribonucleoparticle-independent transport of presecretory proteins into the endoplasmic reticulum. Refer to Discussion for details.

pears that the signal peptide interferes with folding of the precursor to the native conformation of the mature part to a certain degree. Therefore, precursor proteins interact with molecular chaperones at some stage of their synthesis. This interaction has to be reversible, however, in order to eventually allow translocation. This may represent the point where ATP hydrolysis and the additional component come into action. Membrane association of the precursor proteins occurs via a putative signal peptide receptor. At this stage the precursor may be in a native-like folding state or in the molten globule state; it may be free or bound to a molecular chaperone.

With the help of the translocase, the signal peptides are then inserted into the membrane, most likely in the form of a loop structure which is made up by the signal peptide plus the amino terminus of the mature part. The ATP hydrolysis at the microsomal level seems to be directly providing the energy for membrane insertion. In order to become inserted, the precursor has to unfold at least partially, starting at its amino terminus. The question is where does the energy for unfolding come from. Practically all precursor proteins carry signal peptides that are cleaved off during or after translocation by signal peptidase. Thus, in principle, the differences between the free energies of precursor versus mature forms of a protein could be sufficient to drive unfolding at the surface. Furthermore, the energy for complete unfolding of a precursor protein may be as low as 10 kcal/mol, i.e. the initial hydrolysis of one ATP could be sufficient to drive such an unfolding reaction.

In order for translocation to progress, the protein on the cis-side has to unfold further. Again, the question is where does the energy for unfolding come from. A possible answer to this question may reside in the recent observation that protein transport into yeast microsomes involves the trans-acting molecular chaperone BiP. However, a similar requirement for BiP in mammalian microsomes has not yet been observed. It is tempting to speculate that binding of the precursor protein in transit to the trans-acting molecular chaperone provides the energy. Alternatively, completion of translocation may be driven by spontaneous refolding on the trans-side of the target membrane.

3.3. Open questions

Even 20 years after the signal hypothesis was first put forward, one of the major open questions is whether the components of the translocase form a pore, i.e. an aqueous channel that the precursor protein in transit passes through or whether the translocase is a set of enzymes that facilitates translocation at a lipid/protein interface.

Acknowledgements

We would like to acknowledge the collaboration with Hans G. Boman and Gudmundur H. Gudmundsson at the University of Stockholm, Günther Kreil and Christa Mollay at the Austrian Academy of Sciences in Salzburg, Hugh R.B. Pelham and Mike J. Lewis at the Medical Research Council in Cambridge, Peter Mayinger and Martin Klingenberg at the University of Munich, Johannes Buchner, Ursula Jakob and Rainer Jaenicke at the University of Regensburg, and William Wickner and Colin Watts at the University of California in Los Angeles. The authors' work on this subject was supported by the Deutsche Forschungsgemeinschaft and by the Fonds der Chemischen Industrie.

References

1. Wiech, H., Klappa, P. and Zimmermann, R. (1991) FEBS Lett. 285, 182–188.
2. Schlenstedt, G., Gudmundsson, G.H., Boman, H.G. and Zimmermann, R. (1990) J. Biol. Chem. 265, 13960–13968.
3. Zimmermann, R., Sagstetter, M. and Schlenstedt, G. (1990) Biochimie 72, 95–101.
4. Klappa, P., Mayinger, P., Pipkorn, R., Zimmermann, M. and Zimmermann, R. (1991) EMBO J. 10, 2795–2803.
5. Zimmermann, R. and Mollay, C. (1986) J. Biol. Chem. 261, 12889–12895.
6. Müller, G. and Zimmermann, R. (1987) EMBO J. 6, 2099–2107.
7. Müller, G. and Zimmermann, R. (1988) EMBO J. 7, 639–648.
8. Schlenstedt, G. and Zimmermann, R. (1987) EMBO J. 6, 699–703.
9. Yu, Y., Sabatini, D. and Kreibich, G. (1990) J. Cell Biol. 111, 1335–1342.
10. Connolly, T. and Gilmore, R. (1986) J. Cell Biol. 103, 2253–2261.
11. Connolly, T. and Gilmore, R. (1989) Cell 57, 599–610.
12. Connolly, T., Rapiejko, P.J. and Gilmore, R. (1991) Science 252, 1171–1173.
13. Römisch, K., Webb, J., Herz, J., Prehn, S., Frank, R., Vingron, M. and Dobberstein, B. (1989) Nature 340, 478–482.
14. Bernstein, H.D., Poritz, M.A., Strub, K., Hoben, P.J., Brenner, S. and Walter, P. (1989) Nature 340, 482–486.
15. Wiech, H., Sagstetter, M., Müller, G. and Zimmermann, R. (1987) EMBO J. 6, 1011–1016.
16. Zimmermann, R., Sagstetter, M., Lewis, J.L. and Pelham H.R.B. (1988) EMBO J. 7, 2875–2880.
17. Malkin, L.I. and Rich, A. (1967) J. Mol. Biol. 26, 329–346.
18. Blobel, G. and Sabatini, D.D. (1970) J. Cell Biol. 45, 130–145.
19. Bernabeu, C. and Lake, J.A. (1982) Proc. Natl. Acad. Sci. USA 79, 3111–3115.
20. von Heijne, G. (1981) Eur. J. Biochem. 116, 419–422.
21. Perlman, D. and Halvorson, H.O. (1983) J. Mol. Biol. 167, 391–409.
22. von Heijne, G. (1984) EMBO J. 3, 2315–2318.
23. Ainger, K.J. and Meyer, D.I. (1986) EMBO J. 5, 951–955.
24. Wiedmann, M., Kurzchalia, T.V., Bielka, H. and Rapoport, T.A. (1987) J. Cell Biol. 104, 201–208.
25. Robinson, A., Kaderbhai, M.A. and Austen, B.M. (1987) Biochem. J. 242, 767–777.
26. Hortsch, M., Avossa, D. and Meyer, D.I. (1986) J. Cell Biol. 103, 241–253.
27. Nicchitta, C.V. and Blobel, G. (1989) J. Cell. Biol. 108, 789–795.
28. Wiedmann, M., Kurzchalia, T.V., Hartmann, E. and Rapoport, T.A.(1987) Nature 328, 830–833.
29. Krieg, U.C., Johnson, A.E. and Walter, P. (1989) J. Cell Biol. 109, 2033–2043.
30. Hartmann, E., Wiedmann, M. and Rapoport, T.A. (1989) EMBO J. 8, 2225–2229.
31. Görlich, D., Prehn, S., Hartmann, E., Herz, J., Otto, A., Kraft, R., Wiedmann, M., Knespel, S., Dobberstein, B. and Rapoport, T. (1990) J. Cell Biol. 111, 2283–2294.

32. Prehn, S., Herz, J., Hartmann, E., Kurzchalia, T.V., Frank, R., Roemisch, K., Dobberstein, B. and Rapoport, T.A. (1990) Eur. J. Biochem. 188, 439–445.
33. Zimmermann, R., Zimmermann, M., Mayinger, P. and Klappa, P.(1991) FEBS Lett. 286, 95–99.
34. Chirico, W.J., Waters, G.M. and Blobel, G. (1988) Nature 332, 805–810.
35. Deshaies, R.J., Koch, B.D., Werner-Washburne, M., Craig, E.A. and Schekman, R. (1988) Nature 332, 800–805.
36. Ribes, V., Dehaux, P. and Tollervey, D. (1988) EMBO J. 7, 231–237.
37. Poritz, M.A., Siegel, V., Hansen, W. and Walter, P. (1988) Proc. Natl. Acad. Sci. USA 85, 4315–4319.
38. Hann, B.C., Poritz, M.A. and Walter, P. (1989) J. Cell Biol. 109, 3223–3230.
39. Sanz, P. and Meyer, D.I. (1989) J. Cell Biol. 108, 2101–2106.
40. Deshaies, R.J. and Schekman, R. (1987) J. Cell Biol. 105, 633–645.
41. Deshaies, R.J. and Schekman, R. (1989) J. Cell Biol. 109, 2653–2664.
42. Sadler, I., Chiang, A., Kurihara, T., Rothblatt, J., Way, J. and Silver, P. (1989) J. Cell Biol. 109, 2665–2675.
43. Toyn, J., Hibbs, A.R., Sanz, P., Crowe, J. and Meyer, D.I. (1988) EMBO J. 7, 4347–4353.
44. Deshaies, R.J., Sanders, S.L., Feldheim, D.A. and Schekman, R. (1991) Nature 349, 806–808.
45. Vogel, J.P., Misra, L.M. and Rose, M.D. (1990) J. Cell Biol. 110, 1885–1895.

Part C

Vacuoles

CHAPTER 13

Mechanism and regulation of import and degradation of cytosolic proteins in the lysosome/vacuole

HUI-LING CHIANG and RANDY SCHEKMAN

Department of Molecular and Cell Biology, Howard Hughes Medical Institute, University of California at Berkeley, 401 Barker Hall, Berkeley, CA 94720, USA

Abstract

Protein degradation is an essential process in cells serving to eliminate abnormal proteins, regulate protein activities and provide amino acids during starvation for the synthesis of critical proteins. Protein degradation is regulated by hormones and nutrients, and increases when cells are deprived of serum. The increase in degradation is accompanied by the formation of autophagic vacuoles containing sequestered cytoplasm and organelles. Biochemical and morphological evidence indicates that starvation induced protein degradation occurs in lysosomes. This process has been studied by red-cell mediated microinjection to introduce radiolabeled proteins into cultured fibroblasts. Degradation of RNase A is enhanced in response to serum deprivation. The sequence required for RNase A degradation has been localized to a short peptide (lys-phe-glu-arg-gln). Antibodies raised against this pentapeptide precipitate 25–30% of cytosolic proteins. These proteins are selectively degraded when cells are deprived of serum. An hsc73 isozyme specifically recognizes the peptide sequence. Degradation of RNase A by isolated lysosomes is stimulated by ATP and hsc73 protein. We have extended our studies from mammalian lysosomes to the yeast vacuole. Catabolic inactivation of fructose 1,6-bisphosphatase (FBPase) is mediated by vacuolar degradation of FBPase in response to glucose. Degradation of FBPase is dependent on the *PEP4* gene whose product is required for the activities of several vacuolar proteases. Immunofluorescence and cell fractionation experiments further indicate that FBPase is redistributed from the cytosol to the vacuole in response to glucose. The identification of such an import process defines a new protein targeting pathway in yeast. Degradation of FBPase requires protein synthesis. Vacuolar targeting of FBPase also requires the early part of the secretory pathway. We propose that a receptor protein is synthesized in response to glucose.

This protein traverses the early part of the secretory pathway en route to the vacuole. The presence of such a factor on the vacuolar membrane is required for import of FBPase to occur.

1. Introduction

Protein degradation serves in the following essential roles:

(a) *Regulation of protein concentration or enzymatic activity.* Most proteins are synthesized and degraded. The steady state concentration of a protein is, therefore, regulated by the balance of the two processes [1–4]. For example, cyclin, an important factor controlling the cell cycle, changes concentration with the cell cycle, but is synthesized at a constant rate. The variation of cyclin concentration is entirely due to protein degradation [3]. In rat liver, glucocorticoid induces several enzymes [4]. The induction of aminotransferase is more rapid than that of glutamine-alanine transaminase [4]. This difference is due to variation in the rate of degradation of these enzymes [4].

(b) *Elimination of abnormal or denatured proteins.* Abnormal proteins synthesized as a consequence of errors in transcription or translation are rapidly degraded. Proteins that are damaged, mislocalized, or not assembled properly are also eliminated from cells by protein degradation [5–7]. This process requires ATP and ubiquitin conjugation [5–7] Ubiquitin is a small protein of high intracellular abundance which is conjugated to protein via its C-terminal glycine to an ϵ-amino group of lysine residues in the target proteins. Ubiquitin is activated by component E1 (ubiquitin activating enzymes) and transferred to component E2 (ubiquitin conjugating enzymes) prior to transfer to a target protein. Some protein substrates require an additional factor, E3 (ubiquitin recognition enzymes), to become ubiquitinated [7]. Protein-ubiquitin conjugates are degraded in a large protease complex, called the proteosome [8]. Activation as well as degradation requires ATP hydrolysis [5–8]. Ubiquitin is involved in strikingly diverse cellular functions such as cell cycle control [9], DNA repair [10], ribosome biosynthesis [11], stress response [12] and ATP-dependent degradation of certain proteins [13–15]. Using mouse ts85 mutant cells containing a defective E1 at the non-permissive temperature, Ciechanover et al. showed that 90% of short-lived and abnormal proteins are degraded by the ubiquitin dependent pathway [13]. In *Saccharomyces cerevisiae*, ubiquitin is also involved in degrading abnormal and short-lived proteins [14,15]. Cells in which *UBC1*, *UBC4* and *UBC5* are deleted are defective in degrading these proteins [14,15].

(c) *Protein degradation is regulated by hormones and changes in external environment.* Several hormones or nutritional factors are important in growth regulation and influence protein synthesis, protein degradation, or both. Examples of regulators of protein degradation include insulin [16,17], glucagon [18,19], growth factors [20,21] and amino acids [22,23]. Insulin is released after a meal and stimulates protein synthesis while inhibiting protein degradation [16,17]. Glucagon is released between meals and increases protein degradation [18,19]. Starvation of amino acids increases protein degradation, while addition of amino acid decreases protein degradation [20,22,23]. Increased protein degradation leads to the mobilization of tissue proteins and provides amino acids to be used directly as energy sources, for the synthesis of critical enzymes, and for gluconeogenesis.

(d) *Protein degradation is important for cell growth and development.* During growth and development, massive remodeling is required. Cells increase synthesis of certain proteins, and at the same time eliminate proteins that are no longer needed. Massive protein degradation has been observed when yeast cells are grown in poor media to induce sporulation [24–26]. In this condition, cells degrade 70–80% of the pre-existing vegetative proteins [24–26]. Degradation occurs in the vacuole, as mutants defective in vacuolar proteinases do not engage in large-scale protein degradation during meiosis and as a consequence do not complete the sporulation process [25,26].

(e) *Altered protein degradation is associated with diseases.* Muscular dystrophy [27], muscle atrophy following denervation [28] or muscle wasting in cancer patients [29] are all associated with abnormal protein degradation. In aged cultured fibroblasts, protein degradation is decreased [30]. Lysosomes filled with subunits of the mitochondrial ATP synthase have been observed in Batten's disease [31]. Accumulation of ubiquitin is also found in the granulovacuolar areas in Alzheimer's disease [32] and certain other neurodegenerative diseases [33].

2. Intracellular protein degradation in mammalian lysosomes

There are several pathways of protein degradation in cells. Ubiquitin dependent degradation has been studied extensively and reviewed [34,35]. Other pathways include mitochondrial [36,37], ER proteolysis [38,39] and calcium dependent proteases [40,41]. This chapter focuses on lysosomal/vacuolar mediated degradation.

In rat liver, the rates of protein degradation vary from 0.3%/h to 4.5%/h depending on the growth conditions [42–45]. Usually, 0.3–1.5%/h is defined as basal degradation, whereas 1.5–4.5%/h is regarded as accelerated or induced protein degradation. The contribution of lysosomes to basal degradation has been controversial, while accelerated degradation is attributed mainly to lysosomes [42–45]. Marzella and Glaumann have estimated that in hepatocytes during starvation, 80% of protein degradation occurs in lysosomes, whereas only less than 20% of protein degradation occurs in the cytosol [45,46]. The role lysosomes play in protein degradation has been elucidated both biochemically and morphologically.

2.1. Microautophagy

Glaumann and colleagues showed that isolated lysosomes are capable of degrading radiolabeled methemoglobin and insulin in vitro [46]. Lysosomes degrade 10% of added protein during 60 min of incubation and chloroquine decreases TCA soluble activity by 50%. Control experiments indicate that rupture of lysosomes during incubation is minimal. When detergent is used to solubilize lysosomes, 40% of exogenously added protein is degraded and that degradation is not affected by chloroquine [46]. Using electron dense Percoll particles to visualize lysosomal degradation, Glaumann et al. found that a cup-like invagination of membrane is formed followed by the fission of the internalized membrane to form intralysosomal vesicles. Subsequent opening of vesicles into the lysosomal matrix permits hydrolysis of the trapped material. This process, called microautophagy, has been postulated as a mechanism by which lysosomes internalize and degrade cytoplasmic proteins under basal conditions [46,47].

2.2. Macroautophagy

Macroautophagy involves the formation of a vacuolar membrane which surrounds a region of cytoplasm that includes cytosol and other organelles [48,49]. The process is completed by fusion of the autophagic vacuole with the lysosomes. Macroautophagy is induced when cells are starved of amino acids or treated with glucagon to increase protein degradation [48–50]. Macroautophagy is suppressed by 3-methyladenine [51–53] which has no effect on protein synthesis, the level of ATP, or degradation of exogenous proteins that are endocytosed and delivered to the lysosome [52,53].

A stepwise pathway of autophagosome maturation and lysosomal fusion has been developed by analysis of membrane antigen in the developing organelle [48,49]. The inner and outer limiting membrane of nascent vacuoles appear to derive from ER as they are decorated by antibodies directed against integral membrane proteins of the ER. Maturation involves fusion with an immature lysosome that lacks the normal hydrolytic enzymes, but contains lysosomal membrane proteins. The fused compartment then becomes acidic and acquires lysosomal hydrolases [48,49].

Fractionation studies reveal a population of autophagic vacuoles that are less dense than lysosomes [54,55]. Glucagon treatment or amino acid deprivation produce these less dense structures which gradually acquire the lysosomal marker, N-acetyl-β-D-glucosaminidase [54,55].

Autophagic degradation is partially blocked by lysosomotropic amines such as ammonia, propylamine, methylamine, and chloroquine [56,57]. Starvation-induced protein degradation is also inhibited by leupeptin (inhibits lysosomal thiol proteases), antipain (thiol and serine protease inhibitors), and bestatin (exopeptidase inhibitor) [58–60].

2.3. Lysosomal protein degradation in cultured fibroblasts

The mechanism of lysosomal protein degradation has been studied in cultured fibroblasts using a red-cell mediated microinjection technique [61,62]. Red cells are lysed in hypotonic solution to release contents and membranes are resealed in a hypertonic solution containing radiolabeled protein. Cells containing trapped radiolabeled proteins are then fused with cultured cells using polyethylene glycol (PEG) or Sendai virus [61,62]. This procedure permits simultaneous injection of radiolabeled proteins in a relatively short period of time. Careful examination indicates that red-cell mediated microinjection does not perturb cell metabolism, cell growth or protein synthesis [61,62].

Several proteins were microinjected into IMR-90 human lung fibroblasts, and their catabolism in the presence and absence of serum was measured. Catabolism of BSA, ovalbumin and lysozyme are not significantly altered by the medium composition, but catabolism of ribonuclease A (RNase A) and certain other proteins are increased upon serum withdrawal [63,64]. The half-life of RNase A in the presence of serum is approximately 100 h. Whereas in the absence of serum, the half-life decreases to 50 h [63,64].

Degradation of RNase occurs in lysosomes. RNase A is randomly distributed throughout the cytoplasm of fibroblasts immediately after injection, and a fraction of injected RNase A is associated with lysosomes at steady state. Microinjected RNase A tagged with [^{3}H]raffinose shows that the radioactive degradation products are completely retained within lysosomes. Since both the lysosomal membrane and plasma membrane are impermeable to raffinose coupled to lysine or small peptides, these results imply that degradation of microinjected RNase A takes place in lysosomes [65]. In addition, degradation of RNase A is inhibited by lysosomotropic agents, and the enhanced degradation in the absence of serum results from an increase in the rate of delivery of this protein to lysosomes [65].

RNase A can be cleaved by subtilisin at residue 20 to generate S-peptide (1–20) and S-protein (21–120). Catabolism of microinjected S-protein is not affected by serum withdrawal, while catabolism of microinjected S-peptide shows a twofold increase upon serum withdrawal. When S-peptide is reconstituted with S-protein, the catabolism of S-protein increases twofold upon serum deprivation [63]. These results suggest that S-peptide contains the information for the enhanced degradation of RNase A.

Covalent linkage of S-peptide to insulin A chain and lysozyme, whose catabolic rates are not affected by serum, causes the degradation of the conjugates to increase twofold upon serum withdrawal [66]. These results indicate that microinjected proteins are selectively degraded and that a recognition sequence for the enhanced catabolism is within the N-terminal 20 amino acids of RNase A. The entire 20 amino acid sequence is not required for the starvation induced degradation since the N-terminal 1–14 and 4–13 derivatives are degraded in a serum dependent fashion, while 1–10 and 2–8 are not [67].

The sequence lys-phe-glu-arg-gln (KFERQ, residues 7–11 of RNase A) has been identified as the crucial region responsible for enhanced degradation of RNase A during serum deprivation. Microinjected KFERQ is degraded twice as rapidly during serum deprivation [67]. Furthermore, co-microinjection of radiolabeled RNase A with increasing amounts of nonradiolabeled KFERQ abolishes the enhanced degradation of RNase A after serum withdrawal, while it has no effect of the degradation of RNase A in the presence of serum [67].

In order to examine whether the KFERQ peptide is of general importance in targeting intracellular proteins to lysosomes during serum deprivation, antibodies directed to the pentapeptide were characterized with regard to binding of cellular proteins. Binding is competitively inhibited by S-peptide, KFERQ, RNase A, and by other proteins containing sequences similar to KFERQ [68]. When cells were radiolabeled and fractionated into nuclear, mitochondria/lysosomes, microsomes and cytosol, KFERQ peptide antibodies recognize a large fraction (25–30%) of cytosolic proteins

Degradation of radiolabeled immunoreactive and nonimmunoreactive cellular

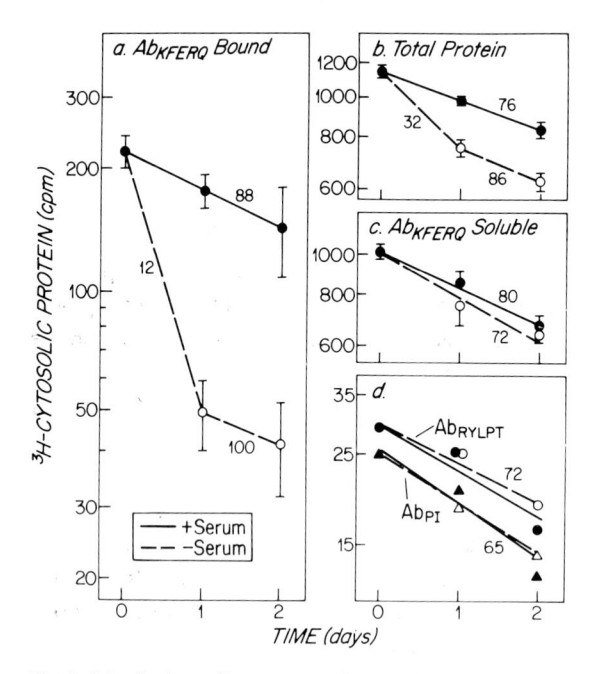

Fig. 1. Selective loss of immunoreactive proteins during serum starvation. Confluent cultures of fibroblasts were radiolabeled and chased in the presence or absence of serum for 1 or 2 days. Cytosolic proteins were fractionated and immunoprecipitated. (a) Radioactivity precipitated with anti-KFERQ IgGs after subtraction of radioactivity associated with pre-immune IgGs. (b) Radioactivity in unfractionated cytosolic proteins. (c) Radioactivity in proteins that are not precipitated with anti-KFERQ IgGs. (d) Radioactivity associated with pre-immune IgGs and affinity purified anti-RYLPT IgGs. Numbers on the figure refer to half-lives in hours. Reprinted with permission of the *Journal of Biological Chemistry* [68].

proteins was examined in serum-deprived and nonstarved cells. Figure 1b shows that unfractionated cytosolic proteins are degraded with an average half-life of approximately 80 h in the presence of serum and that the average half-life is reduced to 32 h after serum deprivation. The enhanced degradation is maintained for 1 day, then degradation returns to basal level. Immunoprecipitated proteins are degraded with a half-life of approximately 88 h in the presence of serum. The half-life dramatically decreases to 12 h during the first day of serum deprivation and returns to basal level after 1 day (Fig. 1a). Cellular proteins that do not interact with antibodies to KFERQ are degraded with a half-life of 80 h both in the presence and absence of serum (Fig. 1c). Based on these results, immunoreactive proteins are estimated to be responsible for 90% of the enhanced protein degradation during serum starvation [68].

As a further control, different antibodies against the pentapeptide, arg-tyr-leu-pro-thr (RYLPT), were raised and affinity purified. Figure 1d shows that anti-RYLPT recognizes a small fraction of cytosolic proteins and the amount of protein precipitated is the same for cells in the presence and absence of serum. In addition, preimmune antibodies precipitate only a small amount of radioactivity and these proteins show no difference in protein degradation in the presence and absence of serum (Fig. 1d).

The binding of cytosolic proteins to anti-KFERQ antibodies competes with proteins that show starvation dependent degradation, such as pyruvate kinase, aspartate aminotransferase, hemoglobin, and triosphosphate isomerase [68]. Sequence analysis of these proteins has been described in detail by Dice [69] and is not discussed here.

As a first step toward understanding of the mechanism of lysosomal degradation, a search for a cytosolic factor(s) that recognizes this peptide sequence was conducted. An S-peptide affinity matrix was incubated with cytosol and eluted with excess S-peptide. A 73 kDa protein (prp73, peptide recognition protein of 73 kDa) from rat liver binds to S-peptide [70].

Amino acid sequence analysis of fragments of prp73 shows 100% identity with chicken hsp70 and rat hsc73. Furthermore, purified prp73 from rat liver and human fibroblasts is recognized by monoclonal antibodies to *Drosophila* hsp70, which recognizes most 70 kDa heat shock proteins from yeast to human.

In vitro experiments demonstrate that prp73 stimulates lysosomal degradation of radiolabeled RNase A and S-peptide both in permeabilized CHO cells and in isolated lysosomes [70]. The stimulatory effect of prp73 is inhibited by chloroquine. In vitro degradation of RNase A is temperature-, ATP-, and Mg^{2+}-dependent. Degradation is not affected by other heat shock proteins, such as grp78 or *E. coli* DNA K protein [70]. It is proposed that prp73, in response to serum deprivation, recognizes proteins containing a KFERQ related sequence, and then binds to an altered conformation of these proteins to facilitate protein incorporation into lysosomes as has been proposed for the role of hsp70 in the translocation of proteins into microsomes and mitochondria [71,72].

3. Protein degradation in the yeast vacuole

In *S. cerevisiae*, protein degradation varies with metabolic conditions. When cells are grown in glucose, proteins are degraded at 0.5–1%/h. Proteins are degraded at the rate of 2%/h when cells are grown in ethanol. Poor carbon sources and nitrogen starvation induce protein degradation at 3.3%/h [73,74]. Protein degradation is inhibited when ATP levels drop below 0.3 mM [73,74].

Yeast vacuoles contain a variety of hydrolytic enzymes enclosed in a single layer of membrane. Proteinase A, proteinase B, and proteinase C (CPY) are the major proteolytic enzymes in the vacuole. These proteins are synthesized in the ER and transported via the Golgi apparatus into the vacuole by a process similar to the mannose 6-phosphate pathway for the biosynthesis of lysosomal proteins in mammalian cells [75,76]. The maturation of proteinase B and proteinase C is dependent on the activity of proteinase A [75,76]. The gene encoding proteinase A (*PEP4*), therefore, is important for the function of all these major proteinases in the vacuole.

Starvation-induced protein degradation occurs in vacuoles [24,25]. In wild type cells, 70–80% of vegetative proteins are degraded when cells are transferred to sporulation medium. Under these conditions, vacuolar proteinases such as proteinase A, proteinase B, and aminopeptidase I are also induced [24,25]. Starvation induced protein degradation is inhibited by energy uncouplers and cycloheximide, indicating that ATP and the synthesis of certain proteins are required [24].

Mutants defective in vacuolar proteolysis have been identified [24,25]. Among these mutants, *pep4-3* shows severe pleiotropic phenotypes. Several vacuolar hydrolases such as proteinase A, proteinase B, proteinase C, vacuolar ribonuclease, nonspecific alkaline phosphatase and aminopeptidase I accumulate as inactive zymogens in *pep4* mutants cells [25]. During logarithmic growth or on sporulation medium, *pep4-3* cells degrade proteins at only 30% of the wild type level [24,25,74]. Proteinase A and B double mutant cells degrade proteins at 15% of the wild type level. Therefore, proteinase A and B are responsible for 70–80% of overall protein degradation that occurs during nitrogen starvation or during growth on a poor carbon supply [24,25,74]. Extracts of mutants lack 86–97% of proteolytic activity with radiolabeled yeast proteins and methylcasein as substrates. Proteinase A and proteinase B deficiency leads to cell death under conditions of nutrient deprivation [74,77], suggesting that these vacuolar proteinases are essential under certain growth conditions.

3.1. Catabolite inactivation

In yeast, several enzymes that are important in maintaining energy production are induced by glucose starvation and are inactivated by glucose addition (Table I). These enzymes include (1) cytosolic proteins that are important in the gluconeogenesis pathway such as phosphoenolpyruvate carboxykinase [78,79], and fructose-1,6-

TABLE I

Catabolite inactivation in yeast

Cytoplasmic proteins	
Phosphoenolpyruvate carboxykinase	Holzer et al. [78]
	Gancedo et al. [79]
Fructose 1,6-bisphosphatase	Gancedo [80]
c-Malate dehydrogenase	Holzer et al. [81,82]
Trehalose-6-phosphate synthase	Hers et al. [89]
Trehalose-6-phosphate phosphatase	Hers et al. [89]
Plasma membrane proteins	
Galactose transporter	Matern and Holzer [83]
Glucose transporter	Bisson and Fraenkel [84]
Maltose permease	Busturia and Lagunas [85]
Peroxisomal proteins	
Alcohol oxidase and catalase in *H. polymorpha*	Veenhuis et al. [86]
in *C. boidinii*	Bormann and Sahm [87]
3-Oxoacyl-CoA thiolase (*S. cerevisiae*)	Chiang and Schekman (unpublished results)

bisphosphatase (FBPase) [80], as well as proteins in the TCA cycle (cytosolic malate dehydrogenase) [81,82]; (2) plasma membrane proteins involved in sugar uptake such as the galactose transporter [83], the high affinity glucose transporter [84] and maltose permease [85]; and (3) peroxisomal proteins. In *Hansenula polymorpha* and *Candida boidinii*, alcohol oxidase and catalase are induced when methanol is used as the sole carbon source. These proteins are inactivated when cells are shifted to glucose [86,87]. In *S. cerevisiae*, peroxisomes are the major organelles involved in beta-oxidation of fatty acids [88]. Enzymes participating in peroxisomal fatty acid metabolism such as 3-oxacyl-CoA thiolase are also induced in response to glucose starvation and degraded when glucose is added to the medium (H.-L. Chiang and R. Schekman, unpublished results).

New members of the family of proteins subject to catabolite inactivation are trehalose-6-phosphate phosphatase and trehalose-6-phosphate synthase, two enzymes involved in the trehalose biosynthesis pathway [89]. The synthesis of 1 mol trehalose from glucose requires 3 mol of ATP. However, no ATP is produced upon hydrolysis of trehalose. The inactivation of trehalose-6-phosphate phosphatase and synthase in response to glucose thus prevents an ATP futile cycle when trehalose is mobilized. Protein synthesis is required for the reappearance of these enzymes, suggesting that protein degradation is involved in the inactivation process [89].

The physiological function of these enzymes during glucose starvation is to maximize the production of energy. This is achieved by an increase in the activities of enzymes participating in the gluconeogenesis, the TCA cycle, the beta-oxidation of fatty acid, and sugar uptake. When glucose is added to trigger glycolysis, these enzymes are no longer needed and are inactivated. This phenomenon, called catabolite inactivation by Holzer, illustrates the importance of protein inactivation in the regulation of cellular metabolism in response to changes of environment. Since catabolite inactivation applies to at least 11 enzymes participating in various pathways of cellular metabolism, it appears to be an important process by which cells transit from one condition to another.

3.2. Vacuolar degradation of FBPase

Catabolic inactivation of FBPase was first described by Gancedo in 1971 [80]. FBPase, a key enzyme in the gluconeogenesis pathway, is induced when cells are grown in glucose limited medium, and is completely inactivated within 1 h following a shift of cells to glucose rich medium [80,90]. Inactivation does not occur when cells are incubated in other carbon sources such as acetate or ethanol [80,90]. Reappearance of FBPase activity requires protein synthesis, suggesting inactivation is followed or completed by protein degradation. Using antibodies against FBPase, Holzer et al. showed that inactivation of FBPase corresponds to the disappearance the polypeptide, indicating that protein degradation is involved in the inactivation process [90].

In order to determine whether catabolite inactivation of FBPase is mediated by vacuolar protein degradation, we followed FBPase degradation in isogenic yeast strains differing only at the *PEP4* locus. We reasoned that if degradation of FBPase takes place in the vacuole, disappearance of FBPase polypeptide is expected in wild type cells following a shift to glucose medium, but not in *pep4* cells. A redistribution of FBPase from the cytosol to the vacuole is expected in *pep4* cells. .

Pulse chase radiolabeling and immunoblot experiments demonstrate that degradation of FBPase is strictly regulated by glucose and is dependent on *PEP4* . Since synthesis of FBPase polypeptide is repressed following a shift to glucose, both immunoblotting and pulse chase experiments indicate that inactivation is mediated by degradation of the pre-existing FBPase [91].

To determine if degradation of FBPase takes place in the vacuole, we followed the distribution of FBPase in fractions of cells transferred to glucose. Extracts from wild type and *pep4* cells were fractionated on a discontinuous gradient to separate the vacuole from the cytosol. Cytosolic and vacuolar fractions were also followed by antibodies against constitutive invertase and dipeptidylaminopeptidase B (DPAPB, a vacuolar membrane marker), respectively. We found that FBPase is redistributed from the cytosol to the vacuole following a shift to glucose in *pep4* mutants. In contrast, wild type cells degrade FBPase completely in 1 h following a shift to glucose.

Protease K protection experiments were performed to determine if FBPase is retained in a membrane sealed compartment when it is associated with the vacuole. We found that FBPase is protected from protease K digestion when it is localized in the vacuole, and becomes sensitive to proteinase K only after detergent is added to solubilize the membrane. In contrast, cytosolic FBPase is accessible to proteinase K digestion in the absence of detergent [91].

The distribution of FBPase was also followed by immunofluorescence using affinity purified anti-FBPase antibodies. Before a shift to glucose, FBPase is localized in the cytosol. Following a shift to glucose, FBPase gradually redistributes to the vacuole or a vacuole-related organelle in *pep4* cells. Interestingly, the profile of FBPase staining does not coincide exactly with the vacuole. Instead, a dotted staining pattern of FBPase within the vacuole is evident as early as 30 min after transfer of cells to glucose. This observation suggests that FBPase is aggregated or contained within vesicles inside the vacuole.

The combination of cell fractionation and immunofluorescence experiments demonstrate that cells respond to glucose by a redistribution of FBPase from the cytosol to the vacuole for degradation [91]. The identification of this novel import process raises may interesting questions. (1) How does glucose trigger FBPase import and degradation? Is import mediated by protein phosphorylation or ubiquitination? (2) Which sequence within the FBPase polypeptide is important for vacuolar import and degradation? (3) What are the factors involved in this import process? (4) Is import and degradation mediated by an autophagic process or by a direct transfer across the vacuolar membrane?

3.3. Mechanism of FBPase degradation

Secretion in yeast is blocked by temperature sensitive *sec* mutants defective at specific stages in the secretory pathway. At the nonpermissive temperature, *sec62* blocks protein translocation into the ER [92], *sec18* blocks protein transport from the ER to the Golgi [93], *sec7* blocks intra-Golgi transport [94], and *sec1* blocks transport of secretory vesicles to the plasma membrane [95]. To determine whether import of FBPase into the vacuole is mediated by the early secretory pathway, as is the case for the biosynthesis of vacuolar proteins, we studied the inactivation and degradation of FBPase in these *sec* mutants. Our results indicate that degradation is blocked in *sec62*, *sec18* and *sec7*. In contrast, *sec1* shows normal degradation. At the permissive temperature, these mutants degrade FBPase at the same rate as wild type cells.

These results can be explained by two mechanisms. FBPase is first imported into the ER, transported to the Golgi and diverted to the vacuole. Alternatively, import of FBPase into the vacuole is dependent on a factor that traverses the early part of the secretory pathway en route to the vacuole. In order to distinguish these possibilities, we performed proteinase K digestion experiments. According to the first possibility, FBPase would stay in the lumen of the ER when transport is blocked in *sec18*. In the second possibility, FBPase would remain in the cytosol. Our results indicate that FBPase remains in a proteinase K accessible compartment (the cytosol) in *sec18*, whereas in the same sample, a glycosylated form of alpha-factor precursor which

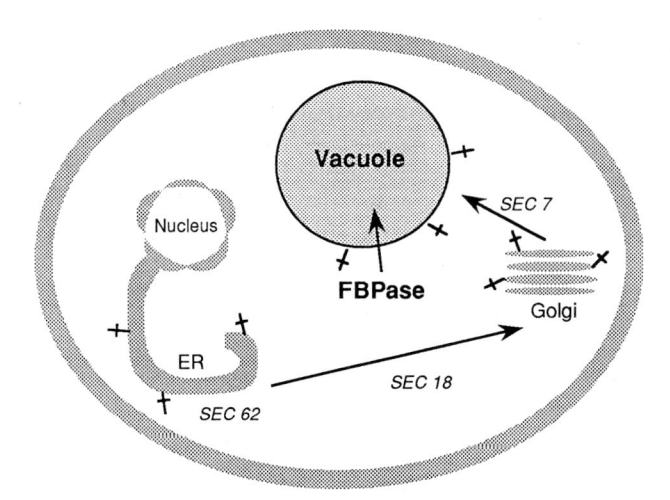

Fig. 2. Pathway of FBPase degradation. FBPase is proposed to translocate from the cytosol into the vacuole by a process that is dependent upon a vacuolar membrane protein (+) whose synthesis is induced when cells are shifted to glucose. The new membrane protein traverses early organelles of the secretory pathway en route to the vacuole. Assembly of the new membrane protein requires translocation into the ER (*SEC62*), transport from the ER to the Golgi (*SEC18*), and transport within the Golgi apparatus (*SEC7*). Reprinted with permission of *Nature* [91].

accumulates in the ER lumen, remains resistant to proteinase K digestion.

These results led to the proposal that cells respond to glucose by synthesizing a new membrane protein that is transported via the Golgi and imported to the vacuole. The presence of such protein, possibly a receptor, on the vacuolar membrane results in the import and degradation of FBPase in the vacuole (Fig. 2).

3.4. Covalent modifications and FBPase degradation

3.4.1. Phosphorylation
Catabolite inactivation is accompanied by an increased intracellular level of cAMP shortly after cells are transferred to glucose medium [90,96]. FBPase is phosphorylated on serine 11 by a cAMP-dependent protein kinase [90,97]. Studies by Holzer have suggested that FBPase phosphorylation may trigger FBPase degradation [90,98]. However, Guerritore et al. [99] demonstrated that FBPase phosphorylation can be uncoupled from degradation. Addition of cAMP to an auxotrophic strain growing on a poor carbon source induces FBPase phosphorylation without triggering FBPase degradation. Hence phosphorylation may be necessary but not sufficient to cause FBPase degradation.

In order to address the requirement for phosphorylation in FBPase degradation, we examined *cdc25* (defective in activating adenylate cyclase due to the accumulation of inactive RAS) and *cdc35* (defective in the catalytic subunit of adenylate cyclase) strains transferred to glucose medium and incubated at a restrictive temperature. No defect in FBPase degradation was detected. Similar findings were made in a *bcy* mutant which expresses adenylate cyclase constitutively. These results suggest that phosphorylation is not required in FBPase degradation.

3.4.2. Ubiquitination

Ciechanover et al. have shown that ubiquitin is important for heat- and starvation-induced protein degradation in the lysosome [100]. In wild type cells, the rate of degradation of long-lived proteins is increased 2.5-fold in response to a shift to high temperature. Starvation also induces protein degradation 4-fold in wild type cells [100]. Both the heat- and starvation-induced degradation is inhibited by ammonium chloride and chloroquine, and is defective in mouse mammary ts85 mutants containing a thermolabile ubiquitin activating enzyme [100]. The defect is also evident in ts20 cells (Chinese hamster ovary cells harboring a heat labile ubiquitin activating enzyme E1).

Immunoelectron microscopy has been used to detect ubiquitin–protein conjugates in lysosomes and autophagic vacuoles, in hepatoma cells [101], 3T3 mouse fibroblasts [102] and neutrophils [103]. Although ubiquitin is also found in other subcellular locations, quantitative experiments demonstrate a 12-fold enrichment of ubiquitin in lysosomes in cultured cells [102].

The mechanism by which ubiquitin participates in enhanced protein degradation in the lysosome is unknown. Ubiquitin may be conjugated directly to cellular proteins that are targeted to the lysosome. Alternatively, proteins involved in the formation or maturation of autophagic vacuoles and the fusion of the vacuole to the lysosome may require ubiquitination to become active. Protein ubiquitination may activate a lysosomal surface protein involved in the recognition or uptake of cytosolic proteins.

The role ubiquitin plays in protein degradation in yeast has been studied using mutants defective in ubiquitin conjugation (*ubc* mutants). Nine ubiquitin conjugating enzymes have been isolated from *S. cerevisiae*. They appear to have distinct substrate specificities and mediate different cellular functions. In yeast, *UBC2* is identical to *RAD6* which mediates DNA repair [10]. *UBC3* is identical to *CDC34* and controls the cell cycle [9]. *UBC1*, *UBC4*, and *UBC5* are important in degrading mis-assembled, mis-localized, and short-lived proteins [14,15]. These genes have overlapping functions. Overexpression of *UBC1* gene partially complements the defect of the *ubc4/ubc5* double mutant [14,15]. Deletion of all these genes results in cell death, suggesting that ubiquitin-dependent protein degradation is essential for life [14,15].

Some ubiquitin enzymes require an additional component, E3, for substrate recognition [34,35]. In yeast, the E3 protein recognizes the N-terminal amino acid of protein substrates. Deletion of E3 (*ubr* mutants) in yeast results in the stabilization of proteins that are normally degraded by the N-end rule pathway [104]. However, *ubr* mutants grow normally [104], suggesting that N-end rule pathway is not essential for cell viability.

Interestingly, degradation of FBPase is defective in *ubc1* and in the *ubc4/ubc5* double mutant. Other *ubc* mutants, such as *ubc6* and *ubc7*, show normal FBPase degradation (unpublished results). Furthermore, *ubr* mutants show FBPase degradation indistinguishable from wild type cells, indicating that FBPase degradation is distinct from the N-end rule pathway of proteolysis. FBPase may be ubiquitinated by one of these *UBC* enzymes to serve as a tag for vacuolar protein degradation. Alternatively, protein components involved in the vacuolar import and degradation pathway may be ubiquitinated to become active (or inactive) to stimulate FBPase degradation.

Acknowledgement

We thank F. Dice, S. Terleckey, C. Plant, L. Wuestehube and M. Hoffman for helpful discussion and technical assistance. This work was supported by a senior fellowship from the American Cancer Society, California Division to H.-L. Chiang, NIH GM 26755 and Howard Hughes Medical Foundation to R. Schekman.

References

1. Goldberg, A. and St. John (1976) Annu. Rev. Biochem. 45, 747–803.
2. Hershko, A. and Ciechanover, A. (1982) Annu. Rev. Biochem. 51, 353–364.
3. Evans, T., Rosenthal, T., Youngblom, D., Distel, D. and Hunt, T. (1983) Cell 33, 389–396.
4. Schimke, R. and Doyle, D. (1970) Annu. Rev. Biochem. 39, 929–976.
5. Finley, D. and Varshavsky, A. (1985) Trends Biochem. Sci. 10, 343–346.
6. Dice, J. (1990) Trends Biochem. Sci. 15, 305–309.
7. Jentsch, S., Seufert, W., Sommer, T. and Reins H.-A. (1990) Trends Biochem. Sci. 15, 195–198.
8. Driscoll, J. and Goldberg, A. (1990) J. Biol. Chem. 265, 4789–4792.
9. Goebl, M., Yochem, J., Jentsch, S., McGrath, J. Varshavsky, A. and Byers, B. (1988) Science 241, 1331–1335.
10. Jentsch, S., McGrath, J. and Varshavsky, A. (1987) Nature 329, 131–134.
11. Finley, D., Bartel, B. and Varshavsky, A. (1989) Nature 338, 394–401.
12. Finley, D., Ciechanover, A. and Varshavsky, A. (1987) Cell 44, 1035–1046.
13. Ciechanover, A., Finley, D. and Varshavsky, A. (1984) Cell 37, 57–66.
14. Seufert, W., McGrath, J. and Jentsch, S. (1990) EMBO J. 9, 4535–4541.
15. Seufert, W. and Jentsch, S. (1990) EMBO J. 9, 543–550.
16. Amenta, J. and Brocher, S. (1981) Life Sci. 28, 1195–1208.
17. Hopgood, M., Clark, M. and Ballard, F. (1977) Biochem. J. 164, 399–407.
18. Surmacz, C., Poso, R. and Mortimore, G. (1987) Biochem. J. 242, 453–458.
19. Lardeux, B. and Mortimore, G. (1987) J. Biol. Chem. 262, 14514–14519.
20. Gunn, J., Brochler, J., Knowles, S. and Ballard, F. (1983) Biochem. J. 210, 251–258.
21. Autreri, J., Akada, A., Bachaki, V. and Dice, J. (1983) J. Cell Physiol. 155, 167–174.
22. Mortimore, G. and Poso, A. (1984) Fed. Proc. 43, 1289–1294.

23. Sommercorn, J. and Swick, R. (1981) J. Biol. Chem. 256, 4816–4821.
24. Betz, H. and Weiser, U. (1976) Eur. J. Biochem. 62, 65–76.
25. Zabenko, G. and Jones, E. (1981) Genetics 97, 45–64.
26. Teichert, U. Mechler, B., Muller, H. and Wolf, D. (1989) J. Biol. Chem. 264, 16307–16045.
27. Turner, P., Westwood, T. and Regen, C. and Steinhardt, R. (1988) Nature 335, 735–738.
28. Furuno, K., Goodman, M. and Goldberg, A. (1990) J. Biol. Chem. 265, 8550–8557.
29. Melville, S., McNurlan, M., Calder, A. and Garlick, P. (1990) Cancer Res. 50, 1125–1131.
30. Dice, J. (1989) Exp. Gerontol. 24, 451–459.
31. Hall, N., Lake, B., Dewji, N. and Patrick, A. (1991) Biochem. J. 275, 269–272.
32. Budka, H., Jellinger, K., Seitelberger, F., Grundke-Igbal, K. and Wisniewski, H. (1989) Prog. Clin. Biol. Res. 317, 837–848.
33. Lowe, J. Blanchard, A., Morrell, K., Lennox, G., Billett, G. Landon, M. and Mayer, J. (1988) J. Pathol. 155, 9–15.
34. Jentsch, S., Seufert, W. and Hauser, H. (1991) Biochim. Biophys. Acta 1089, 127–139.
35. Reichsteiner, M. (1988) Ubiquitin, Plenum Press, New York.
36. Desautels, M. and Goldberg, A. (1982) J. Biol. Chem. 257, 11673–11679.
37. Desautels, M. and Goldberg, A. (1985) Biochem. Soc. Trans. 13, 290–293.
38. Bonifacino, J., Cosson, P., Shah, N. and Klausner, R. (1991) EMBO J. 10, 2783–2793.
39. Lippincott-Schwartz, J., Bonifacino, J., Yuan, L. and Klausner, R. (1988) Cell 54, 209–220.
40. Pontremoli, S. and Melloni, E. (1986) Annu. Rev. Biochem. 55, 455–482.
41. Bond, J. and Butler, J. (1987) Annu. Rev. Biochem. 56, 333–364.
42. Poso, A. Wert, J. and Mortimore, G. (1982) J. Biol. Chem. 257, 12114–12120.
43. Mortimore, G. (1983) Nutr. Rev. 40, 1–12
44. Mortimore, G. and Ward, W. (1981) J. Biol. Chem. 256, 7659–7665.
45. Ahlerg, J., Beckenstam, A., Henell, F. and Glaumann, H. (1985) J. Biol. Chem. 260, 5847–5854.
46. Ahlerg, J., Marzella, L. and Glaumann, H. (1982) Lab. Invest. 47, 523–532.
47. Mortimore, G., Lardeux, B. and Adams, C. (1988) J. Biol. Chem. 263, 2506–2512.
48. Dunn Jr., W. (1990) J. Cell Biol. 110, 1923–1933.
49. Dunn Jr. W. (1990) J. Cell Biol. 110, 1935–1945.
50. Lardeux, B. and Mortimore, G. (1987) J. Biol. Chem. 262, 14514–14519.
51. Seglen, P. and Gordon, P. (1982) Proc. Natl. Acad. Sci. 79, 1889–1892.
52. Seglen, P. and Gordon, P. (1984) J. Cell Biol. 99, 435–444.
53. Caro, L., Plomp, P., Wolvetang, E., Kerkhof, C. and Mejer, A. (1988) Eur. J. Biochem. 175, 325–329.
54. Schworer, C., Shiffer, K. and Mortimore, G. (1981) J. Biol. Chem. 256, 7652–7658.
55. Mortimore, G., Hutson, H. and Surmacz, C. (1983) Proc. Natl. Acad. Sci. 80, 2179–2183.
56. Rote, K. and Rechsteiner, M. (1983) J. Cell Physiol. 116, 103–110.
57. Seglen, P., Gordon, P. and Holen, I. (1990) Semin. Cell Biol. 1, 441–448.
58. Kominani, E., Hashida, S., Khairallah, E. and Katunuma, N. (1983) J. Biol. Chem. 258, 6093–6100.
59. Seglen, P. (1983) Methods Enzymol. 96, 737–764.
60. Grinde, B. (1982) Biochim. Biophys. Acta 701, 328–333.
61. McElligott, M. and Dice, J. (1983) Biosci. Rep. (1984) 4, 451–466.
62. Netland, P. and Dice, J. (1985) Anal. Biochem. 150, 214–220.
63. Backer, J., Bourret, L. and Dice, J. (1983) Proc. Natl. Acad. Sci. 80, 2166–2170.
64. Neff, N., Bourret, L., Miao, P. and Dice, J. (1981) J. Cell Biol. 91, 184–194.
65. McElligott, M., Miao, P, and Dice, J. (1985) J. Biol. Chem. 260, 11986–11993.
66. Backer, J. and Dice, J. (1986) Proc. Natl. Acad. Sci. 83, 5839–5834.
67. Dice, J., Chiang, H.-L., Spencer, E. and Backer, J. (1986) J. Biol. Chem. 261, 6853–6859.
68. Chiang, H.-L. and Dice, J. (1988) J. Biol. Chem. 263, 6797–6805.
69. Dice, J. (1989) FASEB 1, 349–357.
70. Chiang, H.-L., Terleckey, S., Plant, C. and Dice, J. (1989) Science 246, 382–385.

71. Deshaies, R., Koch, B., Werner-Washburne, M., Craig, E. and Schekman, R. (1988) Nature 332, 800–804.
72. Chirico, W., Waters, M. and Blobel, G. (1988) Nature 332, 805–810.
73. Lopez, S. and Gancedo, J. (1979) Biochem. J. 178, 769–776.
74. Teichert, U., Mechler, B., Muller, H. and Wolf, D. (1987) Biochem. Soc. Trans. 15, 811–815.
75. Ammerer, G., Hunter, C., Rothman, J., Saari, G., Valls, L. and Stevens, T. (1986) Mol. Cell Biol. 6, 2490–2499.
76. Woolford, C., Daniels, L., Park, F., Jones, E., Arsdell, J. and Innis, M. (1986) Mol. Cell Biol. 6, 2500–2510.
77. Teichert, U., Mechler, B., Muller, H. and Wolf, D. (1989) J. Biol. Chem. 264, 16037–16045.
78. Muller, M., Muller, H. and Holzer, H. (1981) J. Biol. Chem. 256, 723–727.
79. Gancedo, C. and Schwerzmann, K. (1976) Arch. Microbiol. 109, 221–225.
80. Gancedo, C. (1971) J. Bacteriol. 107, 401–405.
81. Ferguson Jr., J., Boll, M. and Holzer, H. (1967) Eur. J. Biochem. 1, 21–25.
82. Witt, I., Kronau, R. and Holzer, H. (1966) Biochim. Biophys. Acta 128, 63–73.
83. Matern, H. and Holzer, H. (1977). J. Biol. Chem. 252, 6399–6402.
84. Bisson, L. and Fraenkel, D. (1983) J. Bacteriol. 155, 995–1000.
85. Busturia, A. and Lagunas, R. (1985) Biochim. Biophys. Acta 820, 324–326.
86. Veenhuis, M., Douma, A., Harder, W. and Osumi, M. (1983) Arch. Microbiol. 134, 193–203.
87. Bormann, C. and Sahm, H. (1978) Arch. Microbiol. 117, 67–72.
88. Thieringer, R., Shio, H., Cohen, G. and Lazarow, P. (1991) Mol. Cell Biol. 11, 510–522.
89. Francois, J. Neves, M.-J. and Hers, H.-G. (1991) Yeast 7, 575–587.
90. Holzer, H. (1976) Trends. Biochem. Sci. 1, 178–181.
91. Chiang, H.-L. and Schekman, R. (1991) Nature 350, 313–318.
92. Deshaies, R. amd Schekman, R. (1989) J. Cell Biol. 109, 2653–2644.
93. Kaiser, C. and Schekman, R. (1990) Cell 61, 723–733.
94. Franzusoff, A. and Schekman, R. (1989) EMBO J. 8, 2695–2702.
95. Novick, P., Ferro, S. and Schekman, R. (1981) Cell 25, 461–469.
96. Toyoda, Y., Fujii, H., Miwa, I., Okuda, J. and Sy, J. (1987) Biochim. Biophys. Res. Commun. 143, 212–217.
97. Rittenhouse, J., Moberly, L. and Marcus, F. (1987) J. Biol. Chem. 262, 10114–10119.
98. Muller, D. and Holzer, H. (1981). Biochim. Biophys. Res. Commun. 103, 926–933.
99. Lamponi, S., Gallassi, C., Tortora, P. and Guerritore, A. (1987) FEBS Lett. 216, 265–269.
100. Gropper, R., Braudt, R., Elias, S., Berer, C., Mayer, A. and Schwartz, A. and Ciechanover, A. (1991) J. Biol. Chem. 266, 3602–3610.
101. Schwartz, A., Ciechanover, A., Braudt, R. and Geuze, H. (1988) EMBO J. 7, 2961–2966.
102. Laszlo, L., Doherty, F., Osborn, N. and Mayer J. (1990) FEBS Lett. 261, 365–368.
103. Laszlo, L., Doherty, F., Watson, A., Self, T., Landon, M., Lowe, J. and Mayer J. (1991) FEBS Lett. 279, 175–178.
104. Bartel, B., Wunning, I. and Varshavsky, A. (1990) EMBO J. 9, 3197–3189.

CHAPTER 14

The sorting of soluble and integral membrane proteins to the yeast vacuole

CHRISTOPHER K. RAYMOND, CAROL A. VATER, STEVEN NOTHWEHR, CHRISTOPHER J. ROBERTS and TOM H. STEVENS

Institute of Molecular Biology, University of Oregon, Eugene, OR 97403, USA

Abstract

Soluble proteins and integral membrane proteins of the yeast vacuole appear to be delivered to the organelle by distinctly different mechanisms. Unsorted soluble proteins that enter the secretory pathway are secreted from the cell in *Saccharomyces cerevisiae*. In contrast, evidence is presented indicating that the default compartment for unsorted integral membrane proteins is the vacuole rather than the plasma membrane. We present the additional finding that the cytoplasmic domain of an integral membrane protein normally found in the Golgi apparatus is required for retention of the protein in this organelle. Soluble vacuolar proteins are actively sorted to the vacuole in a process that requires many gene products. One of these products is encoded by the *VPS1* gene, an 80 kDa, hydrophilic GTPase that appears to associate with the Golgi apparatus. Mutational analysis suggests that Vps1p is composed of two functionally distinct domains. The amino-terminal region carries a GTP-binding domain that is similar in sequence to a growing subfamily of large GTP-binding proteins. The carboxy-terminal domain may be required for the association of the protein with subcellular organelles, or it may direct the protein to specific functional interactions with other components of the sorting apparatus. Loss of the amino-terminal, GTP-binding function in Vps1p results in a novel mutant product that confers a dominant, loss of function phenotype when expressed in wild-type cells.

1. Introduction

The organelles that comprise the secretory pathway in eukaryotic cells are diverse in composition and function, yet their synthesis is initiated through common biosynthetic pathways. Current research efforts are focussed on investigating how protein traffic proceeds in an orderly manner from one organelle to another and how newly synthesized proteins that originate in the endoplasmic reticulum are sorted away from one another and transported to distinct cellular organelles. It is the latter issue that is the subject of this review. The process of protein sorting to the lysosome-like vacuole in the yeast *Saccharomyces cerevisiae* was chosen as a model system [1,2]. The vacuole in yeast is a prominent organelle that serves as a storage compartment for a number of important metabolites and as the site of hydrolysis and breakdown of a variety of biomolecules [3,4]. Most resident vacuolar proteins, whether soluble or integral membrane proteins, are synthesized and translocated into the endoplasmic reticulum, transported to the Golgi apparatus, and separated away from secreted proteins or integral membrane proteins of the plasma membrane in a distal compartment of the Golgi complex [5–7].

Understanding the process of protein sorting has several facets. In a pathway that branches, one branch of the pathway can be comprised of proteins that are actively sorted while the other branch carries proteins along a passive, default pathway. Such is the case in yeast, where soluble vacuolar proteins are actively diverted to the vacuole whereas soluble secreted proteins travel to the cell surface by default; in the absence of active sorting, soluble vacuolar proteins are secreted [1,8–13]. To be actively sorted, a soluble vacuolar protein must possess targeting information. The best characterized example of a vacuolar targeting signal has been derived from studies of the vacuolar glycoprotein carboxypeptidase Y (CPY). This protein is synthesized as a precursor with a 20 amino acid, cleavable ER targeting sequence followed by a 91 amino acid pro region. The pro region is proteolytically removed in the vacuole to yield the mature protein, which is encoded in the carboxy-terminal 421 amino acids [13,14]. Fusion protein studies indicate that the vacuolar targeting information in CPY is largely contained within the first 10 residues of pro sequence, and point mutations that alter any one of four contiguous residues within this region (Gln-Arg-Pro-Leu) greatly diminish the efficiency of sorting [9,15]. Unlike soluble lysosomal protein sorting in animal cell fibroblasts, carbohydrate modifications are not required for the sorting of yeast CPY [2,16–18]. Thus it is probable that a protein determinant within the pro region of CPY is recognized by the vacuolar protein sorting apparatus.

Like soluble proteins, newly synthesized integral membrane proteins of the vacuole and the plasma membrane must be segregated from one another after transport through common compartments of the early secretory pathway [7]. On the assumption that the default destination for unsorted membrane proteins would be the cell surface, as with soluble proteins, we have sought to identify vacuolar targeting information on an integral membrane protein of the yeast vacuole. Remarkably, no single region within the vacuolar membrane protein dipeptidyl aminopeptidase B (DPAP B) [7,19] is required for vacuolar delivery. Furthermore, experiments with an integral membrane protein normally localized to a late Golgi compartment, dipeptidyl aminopeptidase A (DPAP A) [20], indicate that the cytoplasmic domain of this protein is required for retention of this protein in the Golgi complex; deletion of the cytoplasmic tail from DPAP A results in localization of the mutant protein to the vacuole. These data have forced us to consider models in which the default destination for membrane proteins is the yeast vacuole.

While it is clear that there are targeting signals on soluble vacuolar proteins that specify active sorting to the vacuole, the molecular components that recognize these sorting signals and mediate delivery to the vacuole are poorly understood. One approach to identify such components has been to isolate vacuolar protein sorting (*vps*) mutants [1,8,21,22]. These mutants secrete substantial levels of newly synthesized CPY as well as a number of other soluble vacuolar hydrolases. Genetic analysis of the *vps* mutants underscores the complexity of the sorting process; as many as fifty *VPS* genes are necessary for efficient sorting (reviewed in [3,4]). None of the *vps* mutants mislocalize vacuolar membrane proteins to the cell surface [23], further illuminating the distinction between the sorting of soluble vacuolar proteins and integral membrane proteins. Molecular characterization of specific *VPS* genes and their products is beginning to provide insights into the molecular basis of soluble protein sorting. Here we summarize data indicating that the *VPS1* product is a member of a subfamily of GTP-binding proteins that participates in a seemingly diverse variety of activities in different eukaryotic organisms. Vps1p is a GTPase that appears to be localized, at least in part, to the Golgi apparatus. Mutational analysis indicates that the protein is composed of two functionally distinct regions.

2. Results

2.1. No single domain of the vacuolar integral membrane protein DPAP B is required for vacuolar delivery

DPAP B is a type II integral membrane protein [24] that is encoded by the *DAP2* gene [4,19]. DPAP B has three distinct domains consisting of a short, amino-terminal cytoplasmic tail, a hydrophobic transmembrane domain, and a large, glycosylated lumenal region that carries the enzymatic activity (Fig. 1). To test if any single domain of DPAP B carried vacuolar targeting information, a variety of gene fusions and deletion mutations were constructed, as shown in Fig. 1. In most cases, the localization of the corresponding proteins was assessed using indirect immunofluorescence [25]. Fixed cells expressing a given fusion construct were double labelled with anti-DPAP B antibodies and antibodies directed against the vacuolar H^+-ATPase 60 kDa subunit. The latter protein is part of a large, multisubunit proton pumping ATPase that resides on the vacuolar membrane, and the 60 kDa subunit staining pattern provided an unambiguous indication of vacuolar localization.

The lumenal region of DPAP B, when expressed as a soluble protein, did not

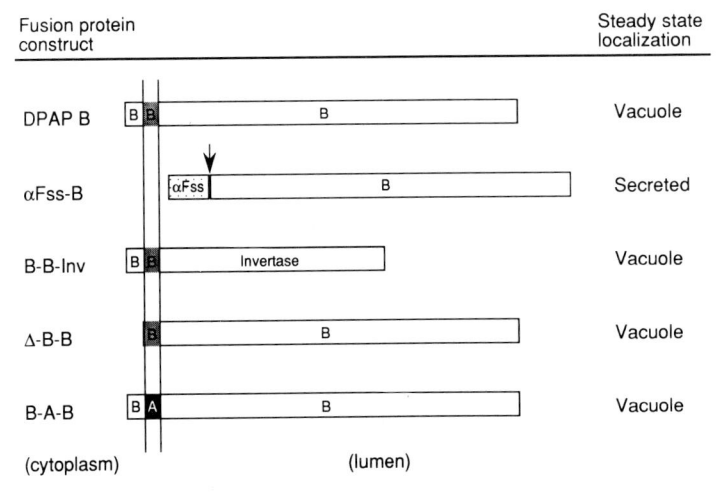

Fusion protein construct		Steady state localization
DPAP B	[diagram: B B — B]	Vacuole
αFss-B	[diagram: αFss — B]	Secreted
B-B-Inv	[diagram: B B — Invertase]	Vacuole
Δ-B-B	[diagram: B — B]	Vacuole
B-A-B	[diagram: B A — B]	Vacuole
(cytoplasm)	(lumen)	

Fig. 1. Constructs used to analyze potential vacuolar targeting signals in the type II vacuolar membrane protein DPAP B. DPAP B and the fusion proteins used in this study are drawn from their amino-termini at the left to their carboxy-termini at the right. The vertical bar represents a lipid bilayer, with the cytoplasm to the left and the organelle lumen to the right of the membrane. The different domains within each fusion construct are denoted by a letter that specifies their origin (A, DPAP A and B, DPAP B). The details of the construction of the fusion proteins are described elsewhere [26; C. Roberts, S. Nothwehr and T. Stevens, unpublished results]. The localization of the soluble form of the DPAP B lumenal domain (αFss-B) was determined by immunoprecipitation of ^{35}S-labelled protein from intracellular and extracellular fractions; the signal sequence cleavage site is denoted by the vertical arrow. The localization of the remaining fusions was determined by indirect immunofluorescence, as described in the text.

possess vacuolar targeting information. The DNA encoding the yeast α-factor signal sequence was fused in frame to the *DAP2* sequences encoding the lumenal region of DPAP B (αFss-B, Fig. 1) [26]. In pulse-chase labelling experiments using [^{35}S]methionine, two distinct populations of the resulting fusion protein were observed (data not shown). Roughly 60% of the labelled material appeared as a broad band on SDS-PAGE, indicative of extensive carbohydrate modification; all of this material was found in the extracellular milieu. The remaining 40% of the protein was found in the intracellular fraction, but migrated as a higher mobility, more homogeneous species. These and other data indicated that a substantial portion of the fusion protein failed to receive Golgi-localized carbohydrate modifications and was instead misfolded material that was retained in the ER [26; C. Roberts, S. Nothwehr and T. Stevens, unpublished results]. None of the material which escaped the ER was localized to the vacuole. Similar published data have shown that the lumenal domain of another vacuolar membrane protein, alkaline phosphatase (ALP), also lacks vacuolar localization information [6,27].

Three other fusion constructs are depicted in Fig. 1, and all of them were localized to the vacuolar membrane when expressed in yeast cells. Hence, a fusion protein consisting of the cytoplasmic and transmembrane regions of DPAP B linked to the normally secreted enzyme invertase (B-B-Inv) is localized to the vacuole. Deletion of the cytoplasmic domain from DPAP B (Δ-B-B) had no effect on vacuolar delivery. While these data would be consistent with the vacuolar targeting information for DPAP B residing in the transmembrane domain, the final fusion argues against this interpretation. When the transmembrane domain of DPAP B was replaced with the transmembrane domain of the Golgi-localized protein DPAP A (see below), the resulting B-A-B fusion was also localized to the vacuole. These data demonstrate that there is no single domain of DPAP B that is necessary for delivery of the protein to the vacuole.

2.2. The cytoplasmic domain of DPAP A is necessary and sufficient for its localization to a late Golgi compartment

DPAP A, which is encoded by the *STE13* gene, is very similar to DPAP B in overall topology and organization (Fig. 2) [7,20,26]. It is a type II integral membrane protein with a 118 amino acid cytoplasmic tail, a short, hydrophobic transmembrane segment, and a lumenal domain of approximately 800 amino acids that shares 48% identity with the lumenal domain of DPAP B over the carboxy-terminal 250 amino acids [7]. Several lines of evidence indicate that DPAP A is localized to a distal

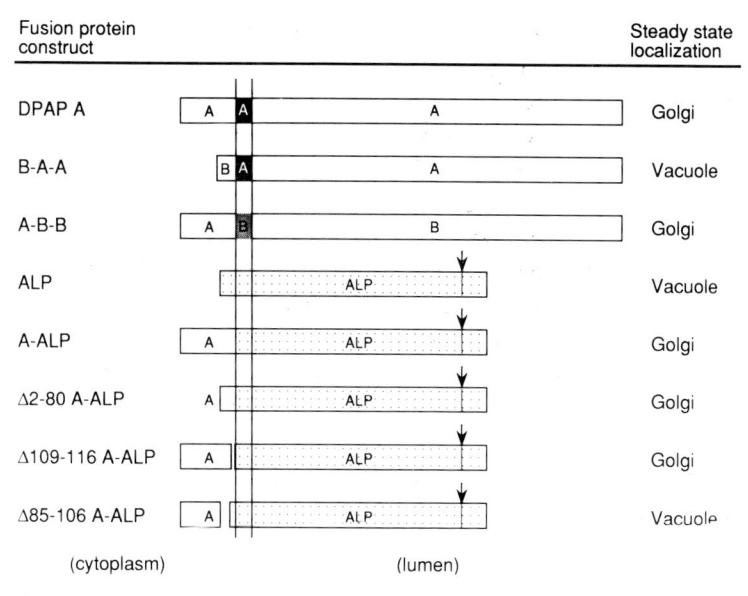

Fusion protein construct		Steady state localization
DPAP A	A | A A	Golgi
B-A-A	B|A A	Vacuole
A-B-B	A |B B	Golgi
ALP	ALP	Vacuole
A-ALP	A | ALP	Golgi
Δ2-80 A-ALP	A | ALP	Golgi
Δ109-116 A-ALP	A | ALP	Golgi
Δ85-106 A-ALP	A | ALP	Vacuole
(cytoplasm)	(lumen)	

Fig. 2. Constructs used to analyze Golgi retention signals on DPAP A. DPAP A, ALP and the fusion constructs used are displayed as in Fig. 1. In the lower four fusions, the cytoplasmic domain of DPAP A or deletion mutants within this region were fused to the transmembrane domain and lumenal segment of ALP. The approximate site of vacuolar proteolytic cleavage in ALP is denoted by the vertical arrow. The localization of the fusion proteins was determined by indirect immunofluorescence.

compartment of the yeast Golgi apparatus. The enzyme is required for proteolytic processing of the mating pheromone α-factor from its precursor species, a process that occurs in a late Golgi compartment [5,20,28]. Furthermore, double label immunofluorescence experiments with cells expressing both Kex2p and DPAP A indicate that the two proteins are localized to the same compartment (S. Nothwehr and T. Stevens, unpublished results). Kex2p, which catalyzes the first step in α-factor processing, has been shown to reside within a distal compartment of the Golgi complex [29–31]. The similarity in structure between DPAP A and DPAP B made them compatible proteins for domain swapping experiments in an effort to identify vacuolar targeting and/or Golgi retention domains.

The cytoplasmic domain of DPAP A was found to be both necessary and sufficient for retention of membrane proteins in the Golgi complex. As shown in Fig. 2, fusion proteins containing the cytoplasmic domain of DPAP A, but the transmembrane domain and lumenal domains of DPAP B (A-B-B), gave indirect immunofluorescence patterns indistinguishable from wild-type DPAP A. Alternatively, substitution of the cytoplasmic domain of DPAP B for the cytoplasmic domain of DPAP A (B-A-A) yielded a fusion construct that was localized to the vacuole. To investigate the region within the cytoplasmic domain of DPAP A required for Golgi retention, this region was fused to the transmembrane and lumenal segments of the type II vacuolar membrane protein alkaline phosphatase (ALP, Fig. 2) [6,32]. ALP is proteolytically processed at its carboxy-terminus upon entry into the vacuolar compartment (Fig. 2), and the cleavage event is readily detected as a mobility shift on SDS-PAGE gels. In addition to indirect immunofluorescence, cleavage of the A-ALP fusion constructs served as a second means to assess the extent of vacuolar versus Golgi localization of the A-ALP fusion constructs. Three deletions spanning the entire 118 residue cytoplasmic segment of DPAP A were investigated in the context of the A-ALP fusion (Fig. 2). Two of these mutants, one spanning from residues 2–80 and the other from residues 109–116, remained localized to the Golgi in an unprocessed form. However, deletion of residues 85–106 gave rise to a mutant protein that was localized to the vacuolar membrane and processed to a significant extent; 60–70% of ^{35}S-labelled fusion protein appeared to be in the cleaved state after a 1-h chase time (S. Nothwehr and T. Stevens, unpublished results). When the same residues are deleted from the cytoplasmic tail of DPAP A, the mutant protein is also mislocalized to the vacuolar membrane. Therefore, mutations that eliminate the retention signals on an integral membrane protein of the yeast Golgi apparatus result in delivery of the mutant protein to the vacuole.

2.3. Vps1p, which is required for the sorting of soluble vacuolar glycoproteins, shares extensive similarity with a subfamily of GTP-binding proteins

VPS1 is a nonessential gene that encodes an 80 kDa, hydrophilic protein. The *VPS1* product, Vps1p, contains a tripartite GTP-binding motif, and it appears to associate

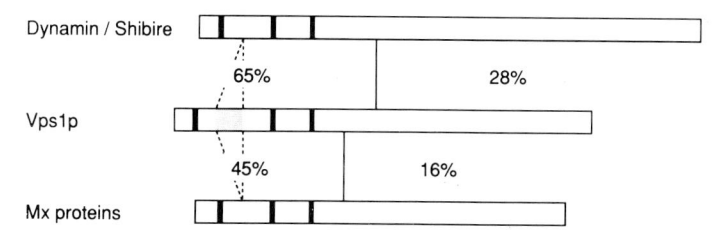

Fig. 3. Sequence similarities within the Vps1p subfamily of GTP-binding proteins. Primary protein sequences are displayed as open horizontal bars that are drawn to scale. The positions of the tripartite GTP-binding motifs [33] within each protein are shown as the three vertical bars. The percent identical amino acid residues between regions in Vps1p and dynamin/*Shibire* protein or the Mx proteins are displayed and the boundaries between highly similar regions and divergent residues are shown as solid vertical lines between the proteins. These data were adapted from published sequence similarities [34,38,40,41]. Dynamin and *Shibire* protein have a unique carboxy terminal segment of roughly 100 residues that is rich in prolines and basic amino acids; this region, not found in Vps1p or the Mx proteins, is shaded. Similarly, Vps1p carries a unique insertion of about 45 residues between elements I and II of the GTP-binding motif that is also shaded and set apart by dashed lines.

with the yeast Golgi complex [33,34]. Furthermore, Vps1p is a member of a growing subfamily of GTP-binding proteins (Fig. 3). Sequence similarity was first detected between the amino-terminal 250 amino acids of Vps1p and the vertebrate Mx proteins [34]. Mx proteins are interferon-inducible proteins that confer resistance to influenza and other viruses via an unknown mechanism [35–37]. The sequencing of the rat cDNA encoding dynamin [38,39], and subsequently the *Drosophilia melanogaster* gene *Shibire* [40,41] revealed that the proteins encoded by these genes are also members of this family. Vps1p and dynamin share 65% sequence identity throughout their amino terminal 300 residues (Fig. 3) [38]. Dynamin, which was purified by its nucleotide-sensitive association with microtubules in crude extracts, was originally thought to be a mechanochemical ATPase that was capable of bundling microtubules and sliding them past one another [39,42,43]. However, the purified protein binds to microtubules in a nucleotide-insensitive manner, and although it does exhibit a GTPase activity that is stimulated many-fold in the presence of microtubules, the purified preparation no longer possesses microtubule sliding activity [38,39,42]. Immunolocalization of dynamin in cultured PC12 cells, coupled with fractionation studies, suggests that dynamin associates with membranous structures in vivo and that these organelles are not arrayed along microtubules; the authors caution that the microtubule association observed in vitro may bear little relevance to the in vivo function of the protein [44]. Unlike other members of this GTP-binding protein subfamily, rat dynamin and the fruit fly *Shibire* protein share significant sequence identity throughout their entire lengths, and they may be the functional homologs of one another in their respective organisms [40,41]. The most prominent phenotype of *shibire*^{ts} mutant flies is a rapid yet reversible block in endocytosis upon shift to the restrictive temperature [45,46]. The process of endocytosis appears to be blocked at the stage in which deeply invaginated coated pits pinch off to become coated vesicles [46].

It is of great interest that Vps1p, which is required for the sorting of soluble vacuolar proteins, and the *Shibire* product, which is necessary for endocytosis, both appear to influence the function of the vacuolar network. The relationship of these proteins to dynamin, coupled with the in vitro association of dynamin with microtubules, suggested that microtubule integrity may be required for Vps1p function in yeast, and thus for vacuolar protein sorting. This possibility was further strengthened when Vps1p proved to be the same protein as Spo15p, a protein encoded by a gene necessary for sporulation [47]. Indeed Spo15p expressed in *Escherichia coli* exhibited nucleotide-insensitive binding to bovine microtubules [48]. In addition, loss of microtubules was reported to influence the morphology of yeast vacuoles [49].

These observations prompted us to examine the effects of the microtubule disrupting drug nocodazole [50] on the efficiency of CPY sorting and on the kinetics of CPY delivery to the vacuolar compartment. Wild-type yeast cultures that had been treated with 15 μg/ml of nocodazole for 2.5 h were pulse-labelled with ^{35}S, and the efficiency and kinetics of CPY sorting were examined (Fig. 4). There was no detectable difference between the treated cells and the control samples. Thorough microscopic examination of the treated cells demonstrated that microtubules were absent, the cells were arrested with a late G2, large-budded cell morphology, and single nuclei were randomly positioned throughout the maternal cell body [23] (data not shown); all of these features are characteristics of yeast cells lacking microtubules [50,51]. Microtubule disruption also had little or no detectable effect on Vps1p localization or vacuolar morphology [23] (data not shown). We conclude from these data that microtubules play no obvious role in vacuolar biogenesis and that Vps1p appears to function independently of the microtubule cytoskeleton.

2.4. Vps1p binds and hydrolyzes GTP

The Vps1p family of proteins all share sequence identity in their amino-termini, where there is a tripartite GTP binding motif (Fig. 3) [34]. We determined that Vps1p was indeed capable of binding GTP by examining the binding of [α-^{32}P]GTP to Vps1p that had been bound to a nitrocellulose blot and subsequently renatured (Fig. 5). Binding of radiolabelled GTP to Vps1p was strongly inhibited by 10-fold molar excess of unlabelled GTP and by a 100-fold molar excess of GDP [23] (data not shown). Addition of unlabelled ATP also inhibited binding, but only at 1000-fold excess over radiolabelled nucleotide. The *VPS1* product was also shown to hydrolyze GTP. Native Vps1p was immunoprecipitated from extracts prepared from either yeast or *E. coli* cells expressing elevated levels of the protein. Immunoprecipitates were also made from cells that lacked the *VPS1* product. [α-^{32}P]GTP was added to these samples and the conversion of GTP to GDP was monitored as a function of time by thin layer chromatography. As is clear in Fig. 5, the immuno-

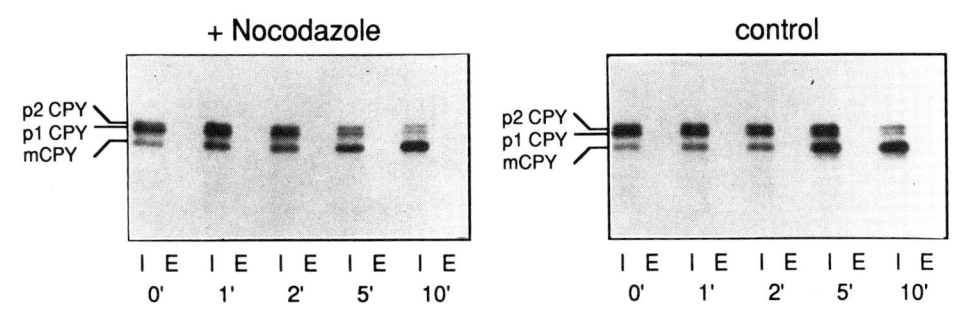

Fig. 4. Microtubules are not required for the sorting of CPY to the yeast vacuole. Yeast cells were treated with 15 µg/ml nocodazole for 2.5 h. One aliquot was examined by microscopy and determined to exhibit the characteristic features of cells lacking microtubules (see text; data not shown). The remaining aliquot and a control culture were pulse-labelled with ^{35}S and chased for various periods of time, as shown. CPY was immunoprecipitated from intracellular (I) and extracellular (E) fractions, separated by SDS-PAGE and visualized by autoradiography. The progression of CPY from precursor ER (p1) and Golgi (p2) forms to the mature vacuolar species (m) is clearly visible as a function of time. No difference in either the efficiency of vacuolar sorting or in the kinetics of vacuolar delivery was detected between the treated sample and the control.

precipitates containing Vps1p possess elevated GTPase activities relative to the control immunoprecipitates. These data show that the *VPS1* protein, as predicted from sequence, binds GTP in a specific manner and possesses an intrinsic GTPase activity.

2.5. Mutational analysis suggests that Vps1p is composed of two functionally distinct domains

Vps1p is one of several GTP-binding proteins that participate in the orderly movement of proteins through the secretory pathway in *S. cerevisiae*. Two other GTP-binding proteins, Ypt1p [52,53] and Sec4p [54,55] are encoded by essential genes, and they appear to mediate vesicle fusion in vesicular traffic between the ER and the Golgi apparatus or the Golgi complex and the plasma membrane, respectively [55–59]. Certain mutant alleles of *YPT1* and *SEC4* confer a dominant-lethal phenotype when expressed in conjunction with nonmutant alleles in wild-type cells, and the corresponding mutations map within the region encoding the third GTP-binding motif [52,55]. These data prompted us to investigate whether point mutations in *VPS1* could produce dominant loss of function mutant phenotypes.

The wild-type *VPS1* gene, carried on a single copy CEN vector, was subjected to in vitro hydroxylamine mutagenesis and then transformed directly into wild-type cells and *vps1-Δ2* mutants. Colonies with a Vps$^-$ phenotype were detected using an assay that detects secretion of CPY [25]. Remarkably, 4% of all wild-type transformants secreted CPY while 8% of the *vps1-Δ2* transformed strains were Vps$^-$. Furthermore, in all cases tested, the Vps$^-$ phenotype in wild-type transformants was plasmid-

A.

Vps1p Western

Vps1p →

Δ C-term →

WT Δ C-term

[α-^{32}P]GTP binding

● 68

● 43

WT Δ C-term

B.

GTPase Activity of VPS1p

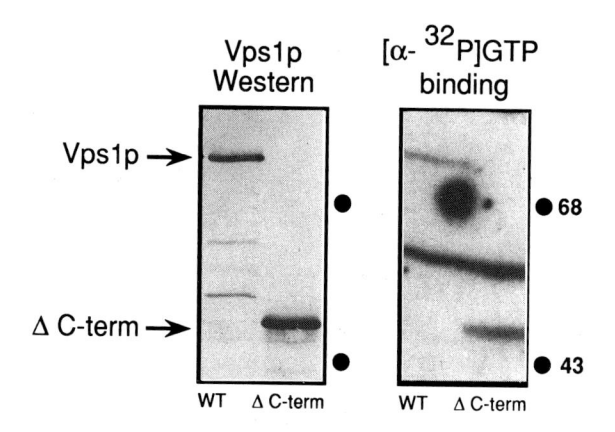

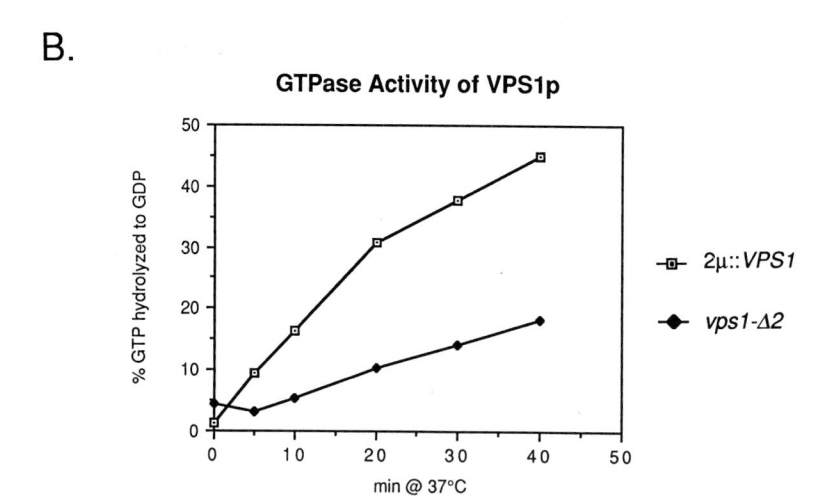

─□─ 2μ::*VPS1*
─◆─ *vps1-Δ2*

x-axis: min @ 37°C
y-axis: % GTP hydrolyzed to GDP

Fig. 5. Vps1p binds and hydrolyzes GTP. (A) Wild-type Vps1p and a mutant form lacking the carboxy-terminal 285 residues (see Fig. 6 C) were expressed in *E. coli*. Protein extracts were separated by SDS-PAGE and duplicate gels were blotted to nitrocellulose. One blot, probed with anti-Vps1p antibody, shows the positions of the two proteins, which migrate as expected relative to molecular weight markers (shown in units of kilodaltons). The other blot was treated and probed with [α-^{32}P]GTP as described elsewhere [23,52]. Both proteins bind GTP. The additional GTP-binding protein present in both lanes is endogenous to *E. coli*. (B) Native Vps1p was immunoprecipitated from yeast cells overexpressing the protein. The conversion of GTP to GDP was assayed in this precipitate relative to an immunoprecipate from yeast cells lacking Vps1p. GTP hydrolysis was monitored by thin layer chromatography and the percent conversion of GTP to GDP was calculated from the quantitative counts of radioactive GDP relative to GTP.

linked, and these plasmid born *vps1* alleles failed to complement *vps1-Δ2* yeast to Vps$^+$. The data indicate that one half of all loss of function mutations in *VPS1* confer a dominant-negative phenotype in wild-type cells. These mutant products somehow interfere with the normal protein sorting process, and therefore these dominant-negative alleles were specifically named dominant-interfering alleles.

The high frequency of dominant-interfering mutations relative to the overall frequency of loss of function mutations suggested that Vps1p might be composed of at least two functionally distinct domains. By this model, mutations that destroy the function of one region leave a separate domain competent to interact with the sorting apparatus; the interaction by such a nonfunctional Vps1p molecule interferes with the function of wild-type Vps1p, thereby causing the observed interference in normal CPY sorting. This model leads to the prediction that dominant-interfering mutations should cluster to a specific region(s) in *VPS1* and that recessive, loss of function mutations should cluster to a different region of the gene. This hypothesis was tested by mapping the position of dominant-interfering mutations within the *VPS1* gene. The approximate location of 12 point mutations that confer this phenotype was determined using a recombinational mapping strategy [60], and the details of this analysis are described more extensively elsewhere [23]. The results of the mapping experiments are shown in Fig. 6. All of the dominant-interfering mutations mapped to the amino-terminal, GTP-binding portion of the Vps1p coding region. In the course of this analysis, we inadvertently isolated an in-frame deletion that spans the GTP-binding region (Fig. 6) [23], and this mutation also confers a dominant-interfering phenotype. The same is true of a larger deletion mutant constructed in vitro (Fig. 6) (C. Vater, C. Raymond and T. Stevens, unpublished results). Hence, mutations and deletions within the GTP-binding portion of Vps1p, the region conserved within the protein family shown in Fig. 3, result in mutant proteins that compromise the wild-type protein sorting apparatus.

Thirty recessive, loss of function *vps1* mutant alleles were also characterized [23]. Protein extracts were prepared from *vps1-Δ2* mutants (which completely lack the *VPS1* coding region [34] that harbored recessive *vps1* alleles on CEN plasmids. These extracts were separated by SDS-PAGE, blotted to nitrocellulose, and probed with anti-Vps1p specific antibodies (data not shown). None of the fully recessive mutants characterized made full length Vps1p product. The majority of mutant genes encoded truncated products ranging in size from approximately 30 to 75 kDa. The reduction in the size of these proteins most likely resulted from nonsense mutations within the *VPS1* ORF. The remaining extracts were devoid of Vps1p specific protein, indicative of mutations that either terminate the protein near the amino-terminus or grossly destabilize the entire protein structure. These data demonstrate that the carboxy-terminal region of Vps1p is necessary for the functional integrity, but not necessarily the stability, of the Vps1p protein. The absence of missense mutations that confer a recessive, loss of function phenotype suggests that the functional integrity of the carboxy-terminal domain is not readily perturbed by single mutations and/or that

A.

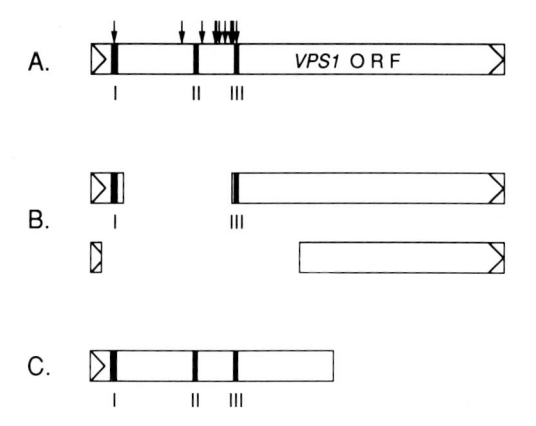

B.

C.

Fig. 6. Point mutations and deletions in *VPS1* that alter Vps1p function. (A) The 704 amino acid *VPS1* protein is displayed as an open rectangle with the amino-terminus on the left. The three GTP-binding elements are designated by vertical bars and I, II and III, respectively. The approximate positions of 12 dominant-interfering point mutations are shown by the vertical arrows above the *VPS1* open reading frame. The map positions of these mutations were determined by recombinational mapping [23,60]. Map positions are only approximations, but the uncertainties in positions was estimated to be no greater than ±50 residues. (B) The two characterized deletion mutations that produce a dominant-interfering mutant product are shown. The top deletion was isolated from in vitro mutagenized plasmids and the extent of the deletion, which is drawn to scale, was determined by extensive restriction mapping. The bottom deletion, also drawn to scale, was constructed by removing the *Mst* II to *Apa* I restriction fragment from the *VPS1* gene and re-ligation of the gene so as to restore the open reading frame. Both mutant products are expressed as stable proteins in vivo. (C) A carboxy-terminal truncation mutant form of Vps1p was generated by creating a frameshift mutation in the *Bam* HI site of the *VPS1* gene. The mutation results in the loss of 285 residues from the carboxy-terminal end of the protein (shown to scale). The mutant product retains the capacity to bind GTP in vitro (Fig. 5A) and it is expressed as a stable protein in vivo. However, this carboxy-truncated Vps1p protein has a recessive, loss of function phenotype.

there are regions in the carboxy-terminal domain that are functionally redundant. Alternatively, the spectrum of mutations induced by hydroxylamine treatment may not include critical residue substitutions that render the carboxy-terminal function inactive.

Characterization of mutant forms of the *VPS1* protein provided further evidence that the protein is composed of two functionally distinct regions. The carboxy-terminal region was not required for the GTP-binding activity of Vps1p; a 419 amino acid truncated form of the protein lacking the carboxy-terminal 285 residues retained the capacity to bind GTP in vitro (Figs. 5A, 6C) [23]. On the other hand, the mutant proteins completely lacking the GTP-binding region (Fig. 6) were expressed as stable proteins in vivo (data not shown), and their expression conferred dominant-interfering phenotypes. This suggests that the carboxy-terminal portion of Vps1p is capable of folding into a stable protein in the absence of the amino-terminal domain. Finally, the carboxy-terminal region of Vps1p is required for a mutant form of the protein to exert dominant interference. In all 12 dominant-interfering point mutation alleles described above, a frame-shift mutation that effectively truncates Vps1p was created at amino acid 420 by filling in and re-ligating a restriction site. These truncated products, while stably expressed in vivo, lost their dominant characteristics and became recessive, loss of function alleles (I. Howald-Stevenson, C. Raymond, and T. Stevens, unpublished results). Taken together, these data support a model in which the carboxy-terminal region initiates and/or maintains some functional interaction necessary for protein sorting. The amino-terminal segment then utilizes GTP hydrolysis to catalyze an event that drives the sorting process forward in one direction. By this model, loss of the carboxy-terminal domain leaves the protein unable to interact with the sorting apparatus and thus eliminates Vps1p function. In contrast, loss of the amino-terminal domain does not eliminate interaction with the sorting apparatus, but the GTP-dependent sorting function is lost. The nonproductive association of the mutant Vps1p with the sorting apparatus poisons the sorting process, leading to the observed dominant-interfering phenotype.

3. Discussion

3.1. Targeting of integral membrane proteins in the secretory pathway of Saccharomyces cerevisiae

Localization studies with mutant forms of the Golgi-localized Kex2p protease [28] or with wild-type Kex2p in a clathrin heavy chain deficient mutant (chc1Δ;) [61] indicated that Golgi membrane proteins that escaped retention were delivered to the plasma membrane. Thus, the default destination for unsorted integral membrane proteins in yeast was thought to be the cell surface. Predicated on this assumption, previous findings indicated that the targeting information for vacuolar membrane proteins was located within their cytoplasmic tails and/or transmembrane regions [7,27]. In our more recent studies, we have found that no single domain of the vacuolar membrane protein DPAP B carries targeting information that is necessary for delivery of the protein to the vacuole [26; C. Roberts, S. Nothwehr, and T. Stevens, unpublished results]. Furthermore, overexpression of the Golgi-localized proteases Kex1p or DPAP A leads to partial localization of these proteins to the vacuole without detectable delivery to the cell surface [26,62]. Finally, mutations in the cytoplasmic tails of DPAP A (see Results; C. Roberts, S. Nothwehr and T. Stevens, unpublished results) or Kex1p (A. Cooper and H. Bussey, unpublished results) result in vacuolar localization of the mutant proteins. These data favor a model in which the vacuole, and not the plasma membrane, is the default compartment for membrane proteins lacking specific targeting information. Furthermore, it may be possible to reconcile the apparent conflict with earlier data. The Kex2p mutant proteins that were delivered to the cell surface lacked portions of their transmembrane domains [28], and thus they may have been secreted from the cell as soluble forms of the *KEX2* protein. The fact that wild-type Kex2p was found at the cell surface in clathrin heavy chain mutants [61] may reflect several roles for clathrin in the function of the secretory pathway; clathrin may be required for both retention of Golgi membrane proteins and for maintaining the integrity of the membrane protein sorting apparatus.

No single domain of a vacuolar membrane protein carries vacuolar targeting information, and mutant Golgi membrane proteins lacking retention signals are delivered to the vacuole. The most simple and conservative interpretation of these data is that the vacuole is the default compartment for unsorted membrane proteins in yeast. For the membrane proteins we have studied, we can rule out the proposal that they are first delivered to the plasma membrane and are then endocytosed. All of the vacuolar membrane protein fusion constructs and the Golgi membrane protein retention mutants that have been examined are delivered to the vacuole in a temperature-shifted *sec1* mutant (S. Nothwehr and T. Stevens, unpublished results). Loss of *SEC1* function blocks protein traffic from the Golgi apparatus to the plasma membrane [63]. Likewise it would seem improbable that mutant membrane proteins lacking targeting or retention signals are delivered to the vacuole by a cryptic garbage pathway. There is no experimental precedent for such a pathway in yeast, and there is mounting evidence that misfolded proteins are retained and degraded in the ER (reviewed in [64,65]). In fact, certain mutant constructs of DPAP A are retained in the ER in an unmodified, enzymatically inactive state [26; C. Roberts, S. Nothwehr and T. Stevens, unpublished results]. In contrast, the membrane proteins that are delivered to the vacuole are enzymatically active and have received carbohydrate modifications indicative of passage through the Golgi apparatus. Thus if such a garbage pathway exists, it cannot be distinguished from the normal route traversed by wild-type vacuolar membrane proteins. Finally, the data are consistent with the more complex hypothesis that both vacuolar and Golgi-localized membrane proteins carry vacuolar targeting information. In the case of Golgi proteins, the retention signal would generally ensure Golgi localization; however, these proteins may also contain vacuolar targeting signals so that proteins that escape from the retention apparatus are sent to the vacuole for degradation. To be fully consistent with the data, such hidden vacuolar targeting signals must be located in the transmembrane domains of these proteins.

The hypothesis that the vacuole is the default compartment for membrane proteins lacking specific targeting or retention information makes some testable predictions. Mutations that abolish the retention signals in membrane proteins normally localized in the ER or in various cisternae of the Golgi apparatus should result in delivery of the mutant proteins to vacuole. As reported here, this prediction has proven to be true of a membrane protein normally localized to a late Golgi compartment. Similarly the vacuole default model makes the prediction that plasma membrane proteins in yeast are actively sorted to the cell surface, that plasma membrane proteins contain positive plasma membrane targeting information, and that mutations in such targeting signals will result in vacuolar delivery of these mutant plasma membrane proteins.

The possibility that the default compartment for integral membrane proteins in *S. cerevisiae* is the vacuole is a surprising departure from studies in mammalian cells, which indicate that membrane proteins are delivered to plasma membrane by default [66–68]. Nonetheless, this possible difference between yeast and animal cells may make biological sense. In animal cells, membrane proteins that do not belong on the plasma membrane may be cleared by extracellular proteases or by endocytosis and subsequent degradation in a lysosomal compartment. In contrast, the plasma membrane of yeast serves as the only impermeant barrier for this unicellular organism, and direct delivery of unsorted membrane proteins to the vacuole may ensure both the integrity of the plasma membrane and the expedient degradation of mislocalized membrane proteins. The mechanism by which membrane proteins are sorted in *S. cerevisiae* and mammalian cells may still be highly conserved, with the major difference being the destination of actively sorted versus passive membrane protein traffic.

3.2. Vps1p, which is a GTPase required for the sorting of soluble vacuolar proteins, is composed of two functionally distinct domains

Vps1p shares sequence identity with a growing subfamily of GTP-binding proteins. The function these proteins perform in vacuolar protein sorting (Vps1p), endocytosis (*Shibire*), or viral resistance (Mx proteins) remains obscure. Tremendous excitement was generated when it appeared that one member of this family, dynamin, was a microtubule-based motor protein [39,42]. Like the kinesin family of motor proteins [69], one inferred hypothesis was that the conserved GTP-binding region in this protein family was a motor domain, and that the divergent regions directed interactions with specific cargo [38,70]. Unfortunately, this appealing notion is increasingly untenable in light of recent experimental evidence. We have observed that loss of microtubules has no detectable influence on vacuolar protein sorting in vivo. Furthermore, dynamin does not appear to associate with microtubules in vivo [44]. Finally, the *Shibire* product, which appears to be the cognate dynamin equivalent in fruit flies, is required specifically for the pinching off of coated pits into coated vesicles [46], and there is no evidence that a microtubule motor protein participates in this process. It is noteworthy, however, that both Vps1p and the *Shibire* product perform functions in the vacuolar network of the secretory pathway.

VPS1 and *SPO15* are the same gene in yeast [23,48]. Spo15p is required for the initial stages of sporulation [47], and recombinantly expressed Spo15p has been shown to sediment with taxol stabilized bovine microtubules [48]. These observations have led to the proposal that Spo15p participates in the microtubule-dependent processes of spindle separation and nuclear division during meiosis. However, vacuolar hydrolases are required for sporulation [71,72] and *vps1* mutants are almost completely deficient for vacuole hydrolase activity [1,73]. Thus the sporulation defect in *vps1/spo15* mutants is likely to be a secondary consequence of vacuole dysfunction.

Random mutagenesis of the *VPS1* gene yielded two classes of *vps1* mutations. One class behaved as recessive, loss of function alleles, and in general the mutations eliminated a portion of the carboxy-terminal region in Vps1p. The other class of mutations, which all contained lesions in the amino-terminal portion of the *VPS1* gene, conferred the unusual property of interfering with protein sorting when the mutant products were expressed in wild-type cells. These and other data described here suggest that Vps1p is composed of two distinct functional regions. The boundary between these regions, as defined by mapping of the various classes of mutations, coincides well with the boundary in sequence similarity between Vps1p and the other members of this protein family (compare Figs. 3 and 6). These data are consistent with the notion that the GTP-binding region in these proteins performs a common function while the divergent carboxy-terminal domain confers specificity.

Biochemical characterization of wild-type and mutant forms of Vps1p has shed some light on the possible functional roles of the two distinct regions in the protein. We have shown that Vps1p binds and hydrolyzes GTP, and, as predicted from sequence, this activity is a property of the amino-terminal, highly conserved region. It seems reasonable to expect that there are cellular factors that modulate binding and/or hydrolysis of GTP by Vps1p. Interestingly, several of the dominant-interfering Vps1p mutant proteins retain the capacity to bind GTP [23]. The amino-terminal mutations in these mutant proteins may affect a region of Vps1p that normally interacts with these hypothetical modulating cellular factors.

The function of the divergent carboxy-terminal region is more obscure. It is clearly an indispensable component for Vps1p function. The fact that expression of this region by itself interferes with protein sorting suggests that it retains the capacity to interact with Vps1p specific factors. This hypothesis is further supported by the observation that increased expression of either wild-type Vps1p or a recently isolated gene *MVP1* (multicopy suppressor of *vps1*-dominant mutation) partially mitigate the interfering phenotype (K. Ekena and T. Stevens, unpublished results). There are at least two models that explain the dominant-interfering influence of the carboxy-terminal region. Vps1p may normally function as part of a multimeric complex, and the carboxy-terminal domain may promote this oligomerization. The dominant-interfering effect would then reflect the consequence of incorporation of nonfunctional subunits into this Vps1p multimer, thereby poisoning the function of the complex. Alternatively, the carboxy-terminal region may interact with other cellular factors, such as Mvp1p, that are present in limited concentration and are required for protein sorting. The dominant-interfering *vps1* protein would thus titrate these factors and thereby compromise protein sorting. Our immediate research goals are to identify factors that functionally interact with Vps1p and to characterize the organellar association of the protein in greater detail.

Acknowledgements

This work was supported by National Institutes of Health predoctoral traineeships to C.K.R. and C.J.R., a Damon Runyon-Walter Winchell Cancer Research Fund Award (DRD 1012) to C.A.V., an American Cancer Society Postdoctoral Fellowship (PF-3608) to S.F.N., grants from the National Institutes of Health (PHS 32448 & 38006) to T.H.S., and an American Cancer Society Faculty Research Award (PRA-337) to T.H.S.

References

1. Rothman, J.H. and Stevens, T.H. (1986) Cell 47, 1041–1051.
2. Stevens, T.H., Esmon, B. and Schekman, R. (1982) Cell 30, 439–448.
3. Klionsky, D.J., Herman, P.K. and Emr, S.D. (1990) Microbiol. Rev. 54, 266–292.
4. Raymond, C.K., Roberts, C.J., Moore, K.E., Howald, I. and Stevens, T.H. (1992) Int. Rev. Cytol. 139, 59–120.
5. Graham, T.R. and Emr, S.D. (1991) J. Cell Biol. 114, 207–218.
6. Klionsky, D.J. and Emr, S.D. (1989) EMBO J. 8, 2241–2250.
7. Roberts, C.J., Pohlig, G., Rothman, J.H. and Stevens, T.H. (1989) J. Cell Biol. 108, 1363–1373.
8. Bankaitis, V.A., Johnson, L.M. and Emr, S.D. (1986) Proc. Natl. Acad. Sci. USA 83, 9075–9079.
9. Johnson, L.M., Bankaitis, V.A. and Emr, S.D. (1987) Cell 48, 875–885.
10. Klionsky, D.J., Banta, L.M. and Emr, S.D. (1988) Mol. Cell. Biol. 8, 2105–2116.
11. Rothman, J.H., Hunter, C.P., Valls, L.A. and Stevens, T.H. (1986) Proc. Natl. Acad. Sci. USA 83, 3248–3252.
12. Stevens, T.H., Rothman, J.H., Payne, G.S. and Schekman, R. (1986) J. Cell Biol. 102, 1551–1557.
13. Valls, L.A., Hunter, C.P., Rothman, J.H. and Stevens, T.H. (1987) Cell 48, 887–897.
14. Blachly-Dyson, E. and Stevens, T.H. (1987) J. Cell Biol. 104, 1183–1191.
15. Valls, L.A., Winther, J.R. and Stevens, T.H. (1990) J. Cell Biol. 111, 361–368.
16. Kornfeld, S. and Mellman, I. (1989) Annu. Rev. Cell Biol. 5, 483–525.
17. Schwaiger, H., Hasilik, A., von Figura, K., Weimken, A. and Tanner, W. (1982) Biochem. Biophys. Res. Commun. 104, 950–956.
18. Winther, J.R. (1991) Eur. J. Biochem. 197, 681–689.
19. Suarez-Rendueles, P. and Wolf, D.H. (1987) J. Bacteriol. 169, 4041–4048.
20. Julius, D., Blair, L., Brake, A., Sprague, G. and Thorner, J. (1983) Cell 32, 839–852.
21. Robinson, J.S., Klionsky, D.J., Banta, L.M. and Emr, S.D. (1988) Mol. Cell. Biol. 8, 4936–4948.
22. Rothman, J.H., Howald, I. and Stevens, T.H. (1989) EMBO J. 8, 2057–2065.
23. Raymond, C.K. (1990) Ph.D. thesis, University of Oregon, Eugene, Oregon.
24. Singer, S.J., Maher, P.A. and Yaffe, M.P. (1987) Proc. Natl. Acad. Sci. USA 84, 1960–1964.
25. Roberts, C.J., Raymond, C.K., Yamashiro, C.T. and Stevens, T.H. (1990) Methods Enzymol. 194, 644–661.
26. Roberts, C.J. (1991) Ph.D. thesis, University of Oregon, Eugene, Oregon.
27. Klionsky, D.J. and Emr, S.D. (1990) J. Biol. Chem. 265, 5349–5352.
28. Fuller, R.S., Brake, A.J. and Thorner, J. (1989) Science (Washington, DC) 246, 482–486.
29. Julius, D., Brake, A., Blair, L., Kunisawa, R. and Thorner, J. (1984) Cell 37, 1075–1089.
30. Franzusoff, A., Redding, K., Crosby, J., Fuller, R.S. and Schekman, R. (1991) J. Cell Biol. 112, 27–37.
31. Redding, K., Holcomb, C. and Fuller, R.S. (1991) J. Cell Biol. 113, 527–538.
32. Kaneko, Y., Hayashi, N., Toh-e, A., Banno, I. and Oshima, Y. (1987) Gene 58, 137–148.

33. Dever, T.E., Glynias, M.J. and Merrick, W.C. (1987) Proc. Natl. Acad. Sci. USA 84, 1814–1818.
34. Rothman, J.H., Raymond, C.K., Gilbert, T., O'Hara, P.J. and Stevens, T.H. (1990) Cell 61, 1063–1074.
35. Arnheiter, H. and Meier, E. (1990) New Biol. 2, 851–857.
36. Pavlovic, J.P., Zurcher, T., Haller, O. and Staehli, P. (1990) J. Virol. 64, 3370–3375.
37. Staehli, P. (1990) Adv. Virus. Res. 38, 147–200.
38. Obar, R.A., Collins, C.A., Hammarback, J.A., Shpetner, H.S. and Vallee, R.B. (1990) Nature (London) 347, 256–261.
39. Shpetner, H.S. and Vallee, R.B. (1989) Cell 59, 421–432.
40. Chen, M.S., Obar, R.A., Schroeder, C.C., Austin, T.W., Poodry, C.A., Wassworth, S.C. and Vallee, R.B. (1991) Nature 351, 583–586.
41. van der Bliek, A.M. and Meyerowitz, E.M. (1991) Nature (London) 351, 411–414.
42. Collins, C.A. (1991) Trends Cell Biol. 1, 57–60.
43. Vallee, R.B. and Shpetner, H.S. (1990) Annu. Rev. Biochem. 59, 909–932.
44. Scaife, R. and Margolis, R.L. (1990) J. Cell Biol. 111, 3023–3033.
45. Poodry, C.A. (1990) Dev. Biol. 138, 464–472.
46. Kessell, I., Holst, B.D. and Roth, T.F. (1989) Proc. Natl. Acad. Sci. USA 86, 4968–4972.
47. Yeh, E., Carbon, J. and Bloom, K. (1986) Mol. Cell. Biol. 6, 158–167.
48. Yeh, E., Driscoll, R., Coltera, M., Olins, A. and Bloom, K. (1991) Nature (London) 349, 713–714.
49. Guthrie, B. and Wickner, W. (1988) J. Cell Biol. 107, 115–120.
50. Jacobs, C.W., Adams, A.E.M., Szaniszlo, P.J. and Pringle, J.R. (1988) J. Cell. Biol. 107, 1409–1426.
51. Huffaker, T.C., Thomas, J.H. and Botstein, D. (1988) J. Cell Biol. 107, 1997–2010.
52. Schmitt, H.D., Wagner, P., Pfaff, E. and Gallwitz, D. (1986) Cell 47, 401–412.
53. Segev, N., Mullholland, J. and Botstein, D. (1988) Cell 52, 915–924.
54. Salminen, A. and Novick, P.J. (1987) Cell 49, 527–536.
55. Walworth, N.C., Goud, B., Kastan-Kabcenell and Novick, P.J. (1989) EMBO J. 8, 1685–1693.
56. Bacon, R.A., Salminen, A., Ruohola, H., Novick, P. and Ferro-Novick, S. (1989) J. Cell Biol. 109, 1015–1022.
57. Goud, B., Salminen, A., Walworth, N.C. and Novick, P.J. (1988) Cell 53, 753–768.
58. Groesch, M.E., Ruohola, H., Bacon, R., Rossi, G. and Ferro-Novick, S. (1990) J. Cell Biol. 111, 45–53.
59. Rexach, M.F. and Schekman, R.W. (1991) J. Cell Biol. 114, 219–230.
60. Kunes, S., Ma, H., Overbye, K., Fox, M.S. and Botstein, D. (1987) Genetics. 115, 73–81.
61. Payne, G.S. and Schekman, R. (1989) Science (Washington, DC) 245, 1358–1365.
62. Cooper, A. (1990) Ph.D. thesis. McGill University, Montreal, Quebec, Canada.
63. Novick, P., Ferro, S. and Schekman, R. (1981) Cell 25, 461–469.
64. Pfeffer, S.R. and Rothman, J.E. (1987) Annu. Rev. Biochem. 56, 829–852.
65. Stafford, F.J. and Bonifacino, J.S. (1991) J. Cell Biol. 115, 1225–1236.
66. Jackson, M.R., Nilsson, T. and Peterson, P. (1990) EMBO J. 9, 3153–3162.
67. Machamer, C.E. and Rose, J.K. (1987) J. Cell Biol. 105, 1205–1214.
68. Williams, M.A. and Fukuda, M. (1990) J. Cell Biol. 111, 955–966.
69. Goldstein, L.S.B. (1991) Trends Cell Biol. 1, 93–98.
70. Hollenbeck, P.J. (1990) Nature 347, 229.
71. Kaneko, Y., Toh-e, A. and Oshima, Y. (1982) Mol. Cell. Biol. 5, 248–252.
72. Zubenko, G.S. and Jones, E.W. (1981) Genetics 97, 45–64.
73. Rothman, J.H. (1988) Ph.D. thesis, University of Oregon, Eugene, Oregon.

Part D

Peroxisomes

CHAPTER 15

Defining components required for peroxisome assembly in *Saccharomyces cerevisiae*

JÖRG HÖHFELD, DAPHNE MERTENS, FRANZISKA F. WIEBEL and WOLF-H. KUNAU

Institut für Physiologische Chemie, Medizinische Fakultät der Ruhr-Universität Bochum, W-4630 Bochum, Germany

1. Introduction

Peroxisomes are nearly ubiquitous organelles of eukaryotic cells (for reviews see [1–5]. The original biochemical characterization led to the discovery of their H_2O_2 metabolism (H_2O_2-producing oxidases and catalase) and this in turn was used as a definition of these subcellular structures [6]. Peroxisomes are the major subgroup of the morphologically defined microbodies. Current knowledge has revealed that the H_2O_2 metabolism is just one aspect of the metabolic functions associated with peroxisomes [5]. Due to the diversity of enzymatic reactions discovered in peroxisomes, they are now viewed as multi-purpose organelles [7]. However, despite their remarkable biochemical heterogeneity, all peroxisomes share the capability to degrade fatty acids via the β-oxidation pathway [5,8,9]. Another notable property is the inducibility of peroxisomal proliferation and peroxisomal metabolic pathways [2,5]. Both diversity of metabolic functions and inducibility of peroxisomal proliferation and metabolism are especially pronounced in eukaryotic microorganisms such as yeasts and filamentous fungi [10–14]. In these microorganisms, the final size, volume fraction and enzyme pattern of peroxisomes can be drastically changed by varying growth conditions. Proliferation of fungal peroxisomes is under the control of glucose repression and de-repression as well as induction by distinct carbon and nitrogen sources. Their properties make peroxisomes rather unique among organelles which in general have retained one major and mostly constitutive function throughout evolution.

Over the last two decades the yeast *Saccharomyces cerevisiae* has been more and more frequently used as a eukaryotic model to study basic functions and structures of eukaryotic cells [15]. Its genetic and biochemical attributes make it an attractive organism for analyzing complex processes. Particularly successful examples are the investigations of cell cycle [16,17] and morphogenesis ([18], the transport of proteins through the secretion pathway [19–21] and the biogenesis of mitochondria [22,23]. The formation of peroxisomes is another process which seems feasible to investigate by exploiting classical and molecular genetics as well as the biochemical potential of this yeast.

The peroxisomes originally demonstrated in *S. cerevisiae* were very rare and small [24]. To date, this is easily explained by the fact that peroxisomal proliferation is also strongly influenced in this yeast by growth conditions [25]. Induction of peroxisomal proliferation in *S. cerevisiae*, using oleic acid as the sole carbon source, was only recently established. It opened the way for the development of a genetic screen to isolate peroxisomal mutants [26]. This approach was initially designed to potentially identify membrane components of the peroxisomal import machinery. However, it also yields mutants defective in a much broader array of peroxisomal gene products. As a first step we isolated strains unable to grow on oleic acid (oleic acid non-utilizing (onu) phenotype). Replica plating on acetate, ethanol or glycerol and oleic acid plates discriminates between cells which cannot utilize oleic acid and those which have mitochondrial defects. This screen enriched cells with an impaired β-oxidation pathway due to either (i) defects in structural genes of β-oxidation enzymes, (ii) defects in peroxisomal protein import or (iii) other peroxisomal defects. To find these peroxisomal mutants among the onu strains, we relied on biochemical methods. Determination of peroxisomal enzyme activities in either whole cell lysates or subcellular fractions obtained from lysates of spheroplasts allowed detection of deficiencies of individual enzyme activities as well as the detection of all or only distinct peroxisomal matrix enzymes in the cytosol rather than in peroxisomes. The mislocalization of peroxisomal enzymes was taken as an indication that peroxisomes are absent or import incompetent. These conclusions were subsequently further verified by electron microscopy and immunocytochemistry which were carried out in collaboration with M. Veenhuis (University of Groningen, The Netherlands). A screen for peroxisomal mutants does not require a search for temperature-sensitive phenotypes because peroxisomes are not essential for the viability of *S. cerevisiae* cells [27,28]. This is in marked contrast to mitochondria [29]. However, the isolation of onu strains on oleic acid plates does allow the selection of temperature-sensitive mutants.

The further development of classical and molecular genetics of other yeasts, e.g. *Pichia pastoris* [30] and *Hansenula polymorpha* [30,31] has recently increased the number of organisms suitable for genetic dissection of peroxisome biogenesis. This is emphasized by the isolation of peroxisomal mutants of *H. polymorpha* [31,32] and *P. pastoris* (Cregg and Veenhuis, pers. commun.). In addition, peroxisomal mutants of Chinese hamster ovary cells [33–35] provide a model system in higher eukaryotes.

Interestingly, the biochemical and morphological phenotypes of the isolated peroxisomal mutants of *S. cerevisiae* resemble those found in fibroblasts of patients with peroxisomal disorders [5,36,37]. This is a recently discovered class of human inborn errors with severe clinical symptoms. These diseases are mostly lethal. It is assumed that for one subgroup, the primary defect is an impairment in peroxisome biogenesis.

In this chapter, we describe the different types of peroxisomal mutants of *S. cerevisiae* isolated by the above mentioned screening procedure and summarize our current knowledge about genes essential for peroxisome assembly in this yeast.

2. Results

2.1. Peroxisomal mutants of Saccharomyces cerevisiae

The devised screen for peroxisomal mutants of *S. cerevisiae* includes colony screening, biochemical characterization of the identified oleic acid non-utilizing (onu) mutant strains (Fig. 1) and morphological analysis [26]. Based on this combined screening procedure, two principally different classes of peroxisomal mutants could be isolated, *fox* and *pas* mutants.

The *fox* mutants (fatty acid oxidation) are characterized by a deficiency of one of the individual β-oxidation proteins. Thus, three complementation groups defining the structural genes of these proteins were identified: *FOX1*, acyl-CoA-oxidase; *FOX2*, multi-functional protein; *FOX3*, 3-oxoacyl-CoA thiolase. The detailed characterization of these mutants is beyond the scope of this study and will be described elsewhere (Erdmann and Kunau, unpublished results).

The second group comprises mutants affected in peroxisome assembly (*pas* mu-

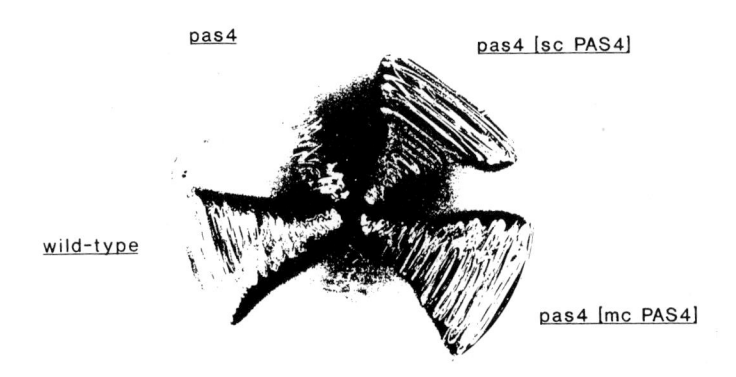

Fig. 1. Screening for growth on oleic acid as sole carbon source. In contrast to the wild-type strain, two *pas4* mutants are not able to grow on oleic acid minimal medium plates even after incubation for 14 days at 30°C (*pas4*, original mutant strain; *pas4::LEU2*, mutant constructed by gene deletion). Transformation of the mutant strains with a single or multi-copy plasmid carrying a wild-type copy of the *PAS4* gene led to transformants which regained the ability to utilize oleic acid.

tants) [26]. Although the β-oxidation enzymes are detectable and can be induced to wild-type levels, these mutant strains also do not grow on oleic acid medium suggesting that effective functioning of fatty acid degradation requires the compartmentation of the β-oxidation pathway and most likely the proper spatial assembly of the β-oxidation complex in peroxisomes. Our screening procedure allowed us to discriminate subgroups among the *pas* mutants according to their different phenotypes. These subgroups were designated type I to type III *pas* mutants.

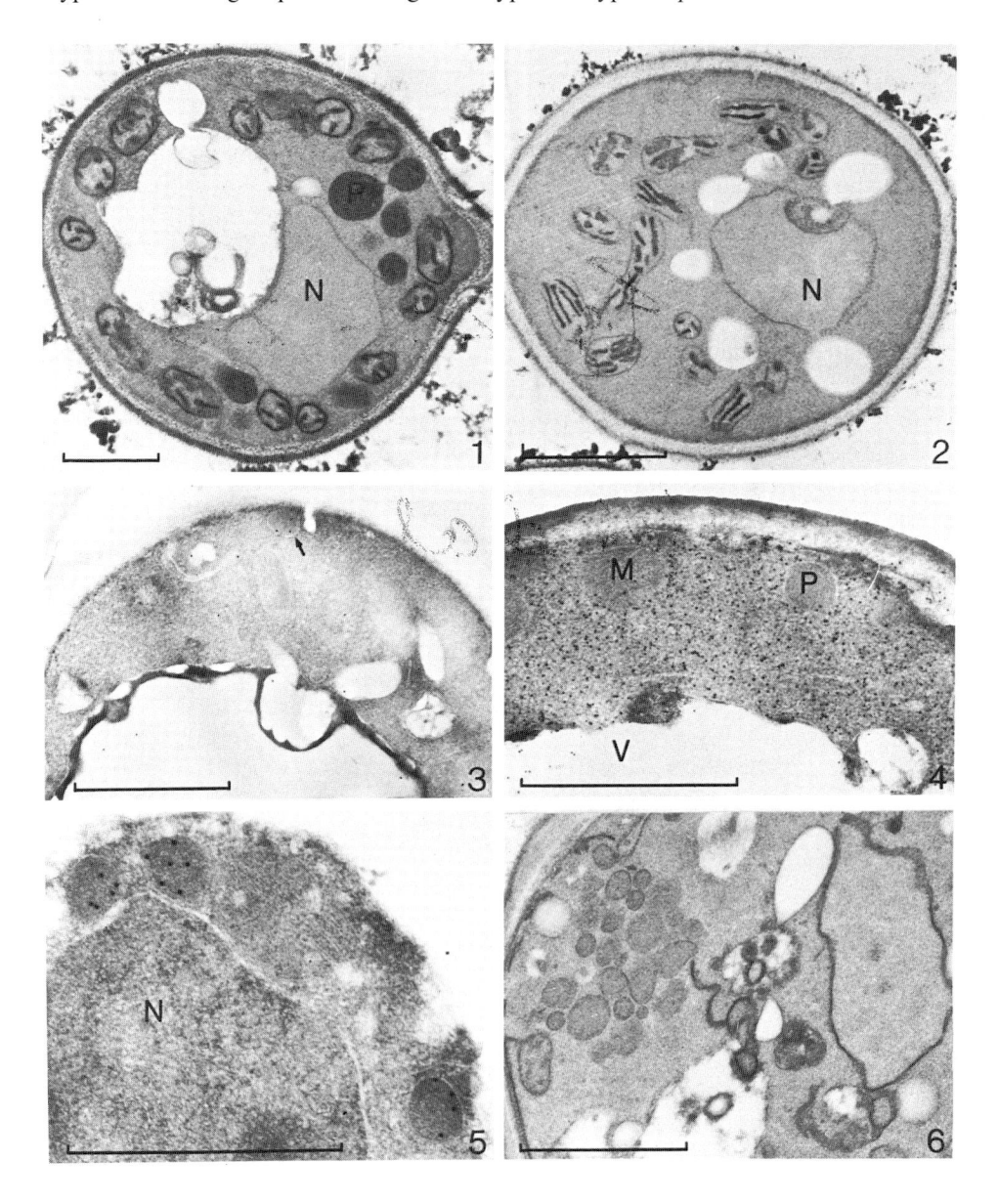

2.1.1. Defects in peroxisome formation (type I pas mutants: pas1, pas2, pas3, pas5)

The phenotype of type I *pas* mutants is characterized by a lack of morphologically detectable peroxisomes. Even after incubation in oleic acid medium, no peroxisomes could be detected as was confirmed by serial sections and a cytochemical staining for catalase activity (Fig. 2). Peroxisomal matrix enzymes can be induced to wild-type levels, but were mislocalized to the cytosol of the mutant strains as was demonstrated by immunocytochemistry and subcellular fractionation experiments (Table I). This phenotype indicates that a gene is affected which encodes a protein essential

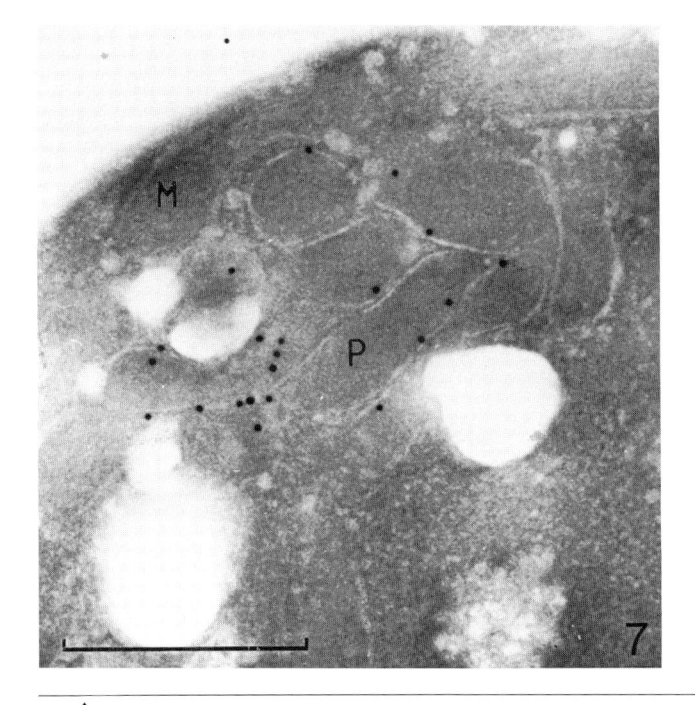

← ↑

Fig. 2. Morphological analysis of *S. cerevisiae* wild-type strain and different *pas* mutants after growth on oleate as sole carbon source. (1) Growth of *S. cerevisiae* wild-type on oleate results in proliferation of peroxisomes (densely stained due to a cytochemical catalase activity staining with diaminobenzidine (DAB)). (2) In contrast, peroxisomes were not detectable in DAB-stained sections of *pas1* mutant cells grown under the same conditions. (3) Immunocytochemical detection of a single small peroxisome present in the constructed *pas4* null mutant after growth on oleate (arrow). The peroxisome was labeled using antithiolase antibodies. (4) Overexpression of thiolase in the thiolase-import mutant (*pas7*). Immunocytochemical staining using antithiolase antibodies demonstrates the cytosolic localization of the overexpressed protein whereas detectable peroxisomes are not labeled. (5) Labeling of the peroxisomal matrix in ultrathin sections of oleate-grown wild-type cells after incubation with antithiolase antibodies. (6) Overexpression of *PAS4* in a wild-type strain based on transformation with a multi-copy plasmid results in a marked proliferation of peroxisomes (cf. non-transformed wild-type cells grown under the same conditions (1)). (7) After expression of a Pas3p-β-galactosidase fusion protein in wild-type cells, labeling of the peroxisomal membrane was observed using anti-β-galactosidase antibodies. M, mitochondrion; N, nucleus; P, peroxisome; V, vacuole; bar, 1 μm.

TABLE I

Distribution pattern of peroxisomal and mitochondrial marker enzymes in the 25 000 × g supernatant and pellet fractions of cell lysates from wild-type strain and different *pas* mutants[a]

Strain	Activity in 25 000 × g		
	a1 Supernatant fraction	a2 Pellet fraction	a1:a2
Wild type			
Catalase	1.3×10^4	6.3×10^4	0.2
Epimerase	15.2	93.2	0.2
Thiolase	24.5	73.8	0.3
Fumarase	17.5	58.3	0.3
pas3–12			
Catalase	23.2×10^4	0.5×10^4	43.3
Epimerase	216.2	68.3	3.2
Thiolase	182.8	28.5	6.4
Fumarase	95.8	173.5	0.6
pas4			
Catalase	2.5×10^4	0.03×10^4	83.3
Epimerase	83.3	21.5	3.9
Thiolase	78.8	10.2	7.7
Fumarase	33.4	95.2	0.4
pas4 [sc-PAS4]			
Catalase	1.4×10^4	3.2×10^4	0.4
Epimerase	12.7	24.8	0.5
Thiolase	6.5	30.3	0.2
Fumarase	3.8	7.5	0.5
pas7			
Catalase	0.8×10^4	1.6×10^4	0.5
Epimerase	10.0	23.3	0.4
Thiolase	10.0	1.7	5.9
Fumarase	50.0	86.7	0.6

[a] The table includes a *pas4* mutant strain transformed with a wild-type copy of the PAS4 gene (*pas4 (sc-PAS4)*). A ratio a1:a2 below 1 indicates a particulate localization, whereas a ratio above 1 implies a cytosolic mislocalization of peroxisomal enzymes. The enzyme activities are expressed as nkat per fraction.

for peroxisome formation. Moreover, the described phenotype resembles that of fibroblasts of Zellweger patients, a peroxisomal disease in man for which a defect in the peroxisomal import machinery is discussed (for review see [5]). It was recently demonstrated that Zellweger fibroblasts contain empty peroxisomal membrane structures (peroxisomal ghosts) [38–40]. However, similar structures have not yet be detected in the *pas* mutants of this subgroup (see Discussion).

2.1.2. Defects in peroxisome proliferation (type II pas mutants: pas4 and pas6)
By morphological analysis it could be demonstrated that *pas4* and *pas6* mutant cells contain very few small peroxisomes after induction on oleic acid medium (Fig. 2). Unlike wild-type cells where these growth conditions result in peroxisome proliferation, the mutant cells rather resemble those of a glucose-repressed state. In contrast to the degree of proliferation, induction rates of β-oxidation enzymes are not affected. Although peroxisomes in reduced numbers are present, the bulk of peroxisomal proteins are mislocalized to the cytosol thereby resembling the *pas* mutants type I (Table I). Based on the phenotype of a null mutant, it could be excluded that the presence of peroxisomes in the *pas4* mutant is due to a leaky mutation of a gene essential for peroxisome formation (type I *PAS* gene). *PAS4* rather seems to be involved in the process of induced proliferation (Mertens and Kunau, unpublished results).

2.1.3. Defects in import of 3-oxoacyl-CoA thiolase (type III pas mutant: pas7)
In case of the *pas7* mutant, a cytosolic mislocalization of the peroxisomal matrix enzyme thiolase only was observed. In contrast to other matrix enzymes, thiolase activity was predominantly detected in the $25\,000 \times g$ supernatant fraction after subcellular fractionation (Table I). This finding was confirmed by morphological data. Although peroxisomes were clearly detectable, it was not possible to label them immunocytochemically using antithiolase antibodies (Fig. 2). To investigate whether this phenotype is caused by a mutation in the thiolase structural gene (e.g. peroxisomal targeting signal), the *pas7* mutant was transformed with a wild-type copy of the thiolase gene (*FOX3*). The transformation led neither to import of thiolase into peroxisomes nor to functional complementation indicating that a protein other than the thiolase itself is affected in this mutant strain. This as yet unidentified protein seems to be essential for the import of thiolase into peroxisomes providing first experimental evidence for the existence of at least two peroxisomal import pathways (Marzioch, Erdmann and Kunau, unpublished results).

2.2. Cloning of peroxisomal genes

The isolation of peroxisomal mutants of *S. cerevisiae* enabled us to clone the affected genes by functional complementation [41]. For this purpose, mutant strains were transformed with a genomic bank of *S. cerevisiae* and the resultant transformants were subsequently screened for their ability to grow on oleic acid minimal medium. Transformants which regained this ability were further analyzed by subcellular fractionation studies and electron microscopy. Based on this approach, the following genes have been cloned: *FOX2*, *FOX3* and *PAS1* to *PAS7* (see Table II). The authenticity of the cloned genes was checked by construction of null mutants and subsequent backcross to the original mutant strains.

TABLE II

Characterization of the genes cloned by functional complementation of peroxisomal mutants of *S. cerevisiae*

Gene	Size of ORF (bp)	M_r of the deduced gene product (kDa)	Properties of the gene product
FOX2	2696	98.4	Trifunctional enzyme
FOX3	1251	44.7	3-Oxoacyl-CoA-thiolase
PAS1	3129	117.3	ATP-binding protein[a]
PAS2	549	21.1	Ubiquitin-conjugating protein[a]
PAS3	1323	50.6	48 kDa integral PMP
PAS4	1011	39.1	Zinc finger-like motif
PAS5	812	30.8	Zinc finger-like motif
PAS6	1767	68.2	Carboxyl-terminal SKL
PAS7	1128	42.3	No obvious membrane span[a]

[a] Based on sequence analysis

2.3. Sequence analysis

The genomic fragments of the plasmids complementing the respective mutant strains were sequenced and the open reading frames of the genes identified. The deduced primary sequences of the gene products were compared to known protein sequences of databases. None of the cloned genes was found to be identical to known sequences. For some gene products (Pas1p, Pas2p and Pas4p), sequence similarities to known protein sequences were observed providing a first insight into a possible function of the putative proteins.

2.3.1. PAS1

Pas1p displayed strong partial sequence similarity to Sec18p/NSF, proteins essential for fusion steps of the secretory pathway [42,43], VCP/Cdc48p/p97, of which p97 was shown to have ATPase activity [44–46], and TBP-1, which binds to the tat-protein of the human immunodeficiency virus (HIV) [47] (Fig. 3a). Similarity is restricted to a region of about 180 amino acids, tentatively called Pas1 box which always includes a consensus motif for ATP binding. This box was found twice well conserved in VCP/p97. However, closer investigation revealed that Pas1p and Sec18p/NSF also contain a second Pas1 box. In these cases, similarity is less pronounced for one of the boxes, namely the first one in Pas1p and the second one in Sec18p/NSF. TBP-1, which is half the length of the other proteins, contains only one Pas1 box.

These findings led to the definition of a previously unrecognized family of putative ATPases [27]. It can be assumed that the proteins of this family share a specific common function residing in the Pas1 box which remains to be ascribed.

(a)

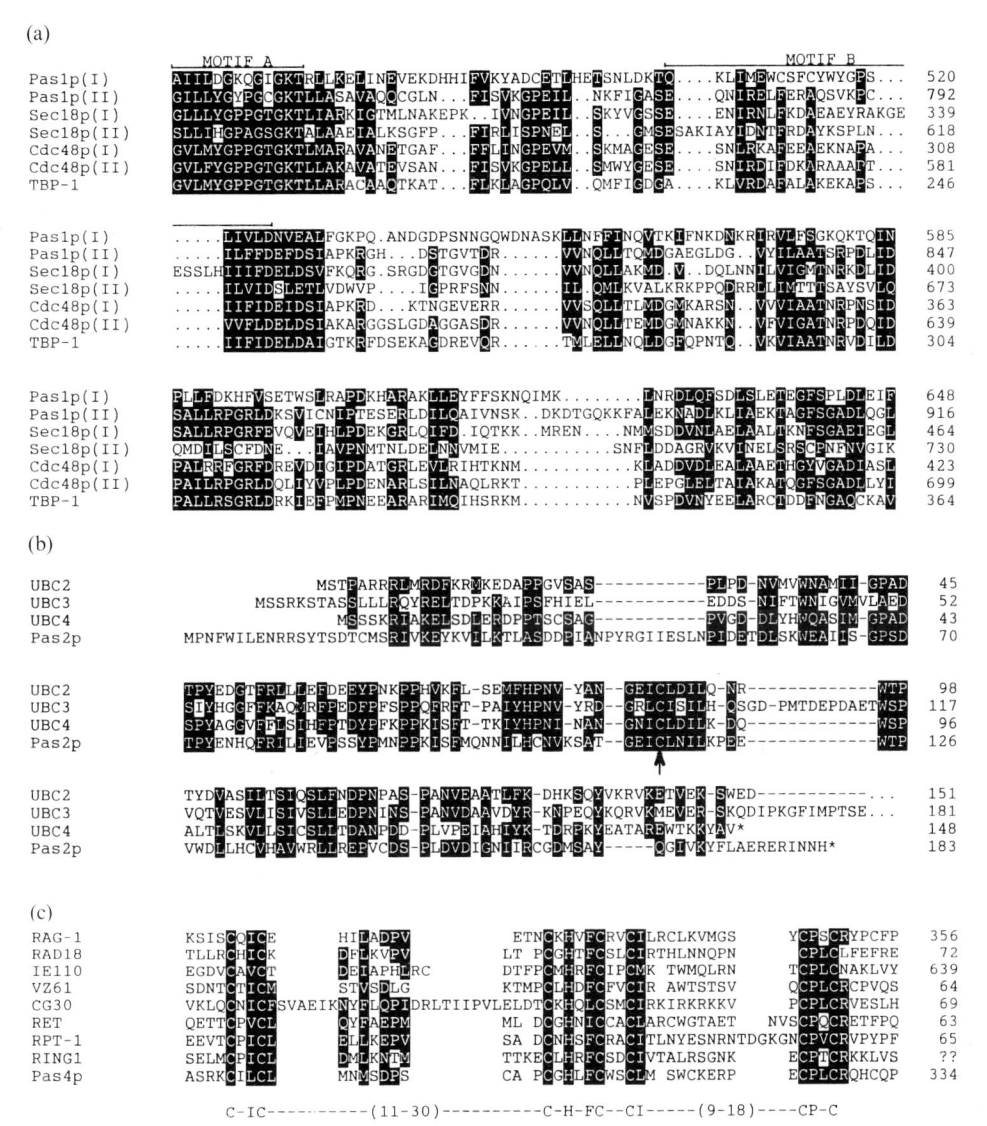

```
              MOTIF A                                              MOTIF B
Pas1p(I)    AIILDGKQGIGKTRLLKELINEVEKDHHIFVKYADCETLHETSNLDKTQ....KLIMEWCSFCYWYGES...  520
Pas1p(II)   GILLYGYPCGKTLLASAVAQQCGLN...FISVKGPEIL..NKFIGASE....QNIRDLFERAQSVKFC...   792
Sec18p(I)   GLLLYGPPGTGKTLIARKIGTMLNAKEPK..IVNGPEIL..SKYVGSSE....ENIRNLFKDAEAEYRAKGE  339
Sec18p(II)  SLLIHGPAGSGKTALAAEIALKSGFP...FIRLISPNEL..S...GMSESAKIAYIDNTFRDAYKSPLN...  618
Cdc48p(I)   GVLMYGPPGTGKTLMARAVANETGAF...FFLINGPEVM..SKMAGESE....SNLRKAFEEAEKNABA...  308
Cdc48p(II)  GVLFYGPPGTGKTLLAKAVATEVSAN...FISVKGPELL..SMWYGESE....SNIRDIFDKARAAART...  581
TBP-1       GVLMYGPPGTGKTLLARACAAQTKAT...FLKLAGPQLV..QMFIGDGA....KLVRDAFALAKEKAES...  246

Pas1p(I)    .....LIVLDNVEALFGKPQ.ANDGDPSNNGQWDNASKLLNFFINQVTKIFNKDNKRIRVLFSGKQKTQIN   585
Pas1p(II)   .....ILFFDEFDSIAPKRGH...DSTGVTDR......VVNQLLTQMDGAEGLDG..VYILAATSRPDLID   847
Sec18p(I)   ESSLHIIFDELDSVFKQRG.SRGDGTGVGDN......VVNQLLAKMD.V..DQLNNILVIGMTNRKDLID   400
Sec18p(II)  .....ILVIDSLETLVDWVP....IGPRFSNN......IL.QMLKVALKRKPPQDRRLLIMTTISAYSVLQ   673
Cdc48p(I)   .....IIFDEIDEIAPKRD...KTNGEVERR......VVSQLLTLMDGMKARSN...VVVIAATNRPNSID  363
Cdc48p(II)  .....VVFLDELDSIAKARGGSLGDAGGASDR......VVNQLLTEMDGMNAKKN...VKFVIGATNRPDQID  639
TBP-1       .....IIFIDELDAIGTKRFFDSEKAGDREVQR......TMLELLNQLDGFQPNTQ..VKVVIAATNRVDILD   304

Pas1p(I)    PLLFDKHFVSETWSLRAPDKHARAKLLEYFFSKNQIMK........LNRDLQFSDLSLETEGFSPLDLEIF   648
Pas1p(II)   SALLRPGRLDKSVICNIPTESERLDILQAIVNSK..DKDTGQKKFALEKNADLKLIAEKTAGFSGADLQGL   916
Sec18p(I)   SALLRPGRFEVQVETHLPDEKGRLQIFD.IQTKK..MREN...NMMSDDVNLAELAALITKNFSGAEIEGL   464
Sec18p(II)  QMDILSCFDNE...IAVPNMTNLDELNMVMIE..........SNFLDDAGRVKVINELSRSCPNFNVGIK   730
Cdc48p(I)   PALRRFGRFDREVDIGIPDATGRLEVLRIHTKNM.......KLADDVDLEALAAEIHGYVGADIASL     423
Cdc48p(II)  PAILRPGRLDQLIYVPLPDENARLSILNAQLRKT.........PLEPGLELTAIAKATQGFSGADLLYI  699
TBP-1       PALLRSGRLDRKIEFPMPNEEARARIMQIHSRKM.........NVSPDVNYEELARCTDDFNGAQCKAV  364
```

(b)

```
UBC2    MST.PARRRLMRDFKRMKEDAPFGVSAS----------PLPD-NVMVWNAMII-GPAD        45
UBC3    MSSRKSTASSLLLRQYRELTDPKKAIPSFHIEL----------EDDS-NIFTWNIGVMVLAED   52
UBC4    MSSSKRIAKELSDLERDPPTSCSAG----------PVGD-DLYHWQASIM-GPAD           43
Pas2p   MPNFWILENRRSYTSDTCMSRIVKEYKVILKTLASDDEIANPYRGIIESLNPIDETDLSKVEAIIS-GPSD 70

UBC2    TPYEDGTFRLLLEFDEEYPNKPPHVKEL-SEMFHPNV-YAN--GEICLDILQ-NR-----------WTP   98
UBC3    SIYHGGFFKAQLRFPEDFPFSPQFRET-PAIYHPNV-YRD--GRLCISILH-QSGD-PMTDEPDAETWSP 117
UBC4    SPYAGGVFFLSHFPTDYFKPKISET-TKIYHPNI-NAN--GNICLDILK-DQ-----------WSP     96
Pas2p   TPYENHQFRILLEVPSSYPMNPPKISFMQNNILHCNVKSAT--GEICLNILKPEE----------WTP  126

UBC2    TYDVASIITSIQSLFNDPNPAS-PANVEAATLFK-DHKSQYVKRVKETVEK-SWED----------.. 151
UBC3    VQTVESVLISIVSLLEDPNINS-PANVDAAVDYR-KNPEQYKQYKMEYER-SKQDIPKGFIMPTSE... 181
UBC4    ALTLSKVLLSICSLLTDANPDD-PLVFEIAHIYK-TDRPKYEATAREWTKKYAV*                148
Pas2p   VWDLLHCVHAVWRLLREPVCDS-PLDVDIGNIIRCGDMSAY-----QGIVKYFLAERERINNH*     183
```

(c)

```
RAG-1   KSISCQICE     HILADPV      ETNCKEVFCRVCILRCLKVMGS   YCPSCRYPCFP   356
RAD18   TLLRCHICK     DFLKVPV      LT PCGHTFCSLCIRTHLNNQPN  CPLCLFEFRE    72
IE110   EGDVCAVCT     DEIAPHLRC    DTFPCMHRFCCIFCMK TWMQLRN  TCPLCNAKLVY   639
VZ61    SDNTCTICM     STVSDLG      KTMPCLFDFCFVCIR AWTSTSV   QCPLCRCPVQS   64
CG30    VKLQCNICFSVAEIKNYFLQPIDRLTIIPVLEDKCHQLCSMCIRKIRKRKKV   PCPLCRVESLH   69
RET     QETTCPVCL     QYFAEPM      ML DCGHNICCACLARCWGTAET   NVSCPQCRETFPQ 63
RPT-1   EEVTCPICL     ELLKEPV      SA DCNHSFCRACITLNYESNRNTDGKGNCPVCRVPYPF 65
RING1   SELMCPICL     DMLKNTM      TTKECLFRFCSDCIVTALRSGNK   ECPTCRKKLVS   ??
Pas4p   ASRKCILCL     MNVSDPS      CA PCGHLFCWSCLM SWCKERP   ECPLCRQHCCP  334

        C-IC----------(11-30)---------C-H-FC--CI-----(9-18)----CP-C
```

Fig. 3. Alignments of the deduced primary sequence of different Pas proteins with a selection of proteins with which they display sequence similarity. Sequence similarities were initially revealed by a search of databases (EMBL Nucleotide Sequence Database and SwissProt Protein Sequence Database, EMBL, Heidelberg). Numbers refer to amino acid positions. (a) Alignment of Pas1 boxes of TBP-1 and yeast members belonging to the proposed ATPase family (see Results). Motifs A and B of the consensus sequence for ATP binding are indicated. Five or more identical or similar amino acid residues are marked. (b) Alignment of Pas2p to UBC proteins. The conserved cystein residue is pointed out by an arrow. Stars mark ends of sequences, dots indicate where complete sequences are not shown. Three or more identical or similar amino acid residues are marked. (c) Alignment of Pas4p to a subgroup of zinc-finger proteins as defined by Freemont et al. (see Results) in the area of the consensus motif (shown below). Six or more identical amino acid residues are marked.

2.3.2. PAS2

Pas2p shows strong sequence similarity to the UBC family of proteins [48] which conjugate ubiquitin to substrates in the ubiquitin conjugation cascade [49,50] (Fig. 3b). Similarity is in the range of 45–60% and thus comparable to that between known UBC proteins. Moreover, a cystein residue in a specific context which is conserved in all UBC proteins [51] is also present in Pas2p. Therefore, Pas2p can be expected to be a ubiquitin conjugating protein itself (Wiebel and Kunau, unpublished results).

2.3.3. PAS4

Sequence comparison revealed that the deduced Pas4p contains a perfect match with a novel cystein-rich motif at its carboxyl-terminal end (Fig. 3c). This sequence, which resembles the well known zinc-finger motif, was recently found in eight proteins most of which are supposed to interact with DNA because of their putative function [52].

The deduced primary sequences of the *PAS* genes were also analyzed with respect to the presence of putative membrane spanning domains using the algorithm of Kyte and Doolittle [53] with a window size of 19 amino acids. Only in the case of the Pas3 primary sequence did this approach lead to the identification of two hydrophobic regions (Fig. 4). In particular, the amino-terminal sequence (amino acids 18–39) fulfills the requirements for a transmembrane span, a length of at least 19 amino acids with a hydrophobicity score higher than 1.6 [53] suggesting the putative Pas3p to be an integral membrane protein [28].

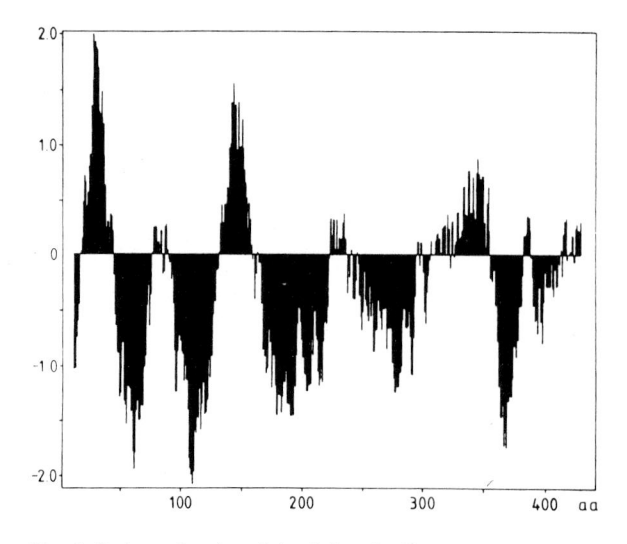

Fig. 4. Hydropathy plot of the deduced primary sequence of Pas3p according to Kyte and Doolittle [53] with a window size of 19 amino acids.

2.4. Identification and characterization of the gene products

To identify the putative gene products, antisera against Pas1, Pas2 and Pas3 protein have been raised so far. For this purpose, a major portion of the respective open reading frame was subcloned in an *E. coli* expression system. In the case of *PAS1*, the gene was expressed as part of a fusion behind the *lacZ* gene (*pUR* vector [54], whereas *PAS2* and *PAS3* were expressed as truncated forms containing only very few foreign amino acids at the amino-terminus encoded by the vector system (*pEXP* vectors [55]. Even using affinity-purified antibodies, it was not possible to detect Pas1p or Pas2p in crude extracts of oleic acid-induced wild-type cells. Only when the respective gene was overexpressed (on a multi-copy plasmid or under regulation of the Gal1 promoter) could polypeptides with an apparent molecular weight of 117 kDa (Pas1p) and 26 kDa (Pas2p) be identified, respectively. This might indicate that Pas1p and Pas2p are very low abundant proteins even in oleic acid-induced wild-type cells.

In contrast to these findings, the affinity purified anti-Pas3p antibodies specifically recognized a 48 kDa protein in crude extracts of wild-type cells [28]. This oleic acid inducible protein could be identified as a peroxisomal protein by sucrose density gradient centrifugation (Fig. 5). Moreover, judged by the carbonate extraction

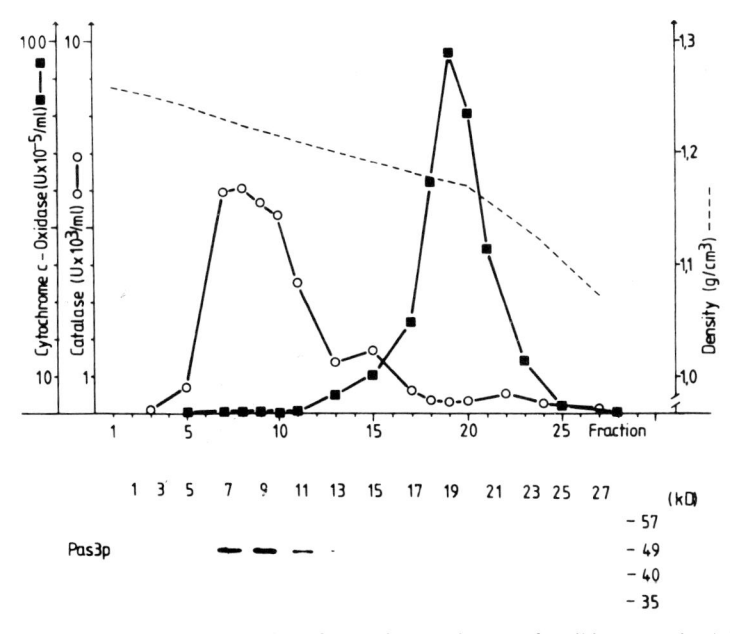

Fig. 5. Immunological detection of Pas3p in peroxisomes of a wild-type strain. A 25 000 × g pellet fraction prepared from spheroplasts was loaded onto a continuous 32–54% sucrose density gradient. After centrifugation, peroxisomes (1.22 g/cm^3) were separated from mitochondria (1.18 g/cm^3) as indicated by the profiles of catalase and cytochrome *c* oxidase activity. The Pas3p was clearly detected in the peroxisomal peak fractions (7–10) by Western blot analysis.

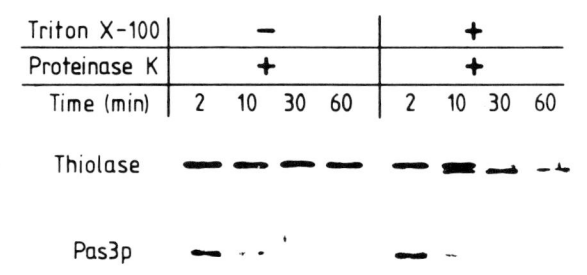

Triton X-100	–				+			
Proteinase K	+				+			
Time (min)	2	10	30	60	2	10	30	60
Thiolase	▬	▬	▬	▬	▬	▬	▬	▬
Pas3p	▬	.	.		▬	▬		

Fig. 6. Western blot analysis using antibodies against thiolase and Pas3p after proteinase K treatment of a crude organellar pellet. Two samples of a 25 000 × g pellet fraction were incubated in the absence (−) or presence (+) of 25 mg/ml agarose-coupled proteinase K and 0.1% Triton X-100 at 30°C. After 2, 10, 30 and 60 min, aliquots of each sample were precipitated in 10% TCA and prepared for SDS-PAGE. The peroxisomal matrix enzyme thiolase was protected from proteolytic attack by the intactness of the peroxisomal membrane. Degradation was only observed when the organelles were lysed by the addition of detergent. In contrast, Pas3p was accessible to the added protease in the presence as well as in the absence of Triton X-100. This indicates that the major portion of Pas3p protrudes into the cytosol.

method [56], Pas3p could be characterized as an integral peroxisomal membrane protein supporting the conclusion drawn from hydropathy analysis (see above). The topology of Pas3p within the peroxisomal membrane was analyzed by proteinase K treatment of intact peroxisomes (Fig. 6). It could be shown that Pas3p is not protected from proteolytic attack by the peroxisomal membrane. This indicates that the large carboxyl-terminal hydrophilic portion of the protein protrudes into the cytosol [28].

2.5. Analysis of the function of the cloned genes for peroxisome biogenesis

It still remains difficult to elucidate the true function of the *PAS* gene products for peroxisome biogenesis. These investigations are especially hampered by the fact that a general and reliable in vitro import system for peroxisomes in yeast is not yet available. Nevertheless, we have started different approaches to gain further insight into putative functions of the identified gene products. These included: (i) the analysis of effects of overexpression of the cloned genes on the morphology of a wild-type strain; (ii) site-directed mutagenesis; (iii) construction of conditional mutants and their analysis in continuous cultures.

2.5.1. Overexpression of PAS4
So far, only *PAS4* overexpression has resulted in a notable change of the morphological phenotype (see Discussion). The overexpression of *PAS4* was achieved by subcloning the gene into a multi-copy plasmid (YEp352 [57] and subsequent transformation of the resultant construct into a *PAS4* null mutant or a wild-type strain of *S. cerevisiae*. This overexpression led to a massive proliferation of peroxisomes in oleic acid-induced cells resulting in a much higher number of organelles than observed for non-transformed wild-type cells (Fig. 2).

2.5.2. Site directed mutagenesis
The amino acid sequence of Pas1p displays two putative ATP-binding sites each imbedded in a copy of the Pas1 box (see above and [27]). Both sites contain the two motifs A and B often found in ATP-binding proteins [58,59]. Site directed mutagenesis of a strictly conserved lysine residue in motif A (GXXGXGKT) of the consensus sequence for ATP-binding was carried out for both sites independently, changing either Lys-467 or Lys-744 to a glutamic acid residue. To test the significance of the corresponding sites for Pas1p function, the two different mutated genes were transformed to a *PAS1* deletion mutant on both single and multi-copy plasmids. The resulting phenotype was determined by testing the ability of growth on oleic acid and subcellular localization of enzyme activity. The gene carrying a mutation in the first putative ATP-binding site could still complement the *pas1* mutant in both single and multi-copy versions. In contrast, transformants containing an allele mutagenized in the second consensus sequence for ATP-binding displayed mutant phenotype (Krause, Erdmann and Kunau, unpublished results). This suggests that only the second consensus sequence, which is located in the well conserved Pas1 box, is essential for Pas1p function.

Pas2p is suggested to be a member of the UBC family of proteins [48]. UBC proteins share an essential cystein residue imbedded in a highly conserved context which was shown to be responsible for the formation of a thiolester bond with ubiquitin during ubiquitin conjugation [51]. To confirm the functional homology of Pas2p to UBC proteins, Cys-115 of Pas2p was altered to either serine or alanine and the mutated genes tested for their ability to complement the null mutant of *PAS2* as judged by growth on oleic acid and subcellular localization of peroxisomal enzymes. Both versions of the mutated *PAS2* gene were unable to complement the *pas2* defect even on a multi-copy plasmid, indicating the essential role of Cys-115 for Pas2p function (Wiebel and Kunau, unpublished results). This supports the idea of Pas2p being a UBC protein.

2.5.3. Conditional peroxisomal mutants
Conditional *pas* mutants were constructed by subcloning the open reading frame of individual *PAS* genes behind a galactose-inducible promoter (Gal1 promoter) and subsequent transformation of the resulting constructs into the corresponding mutant strains. This enabled us to regulate the expression of the *PAS* genes in *pas* mutants by growth on glucose (no expression) or on increasing amounts of galactose (progressive expression) as carbon sources. For the analysis of the effects on the expression of distinct *PAS* genes, the conditional mutants were cultivated in continuous cultures [60], a procedure that guarantees highly reproducible experimental conditions. Very recently, we demonstrated that induction of peroxisome proliferation by oleic acid can be performed in a glucose-limited chemostat culture of a *S. cerevisiae* wild-type strain [61]. Based on this finding, a glucose/oleic acid adapted chemostat culture was also used for the analysis of a conditional *pas1* mutant. In a

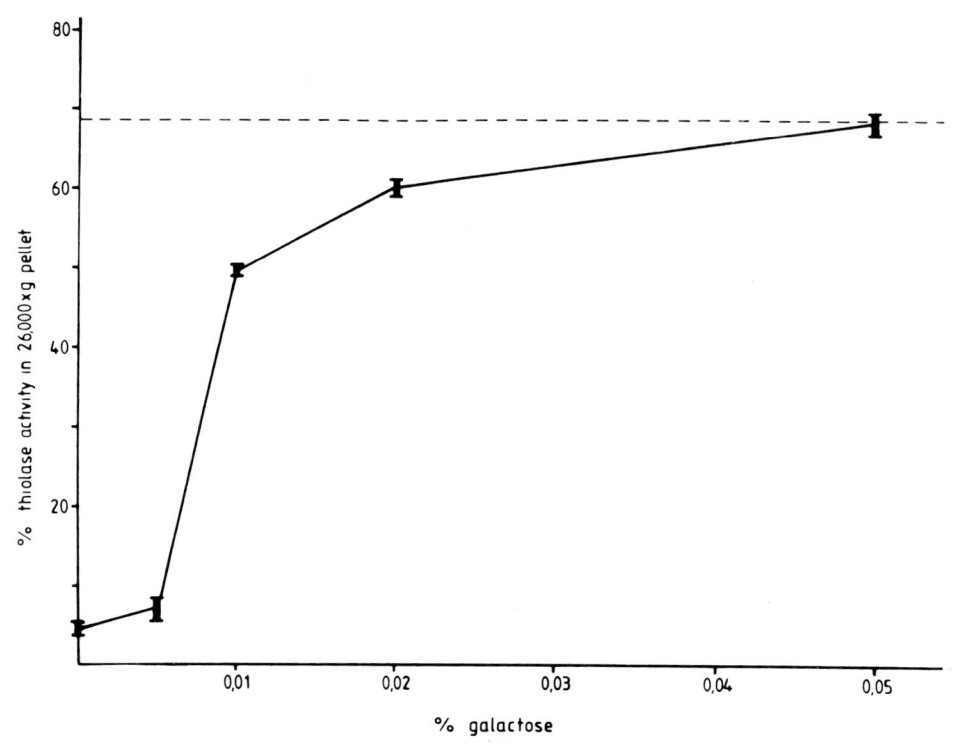

Fig. 7. Dependency of the percentage thiolase activity detected in the 26 000 × g pellet fraction on the galactose concentration. Changes of the phenotype of a conditional *pas1* mutant (PAS1 gene under control of Gal1 promoter) were analyzed by subcellular fractionation after adaptation on different concentrations of galactose. The dashed line gives the average value for the mitochondrial marker enzyme fumarase.

series of experiments, different concentrations of galactose were added to such a culture and the changes of the phenotype were investigated under steady state conditions. Centrifugation (26 000 × g) of spheroplast lysates revealed a correlation between the galactose concentration and the percentage of thiolase activity detected in the particulate fraction (Fig. 7). This particulate fraction contains organelles including newly formed peroxisomes. A concentration as low as 0.05% galactose was sufficient to regain a wild-type phenotype, regarding the distribution of thiolase activity between pellet and supernatant. Based on the morphological analysis, a galactose-dependent increase in the total peroxisomal volume was found (Table III). It seems that the expression levels of *PAS1* have a direct influence on peroxisome formation. The detailed data (Höhfeld, Veenhuis and Kunau, unpublished results) imply that with this system it is possible to directly induce intermediate states between the original *pas* and the wild-type phenotype creating new possibilities to study peroxisome biogenesis.

TABLE III

Morphological analysis of the phenotype of a conditional *pas1* mutant after adaptation of the chemostat culture on different concentrations of galactose[a]

% Galactose	% Peroxisomal volume of total cytosolic volume
0	Non-detectable
0.0005	0.2
0.010	4.4
0.050	4.9
wild-type	11.6

[a] Increasing concentrations of galactose led to an increase of the total peroxisomal volume most likely due to the different expression levels of Pas1p (the *PAS1* gene was under control of the Gal1 promoter).

2.6. Fusion proteins as tools for further investigations

The construction of fusion proteins containing a well detectable reporter-protein moiety (e.g. β-galactosidase) is a frequently used method to characterize newly cloned genes and their corresponding gene products [62]. This includes analysis of gene expression (e.g. [63]), identification of targeting signals (e.g. [64]) and analysis of protein topology (e.g. [65]). Similar approaches have been started for *PAS1*, *PAS2*, *PAS3* and *PAS4*. Here, the investigations concerning a Pas3p-β-galactosidase fusion protein are described. The Pas3 protein contains a membrane spanning sequence on its extreme amino-terminus anchoring a large hydrophilic domain in the peroxisomal membrane (see above and [28]). Due to the fact that this hydrophilic domain protrudes into the cytosol, it is conceivable that Pas3p is targeted to the peroxisomal membrane by an amino-terminal targeting signal. As a first step to prove this hypothesis, a fusion gene was constructed containing a 900 bp fragment of the *PAS3* gene (the *PAS3* promoter region and the adjacent 50% of the open reading frame) in front of the *lacZ* gene. After transformation of the construct on a single as well as multi-copy plasmid into a wild-type strain, β-galactosidase activity was almost exclusively assayed in the particulate fraction (25 000 × g pellet). By sucrose density gradient centrifugation and subsequent membrane preparation from isolated peroxisomes, it was shown that the Pas3p-β-galactosidase fusion protein behaves like an integral peroxisomal membrane protein. These results were confirmed by immunocytochemistry using polyclonal antibodies against β-galactosidase (Fig. 2) (Höhfeld, Veenhuis and Kunau, unpublished results). These findings indicate that the amino-terminal half of Pas3p is sufficient to target a reporter protein to peroxisomes and anchor it in the peroxisomal membrane.

3. Discussion

According to the current model of peroxisome biogenesis, these organelles arise by growth and division of pre-existing peroxisomes (for reviews see [2–5]) and cannot be formed by budding from the endoplasmatic reticulum as was suggested earlier (reviewed in [2]). Experimental evidence shows that both peroxisomal matrix and membrane proteins are encoded by nuclear genes, synthesized on free polyribosomes in the cytosol and post-translationally imported into the organelle. In these respects, peroxisomes resemble mitochondria and chloroplasts, organelles which are now generally accepted to be of endosymbiotic origin [66,67]. However, in contrast to mitochondria and chloroplasts, peroxisomes do not contain DNA, are surrounded by only a single membrane and most of their proteins are synthesized at their final size (for reviews see [2,5]). Moreover, a characteristic feature of peroxisomes is their pronounced inducibility in size and number by external stimuli [5,11,68]. Despite the apparent lack of an inherent genome, the possibility has been raised that peroxisomes might also originally have entered a protoeukaryote as endosymbionts [2–4,66,69]. A de novo synthesis of peroxisomes is considered extremely unlikely although it cannot be formally excluded. This notion is based on the lack of any experimental evidence that cellular membranes can be formed without a pre-existing one. Based on this current model of biogenesis, during mitosis at least one peroxisome must be transferred in a temporally and spatially concerted fashion to the daughter cell. This basal peroxisome can subsequently function as an initiator organelle in the process of peroxisomal proliferation due to changes of environmental conditions. A schematic view of these two events, inheritance of peroxisomes during cell division and their proliferation within existing cells, is presented in Fig. 8. This view of peroxisomal biogenesis implies that for the formation, maintenance and correct function of peroxisomes within a strain, a number of cellular processes have to be intact including: (i) import of peroxisomal membrane proteins; (ii) import of peroxisomal matrix proteins; (iii) transfer of phospholipids to the peroxisomal membrane; (iv) induction of peroxisomal proteins. In analogy to mitochondria [22,23] and chloroplasts [70–72], first experimental evidence for peroxisomes indicate that import of peroxisomal membrane and matrix proteins is also directed by targeting signals on the proteins to be imported [73,74] and membrane-bound components of a peroxisomal import machinery [28,75,76].

3.1. Peroxisomal mutants as a tool to dissect peroxisome biogenesis

The availability of mutants in *S. cerevisiae* impaired in peroxisomal function allows the genetic dissection of peroxisome biogenesis. For some of these mutants, the defect was shown to be due to the impairment of a metabolic function (*fox* mutants, defective in β-oxidation). The other peroxisomal mutants analyzed so far in our laboratory are characterized by mislocalization of peroxisomal matrix proteins

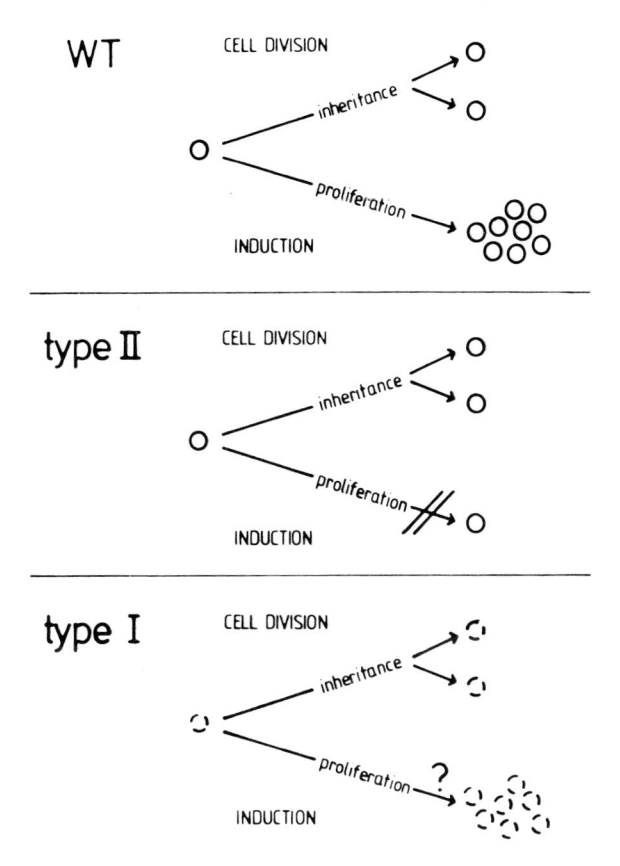

Fig. 8. Hypothetical model of peroxisome biogenesis in a S. *cerevisiae* wild-type strain and different types of *pas* mutants (for explanation see text).

to the cytosol of oleic acid-induced cells. They comprise 13 complementation groups. Five additional complementation groups have been isolated by Tabak and co-workers (van der Leij et al., unpublished results).

The complexity of peroxisome assembly is not only reflected by the genetic diversity of the peroxisomal mutants but also by their different phenotypes. Mislocalization can concern all (type I and type II) or only distinct peroxisomal matrix proteins (type III). Peroxisomes can be normal in size and number (type III), present but non-proliferating (type II) or morphologically not detectable (type I). In all cases, there are no functional peroxisomes resulting in the inability of the cells to grow on oleic acid, but defects are not due to impairments in peroxisomal metabolic enzymes (excluding, e.g. *fox* mutants). This is a new definition of the *S. cerevisiae pas* phenotype in order to include the newly identified mutants (type II and III). Initially, we defined as *pas* mutants only those mutants which lack peroxisomes and exhibit a general mislocalization of matrix proteins (type I).

The *pas* mutants with their distinctly different phenotypes are not only obvious

3.2. Do peroxisomal prestructures exist in type I pas mutants?

The analysis of *pas* mutants re-emphasized the question whether a de novo synthesis of peroxisomes is possible [4]. For genetic analysis, haploid *pas* mutants of type I were crossed yielding diploids which regained wild-type phenotype when different genes were affected [26]. Moreover, we have cloned several *PAS* genes of *S. cerevisiae* by functional complementaion; i.e. transformation of haploid *pas* mutants with a single gene led to wild-type phenotype. If we assume that there is no de novo synthesis of peroxisomes, one has to postulate the presence of peroxisomal prestructures which can serve as the starting point for peroxisome formation. In this respect, it is interesting to note that the phenotype of type I *pas* mutants resembles that of fibroblasts of Zellweger patients (for a review see [5]) and that of peroxisomal mutants of CHO cell lines [33–35]. In both Zellweger fibroblasts as well as CHO mutant cells empty peroxisomal membrane structures (ghosts) have been detected by immunofluorescence studies using antibodies against membrane proteins of rat peroxisomes [38,39,77]. The demonstration of similar structures in *pas* mutants type I of *S. cerevisiae* has long been hampered by the lack of antibodies against peroxisomal membrane proteins of yeast suitable for immunofluorescence or immunocytochemistry. The identification of Pas3p as an integral peroxisomal membrane protein [28] and the observation that a lacZ fusion protein of it (see Results) is specifically targeted to the peroxisomal membrane of wild-type cells (Höhfeld, Veenhuis and Kunau, unpublished results) should open the possibility of identifying peroxisomal prestructures by a combined biochemical and immunocytochemical approach.

We recently observed unusual membrane structures in some of the type I *pas* mutants by electron microscopy which are not detected in wild-type cells. These structures resemble catalase-negative peroxisomal membrane loops of rat liver [78] and membrane structures of oleic acid-induced cells of the methylotrophic yeast *Hansenula polymorpha* [14]. Here again, the Pas3p-β-galactosidase fusion protein should be a helpful tool as a peroxisomal membrane marker to demonstrate the possible peroxisomal nature of these membranes.

In the case where the predicted essential prestructures in the type I *pas* mutants exist, one can also envisage defects that affect the formation or inheritance of these prestructures and thus lead to their complete absence. The corresponding mutant phenotype would then be expected to be complemented neither by transformation nor by reciprocal crossing. They would thus seem to fall into a single complementation group exhibiting type I *pas* phenotype. Moreover, due to the lack of an initiation point for peroxisome formation, none of them should regain the ability to form peroxisomes by uptake of the wild-type copy of the affected gene. Such a class of mutants has not yet been detected. All our current type I complementation groups led to the cloning of the corresponding genes by functional complementation.

3.3. Are type I pas mutants peroxisomal import mutants?

The assumption of peroxisomal prestructures in type I *pas* mutants leads to considerations of which aspect of peroxisome assembly might be affected in these mutant strains. To avoid irreversible loss, the putative prestructures would have to segregate in a similar way as the mature organelle (Fig. 8). Thus, these prestructures have to multiply by growth and division, which should include uptake of phospholipids and specific membrane proteins to define their peroxisomal character. Due to the normal induction of membrane proteins in the type I *pas* mutants the putative membranous prestructures should even proliferate during growth on oleic acid medium (Fig. 8). This in turn would suggest that the absence of morphologically detectable peroxisomes might be due to defects directly or indirectly affecting the import of peroxisomal matrix proteins. Under this assumption, the large number of type I complementation groups reflects a great complexity of the peroxisomal import machinery. The functional diversity of the proteins encoded by the *PAS* genes cloned and sequenced so far seems to support this notion. Pas1p has been proposed to belong to a novel subclass of ATPases of yet unknown specificity. Its members are involved in a variety of processes of no obvious functional relationship (cf. Results). Pas2p is a new member of the UBC family thus linking the ubiquitin conjugation pathway to peroxisomal protein import. Pas3p as an integral peroxisomal membrane protein with the major portion of its molecule protruding into the cytosol, meets the topological properties expected for an import receptor. These three examples point in different directions that cannot be interrelated so far.

3.4. Do more peroxisomal import routes exist other than the SKL-mediated pathway?

The diversity of the peroxisomal import machinery is directly demonstrated by the phenotype of the type III mutant, *pas7*. Analysis of its defect revealed the specific mislocalization of thiolase although this phenotype is not due to a mutation of its structural gene. Interestingly, unlike the peroxisomal β-oxidation multi-functional protein (Fox2p) [79] which is imported in peroxisomes of the *pas7* mutant (see Table I), thiolase (Fox3p) possesses no carboxyl-terminal SKL tripeptide. Such an SKL motif was identified as a peroxisomal targeting signal at the extreme carboxyl-termini of several peroxisomal proteins [73,74]. However, the presence of peroxisomal matrix proteins without an SKL signal suggests the possible existence of an additional peroxisomal targeting signal. Consistent with this notion, there are recent reports indicating amino-terminal targeting signals for two peroxisomal matrix proteins, malate dehydrogenase of watermelon [80] and thiolase of rat liver [81]. The *pas7* mutant now clearly points to the existence of such a second route which might not only be required for thiolase import alone but for that of other peroxisomal proteins lacking a carboxyl-terminal SKL motif. Whether these two peroxisomal import pathways converge and thus are at least partially identical or whether they are entirely different cannot yet be answered. The *PAS7* gene, however, will define the first component specific for the import of thiolase. In contrast, type I *PAS* gene products should be essential for import of matrix proteins in general.

Several lines of evidence indicate that peroxisomal membrane proteins or at least part of them are imported by still another pathway different from those which lead to the peroxisomal uptake of matrix proteins. The same reasoning which argues for the presence of peroxisomal prestructures in type I *pas* mutants in which matrix proteins are mislocalized implies the existence of a specific route for insertion of peroxisomal membrane proteins. In fact, ghosts in fibroblasts of Zellweger patients have been detected and characterized by means of their peroxisomal integral membrane proteins [38–40,82]. Furthermore, none of the peroxisomal integral membrane proteins for which the primary sequence has been deduced possesses the carboxyl-terminal tripeptide SKL [28,83]. In addition, *PAS3* seems to contain its targeting signal in the amino-terminal half of its molecule (see above). Finally, peroxisomal protein import in the *pas7* mutant distinguishes between a matrix protein, thiolase and an integral membrane protein, *PAS3*, both containing their targeting signal not at their carboxyl-termini. Cells of the *pas7* mutant do not import thiolase into their peroxisomes but insert Pas3p into their peroxisomal membranes.

3.5. Are type II pas mutants affected in peroxisome proliferation?

Unlike type I *pas* mutants those of type II (*pas4* and *pas6*) do contain one or a few small peroxisomes under oleic acid induction. In this respect, they resemble glucose-repressed wild-type cells. In addition, these few peroxisomes in a null mutant of *PAS4* could be immunocytochemically labelled for thiolase. Although it thus appears that the principle aspects of peroxisome formation are unaffected, the strain displays onu phenotype. Moreover, despite the presence of import-competent peroxisomes, the bulk of peroxisomal matrix proteins are present in the cytosol. Recently, a most remarkable observation was made when Pas4p was expressed at different rates. Under oleic acid induction, peroxisome proliferation is controlled by the level of Pas4p expression. Comparison of two different cell types, the null mutant of *PAS4* and cells carrying this gene on a multi-copy vector, gives first insights into the processes involved in peroxisome proliferation. While the null mutant shows no proliferation, the overproducer displays a number of peroxisomes much higher than observed in wild-type cells. In contrast to the proliferation rate of peroxisomes, the induction levels of peroxisomal matrix enzymes are not affected by different expression levels of Pas4p. Null mutant and overproducer of *PAS4* possess import-competent peroxisomes, although those of the null mutant seem to be limited in their capacity to import of matrix proteins. However, based on observations in wild-type cells, a limited import capacity should not be the reason for impaired proliferation. It was reported that increased import of matrix proteins leads to an increased volume of individual peroxisomes but not to their proliferation [84,85]. These findings seem to suggest that a step required for peroxisomal membrane biogenesis is the limiting factor.

4. Conclusions

The major progress regarding knowledge about peroxisome biogenesis in *S. cerevisiae* is based on: (i) the discovery that oleic acid triggers peroxisome proliferation in this yeast species; (ii) a screening procedure for peroxisomal mutants; (iii) cloning and characterization of genes essential for peroxisome biogenesis by means of peroxisomal mutants. The combination of these three developments offers in *S. cerevisiae* a unique opportunity to gain insight into the molecular mechanisms of peroxisome biogenesis.

Acknowledgement

We thank A. Hartig for helpful discussion and critical reading of the manuscript and M. Veenhuis for contributing the electron micrographs of this study. We are also grateful to H. Tabak and co-workers for communication of data prior to publication. The original research described in this review was supported in part by the Deutsche Forschungsgemeinschaft (grants Ku 329/11-3 and 11-4). J.H. was supported by a fellowship of the Boehringer Ingelheim Foundation.

References

1. Tolbert, N.E. (1981) Annu. Rev. Biochem. 50, 133–158.
2. Lazarow, P.B. and Fujiki, Y. (1985) Annu. Rev. Cell Biol. 1, 489–530.
3. Borst, P. (1986) Biochim. Biophys. Acta 866, 179–203.
4. Borst, P. (1989) Biochim. Biophys. Acta 1008, 1–13.
5. Lazarow, P. and Moser, H.W. (1989) in: Scriver et al. (Eds.), The Metabolic Basis of Inherited Diseases, McGraw-Hill, New York.
6. de Duve, C. and Baudhuin, P. (1966) Physiol. Rev. 46, 323–357.
7. Opperdoes, F. (1988) Trends Biochem. Sci. 13, 255–260.
8. Kindl, H. and Lazarow, P.B. (1982) Peroxisomes and Glyoxysomes, New York Academy of Sciences, New York.
9. Kunau, W.H., Kionka, C., Ledebur, A., Mateblowski, M., de la Garza, M., Schultz–Borchard, U., Thieringer, R. and Veenhuis, M. (1987) in: C. Fahimi and H. Sies (Eds.), Peroxisomes in Biology and Medicine, Springer-Verlag, Berlin, pp. 128–140.
10. Tanaka, A., Osumi, M. and Fukui, S. (1982) Ann. N.Y. Acad. Sci. 386, 183–199.
11. Veenhuis, M. and Harder, W. (1988) Microbiol. Sci. 5, 347–351.

12. Kunau, W.-H., Bühne, S., del la Garza, M., Kionka, C., Mateblowski, M., Schulz-Borchard, U. and Thieringer, R. (1988) Biochem. Soc. Trans. 16, 418–420.
13. Tanaka, A. and Fukui, S. (1989) in: A.H. Rose and J.S. Harrison (Eds.), The Yeasts, Vol. 3, Academic Press, London, pp. 261–287.
14. Veenhuis, M., Kram, A.M., Kunau, W.-H. and Harder, W. (1990) Yeast 6, 511–519.
15. Roman, H. (1981) in: J.N. Strathern et al. (Eds.), The Molecular Biology of the Yeast Saccharomyces, Cold Spring Harbor, New York, pp. 1–9.
16. Pringle, J.R. and Hartwell, L.H. (1981) in: J.N. Strathern et al. (Eds.), The Molecular Biology of the Yeast Saccharomyces, Cold Spring Harbor, New York, pp. 97–142.
17. Hartwell, L.H. and Weinert, T.A. (1989) Science 246, 629–634.
18. Drubin, D.G. (1991) Cell 65, 1093–1096.
19. Hicke, L. and Schekman, R. (1990) BioEssays 12, 253–258.
20. Newman, A.P. and Ferro-Novick, S. (1990) BioEssays 12, 485–491.
21. Deshaies, R.J., Kepes, F. and Böhni, P.C. (1989) Trends Genet. 5, 87–93.
22. Pfanner, N., Söllner T. and Neupert, W. (1991) Trends Biochem. Sci. 16, 63–67.
23. Glick, B., Wachter, C. and Schatz, G. (1991) Trends Cell Biol. 1, 99–103.
24. Avers, C.J. and Federman, M. (1968) J. Cell Biol. 37, 555–559.
25. Veenhuis, M., Mateblowski, M., Kunau, W.H. and Harder, W. (1987) Yeast 3, 77–84.
26. Erdmann, R., Veenhuis, M., Mertens, D. and Kunau, W.H. (1989) Proc. Natl. Acad. Sci USA 86, 5419–5423.
27. Erdmann, R., Wiebel, F., Flessau, A., Beyer, A., Fröhlich, K.-U. and Kunau, W.-H. (1991) Cell 64, 499–510.
28. Höhfeld, J., Veenhuis, M. and Kunau, W.-H. (1991) J. Cell Biol. 114, 1167–1178.
29. Baker, K.P. and Schatz, G. (1991) Nature 349, 205–208.
30. Gleeson, M.A. and Sudbery, P.E. (1988) Yeast 4, 1–15.
31. Cregg, J.M., van der Klei, I.J., Sulter, G.J., Veenhuis, M. and Harder, W. (1990) Yeast 6, 87–97.
32. Didion, T. and Roggenkamp, R. (1990) Curr. Genet. 17, 113–117.
33. Zoeller, R.A. and Raetz, C.R.H. (1986) Proc. Natl. Acad. Sci. USA 83, 5170–5174.
34. Morand, O.H., Allen, L.-A.H., Zoeller, R.A. and Raetz, C.R.H. (1990) Biochim. Biophys. Acta 1034, 132–141.
35. Tsukamoto, T., Yokota, S. and Fujiki, Y. (1990) J. Cell Biol. 110, 651–660.
36. Schutgens, R.B.H., Heymans, H.S.A., Wanders, R.J.A., van den Bosch, H. and Tager, J.M. (1986) Eur. J. Pediatr. 144, 430–440.
37. Wanders, R.J.A., Heymans, H.S.A., Schutgens, R.B.H., Barth, P.G., van den Bosch, H. and Tager, J.M. (1988) J. Neurol. Sci. 88, 1–39.
38. Santos, M.J., Imanaka, T., Shio, H., Small, G.M. and Lazarow, P.B. (1988) Science 239, 1536–1539.
39. Wiemer, E.A.C., Brul, S., Just, W.W., van Driel, R., Brouwer-Kelder, E., van den Berg, M. Weijers, P.J., Schutgens, R.B.H. van den Bosch, H., Schram, A., Wanders, R.J.A. and Tager, J.M. (1989) Eur. J. Cell Biol. 50, 407–417.
40. Suzuki, Y., Shimozawa, N., Orii, T. and Hashimoto, T. (1989) Pediatr. Res. 26, 150.
41. Struhl, K. (1983) Nature 305, 391–397.
42. Eakle, K.A., Bernstein, M. and Emr, S.D. (1988) Mol. Cell. Biol. 8, 4098–4109.
43. Wilson, D.W., Wilcox, C.A., Flynn, G.C., Ellson, C., Kuang, W.-J., Henzel, W.J., Block, M.R., Ullrich, A. and Rothmann, J.E. (1989) Nature 339, 355–359.
44. Koller, K.J. and Brownstein, M.J. (1987) Nature 325, 542–545.
45. Peters, J.-M., Walsh, M.J. and Franke, W.W. (1990) EMBO J. 9, 1757–1767.
46. Fröhlich, K.-U., Fries, H.-W., Rüdiger, M., Erdmann, R., Botstein, D. and Mecke, D. (1991) J. Cell Biol. 114, 443–453.
47. Nelbock, P., Dillon, P.J., Perkins, A. and Rosen, C.A. (1990) Science 145, 1650–1653.
48. Jentsch, S., Seufert, W., Sommer, T. and Teins, H.-A. (1990) Trends Biochem. Sci. 15, 195–198.

49. Hershko, A. (1991) Trends Biochem. Sci. 16, 265–268.
50. Jentsch, S., Seufert, W. and Hauser, H.-P. (1991) Biochim. Biophys. Acta 1089, 127–139.
51. Seufert, W. and Jentsch, S. (1990) EMBO J. 9, 543–550.
52. Freemont, P.S., Hanson, I.M. and Trowsdale, J. (1991) Cell 64, 483–484.
53. Kyte, J. and Doolittle, R.F. (1982) J. Mol. Biol. 157, 105–132.
54. Rüther, U. and Müller-Hill, B. (1983) EMBO J. 2, 1791–1794.
55. Raymond, C.K., O'Hara, P.J., Eichinger, G., Rothman, J.H. and Stevens, T.H. (1990) J. Cell Biol. 111, 877–892.
56. Fujiki, Y., Hubbard, A.L., Fowler, S. and Lazarow, P.B. (1982) J. Cell Biol. 93, 97–102.
57. Hill, J.E., Myers, A.M., Koerner, T.J. and Tzagoloff, A. (1986) Yeast 2, 163–167.
58. Walker, J.E., Saraste, M., Runswick, M.J. and Gay, N.J. (1982) EMBO J. 1, 945–981.
59. Chin, D.T., Goff, A.S., Webster, T., Smith, T. and Goldberg, A.L. (1988) J. Biol. Chem. 263, 11718–11728.
60. Fiechter, A., Käppeli, O. and Meussdoerffer, F. (1987) in: A.H. Rose and J.S. Harrison (Eds.), The Yeasts, Vol. 2, Academic Press, London, pp. 99–129.
61. Evers, M.E., Höhfeld, J., Kunau, W.H. and Veenhuis, M. (1991) FEMS Microbiol. Lett. 90, 73–78.
62. Silhavy, T.J. and Beckwith, J.R. (1985) Microbiol. Rev. 49, 398–418.
63. Myers, A.M., Tzagoloff, A., Kinney, D.M. and Lusty, C.J. (1986) Gene 45, 299–310.
64. Nakai, M., Hase, T. and Matsubara, H. (1989) J. Biochem. 105, 513–519.
65. Deshaies, R.J. and Schekman, R. (1990) Mol. Cell. Biol. 10, 6024–6035.
66. Cavalier-Smith, T. (1987) Ann. N.Y. Acad. Sci. 503, 55.
67. Gray, M.W. (1988) Biochem. Cell Biol. 66, 325–348.
68. Veenhuis, M. and Harder, W. (1991) in: A.H. Rose and J.S. Harrison (Eds.), The Yeasts, Vol. 4, Academic Press, London, pp. 601–653.
69. de Duve, C. (1982) Ann. N.Y. Acad. Sci. 386, 1–4.
70. Archer, E.K. and Keegstra, K. (1990) J. Bioenerg. Biomembr. 22, 789–810.
71. Smeekens, S. Weisbeek, P. and Robinson, C. (1990) Trends Biochem. Sci. 15, 73–76.
72. Joyard, J., Block, M.A. and Douce, R. (1991) Eur. J. Biochem. 199, 489–509.
73. Gould, S.J., Keller, G.-A., Hosken, N., Wilkinson, J. and Subramani, S. (1989) J. Cell Biol. 108, 1657–1664.
74. Gould, S.J., Keller, G.-A., Schneider, M., Howell, S.H., Garrard, L.J., Goodman, J.M., Distel, B., Tabak, H. and Subramani, S. (1990) EMBO J. 9, 85–90.
75. Kamijo, K., Taketani, S., Yokota, S., Osumi, T. and Hashimoto, T. (1990) J. Biol. Chem. 265, 4534–4540.
76. Tsukamoto, T., Miura, S. and Fujiki, Y. (1991) Nature 350, 77–81.
77. Zoeller, R.A., Allen, L.-A.H., Santos, M.J., Lazarow, P.B., Hashimoto, T., Tartakoff A. M. and Raetz, C.R.H. (1989) J. Biol. Chem. 264, 21872–21878.
78. Baumgart, E., Völkl, A., Hashimoto, T. and Fahimi, H.D. (1989) J. Cell Biol. 108, 2221–2231.
79. Hiltunen, J.K., Wenzel, B., Beyer, A., Fossa, A., Erdmann, R. and Kunau, W.-H. (1992) J. Biol. Chem. 267, 6646–6653.
80. Gietl, C. (1990) Proc. Natl. Acad. Sci. USA 87, 5773–5777.
81. Swinkels, B.W., Gould, S.J., Bodnar, A.G., Rachubinski, R.A. and Subramani, S. (1991) EMBO J. 10, 3255.
82. Tanaka, K. and Coates, P.M. (1990) Fatty Acid Oxidation, Clinical, Biochemical and Molecular Aspects, Alan R. Liss, New York.
83. Kunau, W.-H. and Hartig, A. (1992) Antonie van Leeuwenhoek, in press.
84. Distel, B., van der Leij, I., Veenhuis, M. and Tabak, H.F. (1988) J. Cell Biol. 107, 1669–1675.
85. Roggenkamp, R., Didion, T. and Kowallik, K. (1989) Mol. Cell. Biol. 9, 988–994.
86. Borst, P. and Swinkel, B.W. in: Grundberg-Manago et al. (Eds.), Evolutionary Tinkering in Gene Expression, NATO ASI Series A: Life Sciences, Vol. 169, Plenum Press, New York.

CHAPTER 16

Structure and assembly of peroxisomal membrane proteins

JOEL M. GOODMAN, LISA J. GARRARD*
and MARK T. McCAMMON[†]

Department of Pharmacology, University of Texas Southwestern Medical Center, 5323 Harry Hines Blvd., Dallas, TX 75235-9041, USA

Abstract

We have been characterizing the peroxisomal membrane proteins of the methylotrophic yeast *Candida boidinii* and their importance to the structure, assembly and function of the organelle. The association of newly synthesized PMP20, a peripheral membrane protein, with peroxisomes was much more rapid than that of the matrix enzymes in pulse-chase experiments. During proliferation of peroxisomes on methanol, the synthesis of the peroxisomal membranes and of the integral membrane protein PMP47 preceded synthesis of PMP20 and the major matrix enzymes, alcohol oxidase and dihydroxyacetone synthase. While PMP20 appears to be specific for methanol metabolism, PMP47 and two other integral membrane proteins, PMP32 and PMP31, were abundant components of the peroxisomal membranes when the organelle was induced to proliferate by other growth substrates. Genes encoding PMP20, PMP31 and PMP47 have been isolated, and PMP47 has been expressed in *Saccharomyces cerevisiae*, where it sorts to the peroxisomes and assembles into the membrane with the endogenous membrane proteins. Preliminary results indicate that PMP47 contains a novel, but presently undefined, sorting signal. The importance of the peroxisomal membrane proteins for peroxisomal biogenesis and function are discussed.

---

* *Present address:* Department of Cell Genetics, Genentech, Inc., 460 Point San Bruno Blvd., South San Francisco, CA 94080, USA.
[†] *Present address:* Department of Biochemistry and Molecular Biology, University of Arkansas for Medical Sciences, Little Rock, AR 72205, USA.

1. Introduction

Microbodies, consisting of peroxisomes, glyoxysomes, and glycosomes, perform a variety of essential metabolic functions that vary widely from tissue to tissue and species to species [1,2]. Important functions in plants include the conversion of fat stores to carbohydrate in seedlings and amino acid biosynthesis in leaves. In fungi, they house enzymes of metabolic pathways that permit growth on several carbon sources such as alcohols and fatty acids. Glycosomes of the trypanosomes perform efficient glycolysis to maximize ATP generation. The many important functions of animal peroxisomes include steps in the synthesis of bile acids and plasmalogens and the degradation of long chain fatty acids and purines. These oxidations generate hydrogen peroxide and explain the requirement for catalase, a hallmark enzyme for most types of microbodies. The importance of peroxisomes for humans is clearly demonstrated by Zellweger's syndrome, in which internal (matrix) proteins fail to assemble into the organelle, causing multiple enzyme deficiencies and death at an early age [3,4].

Several aspects of the assembly of microbody matrix proteins are clear. They are synthesized on soluble polysomes and insert post-translationally from the cytosol into the organelle [5–7]. With at least two known exceptions, these proteins do not undergo covalent modifications such as proteolytic cleavage of signal peptides. Many microbody proteins contain the tripeptide sequence serine-lysine-leucine (SKL), or closely related variants at their carboxy-termini [8–10]. This peroxisomal targeting sequence, termed PTS1, is sufficient for targeting a protein to microbodies. 3-Ketoacyl-CoA thiolase [11] and perhaps the microbody isozyme of malate dehydrogenase [12] contain cleaved amino-terminal extensions; these may contain an alternate targeting sequence termed PTS2 [13].

Less is known about the assembly of the peroxisomal membrane. Only a handful of sequences for peroxisomal integral membrane proteins are available (see below), and their functions and assembly are not understood. We have been characterizing this class of proteins in *Candida boidinii*, a methylotrophic yeast in which peroxisomes proliferate to constitute 30–40% of cytoplasmic volume. These organelles can readily be obtained in high purity and good yield [14,15].

2. Results

We have investigated the protein composition of peroxisomes isolated from cultures of *C. boidinii* grown with methanol as a carbon source. The matrix of the organelle is composed almost exclusively of two proteins, alcohol oxidase (AO) and dihydroxyacetone synthase (DHAS) (Fig. 1A) [14,15]. Catalase is also present but in much lower amounts. These enzymes catalyze the first steps of methanol utilization [16]. AO converts methanol to formaldehyde and hydrogen peroxide (subsequently de-

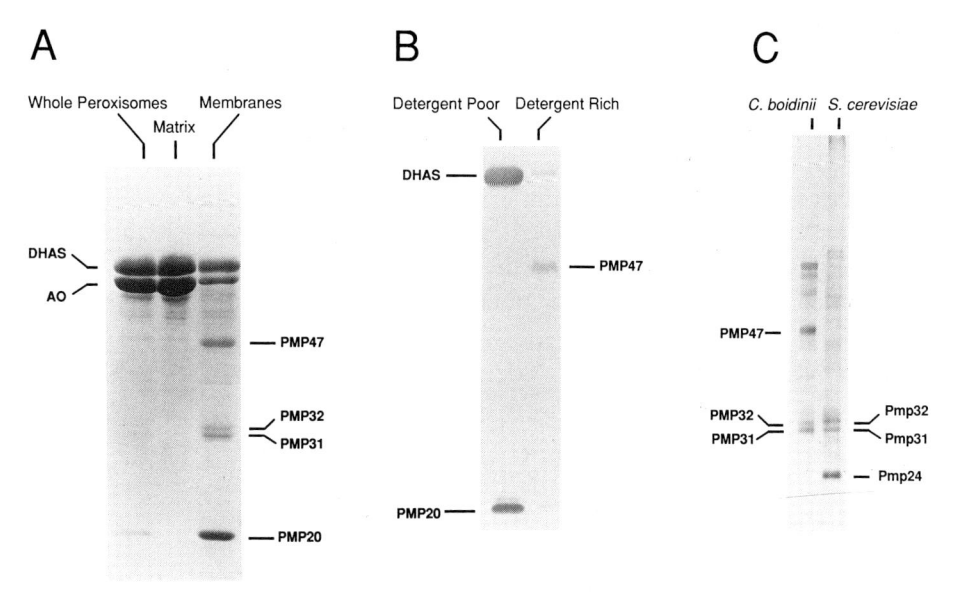

Fig. 1. Protein composition of peroxisomes. (A) Peroxisomes from methanol-grown *C. boidinii* consist primarily of two proteins, alcohol oxidase (AO) and dihydroxyacetone synthase (DHAS) as analyzed by Coomassie-stained SDS polyacrylamide gels. AO and DHAS are located in the soluble matrix of the organelle. The membrane proteins were separated from the bulk of the matrix proteins following lysis in 20 mM Tris–HCl (pH 8.0) by floatation in sucrose gradients. These membranes consist of four major proteins defined as PMP47, PMP32, PMP31, and PMP20. (B) Phase separation of peroxisomal proteins in Triton X-114. A 15 000 × g peroxisome-containing pellet was extracted with the non-ionic detergent Triton X-114 [22]. After phase separation the aqueous (detergent poor) and detergent (detergent rich) phases were subjected to SDS polyacrylamide gel electrophoresis and transferred to nitrocellulose. Antibodies directed against DHAS, PMP20 and PMP47 were used to identify these proteins by an immuno-enzymatic detection system. (C) Peroxisomal membrane proteins from oleate-grown *C. boidinii* and *S. cerevisiae*. Carbonate-insoluble extracts of gradient-purified peroxisomes (see Fig. 3D) from these yeasts were analyzed by equilibrium density gradient centrifugation on 20–40% continuous sucrose gradients. The protein composition of the gradient fractions was analyzed by silver staining after SDS polyacrylamide gel electrophoresis [17,23]. The peak protein fractions of those gradients are shown for comparison, and the major membrane proteins are designated; other bands are primarily due to incomplete extraction by carbonate of the matrix proteins from oleate-induced peroxisomes.

graded by catalase), and DHAS fixes formaldehyde onto ribulose 5-phosphate, generating dihydroxyacetone and glyceraldehyde 3-phosphate. These trioses then enter the cytoplasm for further metabolic interconversions. The peroxisomal membrane composition is also very simple and contains four abundant membrane proteins, termed PMP20, 31, 32, and 47, based on their apparent mass in kilodaltons on SDS polyacrylamide gels (Fig. 1A). Of these proteins, PMP20 is present in greatest abundance.

2.1. Assembly of peroxisomal proteins

We have begun to compare the import pathway of peroxisomal membrane and matrix proteins by determining the kinetics of association of newly synthesized PMP20, AO, and DHAS with peroxisomes in a pulse-chase experiment (Fig. 2). Spheroplasts were incubated for 5 min with [^{35}S]cysteine and then allowed to incubate in the presence of excess unlabeled cysteine. The spheroplasts were then lysed, fractionated into a crude organellar pellet and corresponding supernatant, and the association of these proteins with the organelle was determined by immunoprecipi-

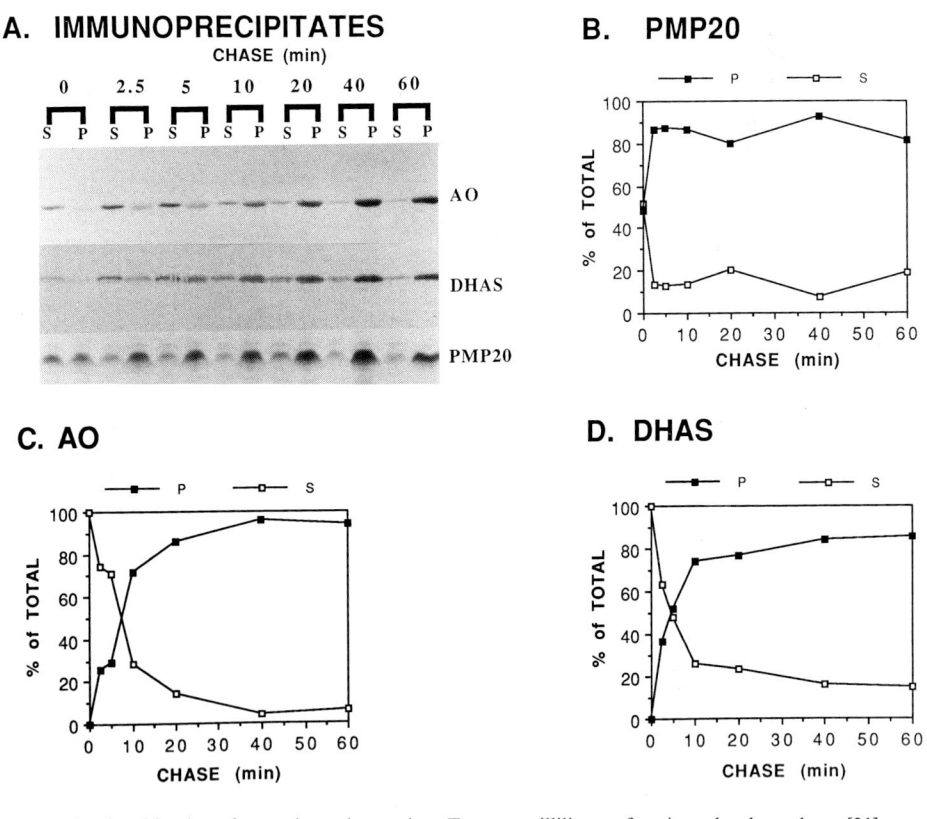

Fig. 2. Sorting kinetics of peroxisomal proteins. Twenty milliliters of activated spheroplasts [31] were labeled with 75 μCi of [^{35}S]cysteine. After 5 min (defined as the zero chase time point), unlabeled cysteine was added to a final concentration of 100 mM. At the indicated times, samples of the spheroplasts were chilled, harvested by centrifugation, lysed, and fractionated into a 15 000 × g supernatant (S) and an organellar pellet (P). Aliquots of these fractions were treated with antisera to immunoprecipitate specifically AO, DHAS or PMP20. (A) Immunoprecipitates of AO, DHAS and PMP20 were separated on SDS polyacrylamide gels and detected by fluorography. (B)–(D) Quantitation by densitometry of PMP20, AO and DHAS, respectively, from immunoprecipitates of supernatant (S) and pellet (P) fractions shown in (A). % of Total is defined as the summation of the signal from the supernatant and pellet from each time point.

tation and SDS polyacrylamide gel electrophoresis. The matrix proteins AO and DHAS existed in the cytosol for several minutes before associating with peroxisomes, confirming previous results [14,15]. In contrast, much of PMP20 was associated with peroxisomes at the beginning of the chase. The amount associated with the pellet at the earliest time points varied from 50% to 100% in different experiments; attempts to maximize the fraction of the cytoplasmic precursor form of PMP20 with shorter labeling times were not successful because completely synthesized PMP20 could not be detected. To localize the newly synthesized PMP20 immediately after the labeling period, the supernatant and pellet were further fractionated. The labeled PMP20 in the supernatant fraction was observed in a 100 000 × g supernatant, and all of the PMP20 in the crude pellet was in gradient purified peroxisomes (data not shown). These results indicate that this newly synthesized membrane-bound protein associated directly with the peroxisomes from the cytoplasm (similar to the matrix proteins) and did not transit any intermediate organellar compartments. Furthermore, PMP20 associated much more rapidly with peroxisomes than did the matrix proteins.

2.2. Proliferation of peroxisomal components

In order to explore the mechanism of peroxisomal proliferation, antibodies against the peripheral and integral membrane proteins, PMP20 and PMP47, respectively (see below), and the matrix proteins AO and DHAS were utilized to detect their expression during peroxisomal proliferation after transferring cells from glucose- to methanol-containing medium [17,18]. Neither PMP20 nor the matrix proteins were detected in cells growing in glucose-containing medium. PMP47 could be detected at low levels by immunoblots and was observed by immunogold labeling on small organelles in thin sections [18]. Thirty minutes after shift into methanol growth medium, proliferation of the peroxisomal membrane could be observed by electron microscopy, and this was accompanied by an increase in PMP47 levels. After 2.5 h of induction by methanol, many small clustered peroxisomes were observed, although the synthesis of PMP20, AO and DHAS was barely detectable [18]. These results indicate that proliferation of peroxisomes in *C. boidinii* was driven by growth of the peroxisomal membrane, rather than by import of matrix proteins. It was also consistent with a role for PMP47 in peroxisomal assembly.

2.3. Structure and composition of peroxisomal membranes

Peroxisomes can be induced to proliferate in *C. boidinii* by several diverse carbon and nitrogen sources [17,19]. To gain insight into the function of the PMPs, we examined the protein composition of the membrane from cells grown in three of these substrates, methanol, oleic acid, and D-alanine. Growth of cells on media containing these carbon sources led to induction of different sets of peroxisomal enzymes to metabolize these substrates, and the different metabolic capacities were reflected by novel patterns of abundant peroxisomal matrix proteins [17]. In contrast, membranes derived from these peroxisomes were very similar, containing high concentrations of PMP31, PMP32, and PMP47. Only cells from methanol cultures, however, expressed PMP20 [17,20,21]. The functions of the substrate-non-specific PMP31, PMP32 and PMP47 are currently not known, but since their abundance and presence supersede specific peroxisomal metabolic pathways, they are most likely related to peroxisomal structure, general function or assembly.

These membrane proteins were resistant to extraction with high or low salt. Treatment of the membranes with sodium carbonate (pH 11.3) extracted contaminating matrix proteins, most of PMP20, but little of the other PMPs [17,20]. Similar results were obtained whether membranes were pelleted or floated after carbonate treatment, further indicating a tight binding of these proteins with the membrane. The hydrophobic nature of PMP47 and PMP20 was investigated further. A peroxisomal fraction was solubilized in Triton X-114. This detergent separates into two phases at elevated temperatures, and integral membrane proteins characteristically partition into the detergent-rich phase [22]. The phases were probed with monoclonal antibodies raised against PMP47, PMP20, and DHAS. DHAS and PMP20 partitioned into the detergent-poor phase, while PMP47 was found exclusively in the detergent-rich phase (Fig. 1B). The location of various peroxisomal proteins were further investigated by immuno-electron microscopy. Gold particles used to detect PMP47 decorated the peroxisomal membranes from cells cultured on a variety of carbon sources; AO and DHAS were localized to the peroxisomal matrix, while PMP20 was observed in both the matrix and at the organellar membrane [17]. Taken together, these results demonstrate that PMP47 is an integral membrane protein while PMP20 is more weakly associated with the organellar membrane.

Another yeast, *Saccharomyces cerevisiae*, also has a simple membrane composition with proteins of molecular mass 32, 31 and 24 kDa, termed Pmp32, Pmp31, and Pmp24 respectively [23] (Fig. 1C). The similarities in apparent molecular mass between the proteins of molecular mass 32 and 31 kDa between the two yeast suggests that these proteins represent corresponding proteins or homologs. Preliminary results have not confirmed this prediction. Affinity purified antibodies against PMP32 and PMP31 of *C. boidinii* recognize neither PMP32 nor PMP31 of *S. cerevisiae* by immunoblot analysis.

2.4. Cloning of genes encoding membrane proteins

We have been isolating the genes encoding the PMPs in order to study the importance of the membrane proteins to peroxisomal function. Two genes were isolated encoding isoforms of PMP20 that are 97% identical at the protein level [24]. PMP20 terminates in the tripeptidyl sequence -AKL, and the last 12 amino acids of the protein containing this sequence is capable of sorting a carrier protein to peroxisomes in mammalian cells [10]. The PMP20 genes are transcribed to different extents by methanol but appear to be specific for methanol metabolism since the proteins are not induced by other peroxisome-proliferating growth substrates as previously described. The precise role of PMP20 in the peroxisomal metabolism of methanol has not yet been identified, but we have speculated based on weak sequence homology to DHAS and glyceraldehyde-3-phosphate dehydrogenase, that it may be involved in the transport of glyceraldehyde 3-phosphate out from the peroxisome [24]. PMP20 has also been found associated with DHAS, although it does not appear to affect the activity of this enzyme in vitro [21].

PMP32 and PMP31 of C. boidinii are also related proteins; tryptic peptides of these proteins clearly have similar sequences. The gene encoding PMP31 has been isolated (unpublished observations), and predicts an integral membrane protein with one to two membrane spanning domains. Screening for the PMP32 gene of C. boidinii and the PMP32-31 of S. cerevisiae are currently in progress. Analysis of tryptic fragments from a 31 kDa membrane protein from peroxisomal fractions of S. cerevisiae yields a peptide sequence identical to mitochondrial porin [25; unpublished results]. Contamination of peroxisomal membranes with this abundant membrane protein cannot easily be ruled out.

Similar to the PMP20s and PMP32-31, PMP47 also appears to be encoded by more than one gene. This conclusion is based on Southern hybridization of one isolated gene to several bands in digests of genomic DNA [26]. Another copy of a gene encoding PMP47 has recently been isolated. The inferred coding sequence closely resembles the PMP47 sequence (unpublished observations), although it is not clear at this time whether it represents a PMP47B gene or a second allele of the original gene. Restriction fragments of S. cerevisiae genomic DNA hybridized to a fragment of C. boidinii PMP47. We have not been able to isolate a PMP47 homolog gene from a S. cerevisiae library, however. Analysis of the inferred C. boidinii protein sequence did not reveal any obvious similarities to proteins in several data bases. PMP47 was predicted to have at least two membrane spanning segments by the algorithm of Kyte and Doolittle [27]. The topology of the protein must be determined experimentally to confirm the models.

2.5. Expression and sorting of PMP47

The gene encoding PMP47 was transformed into S. cerevisiae, a yeast that lacks a close homolog of PMP47 by immunoblot analysis [23]. When peroxisomes were induced to proliferate by oleic acid, PMP47 was observed in peroxisomes by both immuno-electron microscopy and indirect immunofluorescence (data not shown), and was localized exclusively in peroxisomal fractions (Fig. 3). The protein was not solubilized by sodium carbonate (pH 11.3) and co-purified with the endogenous peroxisomal membrane proteins on isopycnic sucrose gradients, indicating that PMP47 was correctly assembling into the peroxisomal membranes of the heterolo-

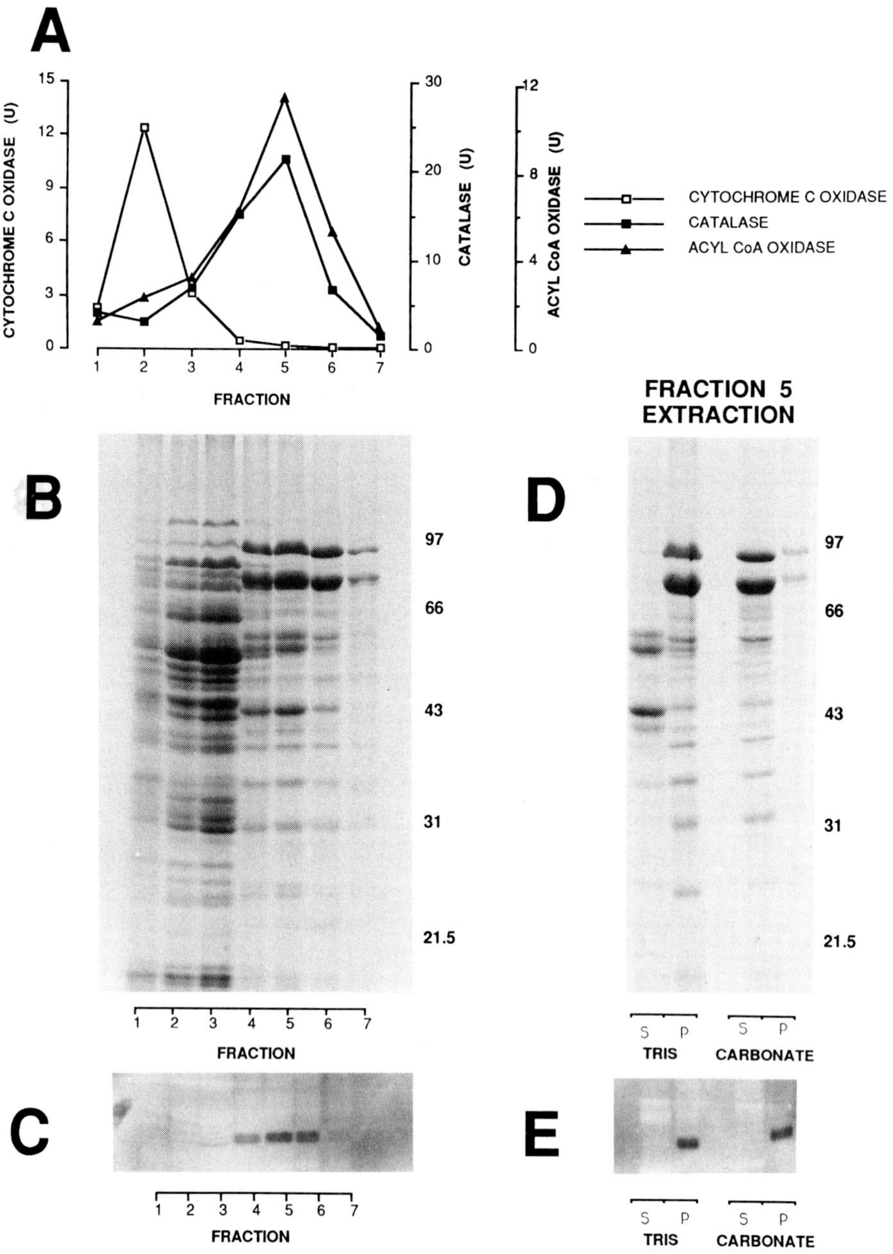
A
CYTOCHROME C OXIDASE (U)
CATALASE (U)
ACYL CoA OXIDASE (U)
CYTOCHROME C OXIDASE
CATALASE
ACYL CoA OXIDASE
FRACTION
B
97
66
43
31
21.5
FRACTION
C
FRACTION
FRACTION 5
EXTRACTION
D
97
66
43
31
21.5
S P
TRIS
S P
CARBONATE
E
S P
TRIS
S P
CARBONATE

gous yeast [23]. We have begun to investigate the sorting of this peroxisomal membrane protein. The carboxy-terminus of PMP47 ends in the sequence -AKE, somewhat similar to the -SKL consensus determined for PTS1, although -AKE did not function as a sorting signal in mammals [9]. The tripeptidyl sequence -SKL- is present at residues 320–322 of PMP47, however. Similar sequences are observed at internal locations of a number of other peroxisomal and non-peroxisomal proteins, and it is not clear whether these internal tripeptidyl sequences can function as peroxisomal targeting sequences. The carboxy-terminal -AKE and the internal -SKL- sequences were altered by substituting alanines for lysines to test whether they were necessary for the sorting of PMP47. Both the single and double mutations did not prevent the altered PMP47 from sorting to peroxisomes (Fig. 4), suggesting that neither of these portions of PMP47 are necessary for sorting. Other alterations of the -SKL- and the -AKE tripeptides confirm these results (unpublished). Further searches for the peroxisomal sorting signal(s) within PMP47 are in progress.

3. Discussion

The simplicity and abundance of peroxisomes in *C. boidinii*, and the ability of these cells to induce peroxisomal proliferation upon incubation in several diverse carbon and nitrogen sources, has allowed us to study the composition, function, and assembly of this organelle. We have found that while the induction of the abundant matrix enzymes is dependent on the growth substrate, three abundant membrane proteins, termed PMP31, PMP32, and PMP47, are present under all inducing conditions. What are the general functions of the peroxisomal membrane that these proteins may be performing? They may have a structural function in maintaining the shape of the organelle or by allowing the shape changes that accompany peroxisomal proliferation and fission. These PMPs are not easily solubilized, even with nonionic detergents, suggesting that they may be a part of a protein network (unpub-

←

Fig. 3. Sorting of PMP47 in *S. cerevisiae*. PMP47 was expressed in *S. cerevisiae* behind the $GAL10_{UAS}$. Cells were pregrown on semisynthetic oleate medium for 20 h before adding 0.05% galactose for 4 h before harvest [26]. Peroxisomes were separated from mitochondria on a discontinuous sucrose gradient [23]. (A) Enzyme distribution in gradient fractions with peaks for peroxisomal enzymes catalase and acyl-CoA oxidase in fraction five. (B) Coomassie-stained SDS polyacrylamide gel of gradient fractions indicating two distinct protein profiles corresponding to mitochondria (fractions 1–3) and peroxisomes (fractions 4–6). (C) Proteins from gel identical to (B) were electrophoretically transferred to nitrocellulose and probed with a monoclonal antibody against PMP47, which was detected exclusively in the peroxisomal fractions. (D) Extraction of fraction 5 with 10 mM Tris–HCl (pH 8), 5 mM EDTA plus proteinase inhibitors (1 mM PMSF, 2 mM benzamidine, and 5 µg/ml of both leupeptin and aprotinin) to yield a soluble fraction (S) and a Tris-insoluble pellet (P); this pellet was subsequently extracted with 0.1 M sodium carbonate (pH 11.3) to yield a supernatant (S) and an insoluble pellet (P). (E) Fractions from (D) above were electroblotted as in (C), and PMP47 was similarly detected.

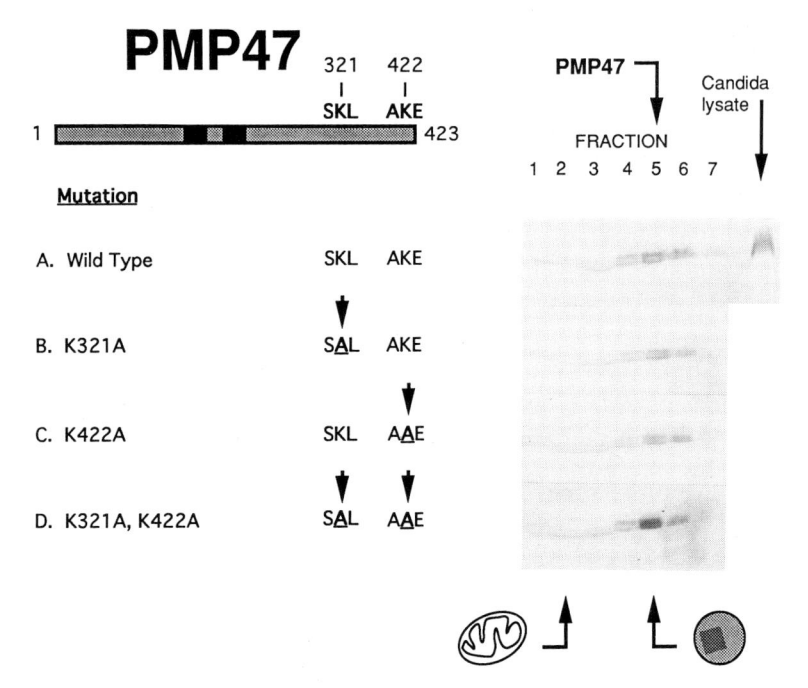

Fig. 4. Neither -SKL- nor -AKE function to sort PMP47 to peroxisomes. The PMP47 protein is shown with two membrane spanning domains (black boxes) and two potential PTS1-like signals (SKL at residues 320–322 and AKE at 421–423). The gene encoding PMP47 was mutagenized by the method of Kunkel [36]. The lysines at residues 321 and 422 of wild type PMP47 (A) were altered to alanines singly (B and C, respectively) and in combination (D). The mutations were expressed in S. cerevisiae, and the sorting of the altered PMP47 protein was monitored by the fractionation system in Fig. 3 above, except that a constitutive construct pRS47/EP was used instead of the $GAL10_{UAS}$ promoter [26]. Shown are the location of the PMP47 mutant proteins from the gradient fractions as detected by immunoblots (as in Fig. 3C). The peaks of mitochondria and peroxisomes are represented by the icons. A lysate of C. boidinii (Candida lysate) is shown as a positive control for the size of PMP47 protein and its expression.

lished results). Another structural function may be in maintaining the clustering of peroxisomes seen in C. boidinii, particularly apparent in cells cultured in methanol [17,18].

Another possible function of the peroxisomal membrane that may be independent of the substrate is as a proteinaceous channel for the exchange of small molecules [28,29]. Indeed, mitochondrial porin may also be a component of the peroxisomes of S. cerevisiae, based on partial sequencing of Pmp31. However, there is no sequence similarity between PMP31 of C. boidinii and known porins. There is evidence that peroxisomes of C. boidinii are acidified [30], and that a proton pump may be required to maintain the acidic environment [31]. Other possible functions for the PMPs are the transport of cofactors such as heme or flavin, or as constituents of a protein import pathway. Indeed, the homology of rat PMP70 with members of the multiple-drug resistant protein family suggests that PMP70 may have both a transport function as well as an ATPase activity [32]. Mutants lacking two other integral membrane proteins, hamster PAF-1 [33] and *S. cerevisiae* Pas3p [34] do not contain recognizable peroxisomes, indicating the importance of these proteins for peroxisomal integrity and/or assembly. However, there is no obvious homology between the PMPs of *C. boidinii* thus far sequenced and these other integral peroxisomal membrane proteins.

We demonstrate here that the association of PMP20 with peroxisomes occurs immediately following its synthesis, unlike the association of the matrix proteins AO and DHAS. We cannot rule out the possibility that much of PMP20 is binding to peroxisomes co-translationally. Since the extreme carboxy-terminus contains a targeting sequence [10], co-translational association would imply the existence of a second targeting signal toward the amino end of the molecule. This could be tested by determining whether a truncated mutant of PMP20 is still capable of sorting. Perhaps a more likely interpretation of our data is that the fully translated and released nascent PMP20 associates with peroxisomes at 0°C during the lysis of the spheroplasts.

Our data with PMP47 suggest that the import pathway of peroxisomal membrane proteins is distinct from that of the matrix proteins. This view is consistent with the findings in Zellweger's syndrome, where membrane proteins can assemble into peroxisomal ghosts while the import of most matrix proteins is blocked [35]. The assembly of PMP47 must occur through a novel and yet unidentified targeting signal, since the alteration of the PTS1-like signals does not affect its sorting. The amino-terminus of PMP47 also does not resemble that of thiolase, which contains PTS2 [13]. Further experiments utilizing fusion proteins and large PMP47 deletions are in progress to identify the targeting element of this protein.

References

1. Tolbert, N.E. (1981) Annu. Rev. Biochem. 50, 133–157.
2. Fahimi, H.D. and Sies, H. (1987) Peroxisomes in Biology and Medicine, Springer-Verlag, Berlin.
3. Schutgens, R.B.H., Wanders, R.J.A., Heymans, H.S.A., Schram, A.W., Tager, J.M., Schrakamp, G. and van den Bosch, H. (1987) J. Inher. Metab. Dis. 10(Suppl. 1), 33–45.
4. Suzuki, Y., Orii, T., Mori, M., Tatibana, M. and Hashimoto, T. (1986) Clin. Chim. Acta 156, 191–196.
5. Lazarow, P.B. and Fujiki, Y. (1985) Annu. Rev. Cell Biol. 1, 489–530.
6. Borst, P. (1986) Biochim. Biophys. Acta 866, 179–203.
7. Borst, P. (1989) Biochim. Biophys. Acta 1008, 1–13.
8. Gould, S.J., Keller, G.-A. and Subramani, S. (1988) J. Cell Biol. 107, 897–905.
9. Gould, S.J., Keller, G.-A., Hosken, N., Wilkinson, J. and Subramani, S. (1989) J. Cell Biol. 108, 1657–1664.
10. Gould, S.J., Keller, G.-A., Schneider, M., Howell, S.H., Garrard, L.J., Goodman, J.M., Distel, B., Tabak, H. and Subramani, S. (1990) EMBO J. 9, 85–90.
11. Hijikata, M., Ishii, N., Kagamiyama, H., Osumi, T. and Hashimoto, T. (1987) J. Biol. Chem. 262, 8151–8158.
12. Gietl, C. (1990) Proc. Natl. Acad. Sci. USA 87, 5773–5777.

13. Swinkles, B.W., Gould, S.J., Bodnar, A.G., Rachubinski, R.A. and Subramani, S. (1991) EMBO J. 10, 3255–3262.
14. Goodman, J.M., Scott, C.W., Donahue, P.N. and Atherton, J.P. (1984) J. Biol. Chem. 259, 8485–8493.
15. Goodman, J.M. (1985) J. Biol. Chem. 260, 7108–7113.
16. Harder, W., Trotsenko, Y.A., Bystrykh, L.V. and Egli, T. (1987) in: H.W. van Verseveld and J.A. Duine (Eds.), Microbial Growth on C1 Compounds, Martinus Nijhoff, Dordrecht, pp. 139–149.
17. Goodman, J.M., Trapp, S.B., Hwang, H. and Veenhuis, M. (1990) J. Cell Sci. 97, 193–204.
18. Veenhuis, M. and Goodman, J.M. (1990) J. Cell Sci. 96, 583–590.
19. Sulter, G.J., Waterman, H.R., Goodman, J.G. and Veenhuis, M. (1990) Arch. Microbiol. 153, 485–489.
20. Goodman, J.M., Maher, J., Silver, P.A., Pacifico, A. and Sanders, D. (1986) J. Biol. Chem. 261, 3464–3468.
21. Garrard, L.J. (1990) Ph.D. Thesis, University of Texas Southwestern Medical Center at Dallas.
22. Bordier, C. (1981) J. Biol. Chem. 256, 1604–1607.
23. McCammon, M.T., Veenhuis, M., Trapp, S.B. and Goodman, J.M. (1990) J. Bacteriol. 172, 5816–5827.
24. Garrard, L.J. and Goodman, J.M. (1989) J. Biol. Chem. 264, 13929–13937.
25. Mihara, K. and Sato, R. (1985) EMBO J. 4, 769–774.
26. McCammon, M.T., Dowds, C.A., Orth, K., Moomaw, C.R., Slaughter, C.A. and Goodman, J.M. (1990) J. Biol. Chem. 265, 20098–20105.
27. Kyte, J. and Doolittle, R.F. (1982) J. Mol. Biol. 157, 105–132.
28. Van Veldhoven, P.P., Just, W.W. and Mannaerts, G.P. (1987) J. Biol. Chem. 262, 4310–4318.
29. Lemmens, M., Verheyden, K., Van Velhaven, P., Vereecke, J., Mannaerts, G.P. and Carmeliet, E. (1989) Biochim. Biophys. Acta 984, 351–359.
30. Waterham, H.R., Keizer-Gunnink, I., Goodman, J.M., Harder, W. and Veenhuis, M. (1990) FEBS Lett. 262, 17–19.
31. Bellion, E. and Goodman, J.M. (1987) Cell 48, 165–173.
32. Kamijo, K., Taketani, S., Yokota, S., Osumi, T. and Hashimoto, T. (1990) J. Biol. Chem. 265, 4534–4540.
33. Tsukamoto, T., Miura, S. and Fujiki, Y. (1991) Nature 350, 77–81.
34. Hohfeld, J., Veenhuis, M. and Kunau, W.-H. (1991) J. Cell Biol. 114, 1167–1178.
35. Santos, M.J., Imanaka, T., Shio, H. and Lazarow, P.B. (1988) J. Biol. Chem. 263, 10502–10509.
36. Kunkel, T.A. (1985) Proc. Natl. Acad. Sci. USA 82, 488–492.

CHAPTER 17

Mechanisms of transport of proteins into microbodies

SURESH SUBRAMANI

Department of Biology, 0322 Bonner Hall, University of California, San Diego, La Jolla, CA 92093, USA

Abstract

The microbodies are a class of electron-dense, single-membrane-bound organelles. They include evolutionarily related subcellular compartments such as peroxisomes, glyoxysomes and glycosomes. Proteins destined to reside in the matrix or membrane of these organelles are synthesized on free cytoplasmic polysomes and transported post-translationally, in a targeting signal-dependent manner, to these compartments. Our work has led to the discovery of two types of signals that transport proteins into the matrix of these organelles. The first (PTS-1) is a conserved, consensus, C-terminal tripeptide which serves as a general microbody targeting signal (McTS). This signal does not need to be part of the primary amino acid sequence of the protein in order to function. The second signal, PTS-2, is an 11 amino acid sequence which functions at the amino-terminus and at least in one internal location to target proteins to peroxisomes. Its ability to function as a general microbody targeting signal remains to be tested. While the PTS-1 signal is found in numerous microbody proteins, the PTS-2 signal has been identified in only one protein, the peroxisomal 3-ketoacyl CoA thiolase. Cells from patients with a peroxisomal deficiency known as Zellweger's syndrome are deficient in the transport of proteins containing PTS-1 but not of those possessing PTS-2. Genetic evidence in *Saccharomyces cerevisiae* also supports the idea that there may be at least two distinct mechanisms for protein transport into peroxisomes.

1. Introduction

Microbodies were first described in mouse kidney tubules as small, single-membrane-bound vesicles surrounding a granular, electron-dense matrix [1]. In the

1960s, the term microbodies was replaced by peroxisomes, to reflect the biochemical property that these organelles contained enzymes that produced and degraded H_2O_2 [2]. However, the word microbody is currently used generically to describe evolutionarily related subcellular compartments such as peroxisomes, glyoxysomes and glycosomes.

Peroxisomes, the most ubiquitous of the microbodies, are found in all eukaryotic organisms except archezoa [3]. Although they were initially viewed as the compartment involved in the production and detoxification of H_2O_2, there is now a large compendium of proteins that claims residence in peroxisomes. Many of these are enzymes that participate in key biochemical pathways which endow peroxisomes with a panoply of important biochemical functions (Table I). It is therefore not terribly surprising that human diseases compromising peroxisomal functions are often lethal [4].

Glyoxysomes were first described in germinating castor bean seedlings [5] and were so named because the five enzymes of the glyoxylate pathway were housed in this compartment. These enzymes play a role in the conversion of fat to carbohydrates. Glyoxysomes are akin to peroxisomes in that they also contain oxidases that generate H_2O_2, catalase to degrade H_2O_2 and β-oxidation enzymes for the oxidation of fatty acids [6].

The third type of microbody, the glycosome, was first discovered in trypanosomes and derives its name from the fact that many glycolytic enzymes are duplicated in this compartment as well as in the cytosol [7]. In addition, however, they contain enzymes involved in diverse pathways such as pyrimidine and glycerol metabolism and purine salvage. The glycosomes of some organisms, such as trypanosomes, do not contain

TABLE I

Biochemical functions of peroxisomes

Function	Reference
Generation and degradation of H_2O_2	De Duve and Baudhuin [2]
β-Oxidation of fatty acids	Lazarow and De Duve [45], Hashimoto [46], Bremer and Norum [47], Singh et al. [48]
Biosynthesis of plasmalogens or alkyl ether phospholipids	Hajra et al. [49], Hajra and Bishop [50]
Biosynthesis of bile acids	Pedersen et al. [51], Krisans et al. [52], Thompson and Krisans [33]
Biosynthesis of cholesterol	Keller et al. [53], Thompson et al. [54], Tsuneoka et al. [55], Keller et al. [56], Thompson and Krisans [33]
Oxidation of L-pipecolic acid	Zaar et al. [57], Wanders et al. [58]
Oxidation of phytanic acid	Draye et al. [59]
Metabolism of glyoxylate	Noguchi and Takada [60]
Prostaglandin metabolism	Diczfalusy et al. [61], Schepers et al. [62]

TABLE II

Few peroxisomal proteins are proteolytically processed

Protein	Primary translation product (kDa)	Processed forms (kDa)	Reference
3-Ketoacyl CoA thiolase	44	41, 3[a]	Fujiki et al. [63]
Acyl-CoA oxidase	72	52, 20	Osumi et al. [64], Miura et al. [65]
Rat sterol carrier protein 2	14.5	13, 1.5[a]	Fujiki et al. [66]

[a] These are presumably degraded because they have not been found in vivo.

catalase. However, related parasites, such as *Crithidia* sp. belonging to the same family, do have catalase [8]. Glycosomes, like peroxisomes, contain enzymes involved in plasmalogen synthesis and in the β-oxidation of fatty acids [8].

All microbodies are believed to arise from pre-existing ones by growth and fission. Since these organelles possess no DNA, all their proteins are encoded by nuclear genes. Almost all the microbody matrix and integral membrane proteins studied so far are synthesized on free polysomes and translated across or into the membrane of the organelle.

Unlike many mitochondrial, endoplasmic reticulum (ER) or chloroplast proteins, most peroxisomal proteins are not proteolytically processed either during or after import into the organelle. The only known exceptions to this rule are shown in Table II. Human diseases involving a general loss of peroxisomal functions fall into at least nine complementation groups [9,10]. In addition, diseases involving loss of individual peroxisomal enzymes are also known to occur. The best studied of the generalized disorders is the cerebro-hepato-renal syndrome of Zellweger [11]. Patients with Zellweger's syndrome fall into at least three different complementation groups. Members of all three groups appear to have aberrant peroxisome ghosts and lack many of the peroxisomal matrix proteins [12–14]. Thus this disease has the hallmarks of a deficiency in protein import into peroxisomes.

2. *A C-terminal tripeptide is a major targeting signal for proteins of the microbody matrix*

A large body of evidence has now been accumulated to demonstrate that variants of a C-terminal tripeptide serve as a ubiquitous McTs. One such variant, consisting of the C-terminal amino acids SKL (in the one letter code), was first shown to be necessary for the targeting of firefly luciferase into peroxisomes of mammalian [15,16] and yeast cells [17]. This tripeptide is also sufficient to direct the transport of a passenger protein, bacterial chloramphenicol acetyltransferase (CAT), into peroxisomes of mammalian cells [16,18] and glycosomes of *Trypanosoma brucei* [19]. The targeting of luciferase to peroxisomes of *S. cerevisiae, Hansenula polymorpha,*

Nicotiana tabacum, *Xenopus laevis*, *Photinus pyralis* and monkey kidney cells and mouse, human and Chinese hamster ovary fibroblast cells provides overwhelming evidence that the targeting signal on luciferase must have been conserved through evolution [20–22; Wendland and Subramani, unpublished data]. Further evidence for the generality of the C-terminal location of the targeting signal in many microbody proteins comes from the demonstration that seven other peroxisomal proteins [18,23,24] and at least one glycosomal protein [19,25] contain sequences at or near their C-termini that are either necessary or sufficient to transport proteins into the appropriate microbody compartment. This implies that transport of these proteins into microbodies would have to be post-translational.

Antibodies directed against the SKL tripeptide specifically recognize by immunoelectron microscopy, peroxisomes in rat liver cells, glyoxysomes in *Pichia pastoris*, *Neurospora crassa* and *Ricinus communis* cells and glycosomes in *T. brucei* [26,27]. In each of these organisms about 15–45% of the total Coomassie-blue stained microbody proteins are recognized in Western blots by the anti-SKL antibody [26,27]. These results prove that the tripeptide McTS is not only conserved in evolution but also that it is used for the targeting of a substantial number of proteins into microbodies [28].

3. Certain variants of the SKL tripeptide can also function as PTS

Site-directed mutagenesis was used to mutate each of the amino acids in the SKL tripeptide in luciferase. The mutant luciferases were then expressed in monkey kidney cells and their subcellular localization was determined by indirect immunofluorescence. It was found that Ser could be substituted by Ala or Cys, Lys could be replaced by Arg or His and the C-terminal Leu could be interchanged with Met, without substantial loss of ability of the tripeptide to function as a PTS [16]. These experiments provide evidence for the existence of a consensus tripeptide microbody targeting signal. Computer searches of the amino acid sequences of peroxisomal proteins reveals the remarkable conservation of the C-terminal tripeptide McTS in at least 27 microbody proteins [28].

4. Peroxisomal protein transport in microinjected mammalian cells

Despite the appearance of a number of publications describing the transport of proteins into rat liver or *Candida tropicalis* peroxisomes in vitro [29–32], the fragility of purified peroxisomes and the reliance on protease protection as the sole criterion for translocation into the organelle have impeded the full exploitation of such assays. The cloning of the cDNA for the peroxisomal thiolases provided an alternative reagent for the study of peroxisomal protein import in vitro. Since this protein is among the few peroxisomal proteins that are cleaved following import, the processing of the ^{35}S-labeled thiolase precursors (44 kDa) to the mature form (41 kDa) serves as a diagnostic hallmark of transport into the organelle. Unfortunately, however, both the thiolases [33] and the proteolytic activity responsible for their cleavage have proved to be notoriously leaky in preparations of rat liver peroxisomes (Shackelford, Wendland, Subramani and Krisans, unpublished data).

To overcome these problems, we turned recently to the use of microinjection to assess peroxisomal import into a variety of mammalian cells [22]. The approach capitalizes upon the absence of irreversible modifications on peroxisomal proteins which renders them ideal molecules for re-transport into the organelle. Microinjection offers two major advantages. First, it allows the assessment of import in an in vivo context while preserving the ease of manipulation of in vitro systems. Second, it can be used in primary cell lines derived from patients with peroxisomal dysfunctions.

Purified luciferase is transported into peroxisomes of mammalian cells in a time- and temperature-dependent fashion, following microinjection of the protein into the cytoplasm. The microinjected protein accumulates in membrane-bound vesicles that are peroxisomes, as judged by the co-localization of bona fide peroxisomal markers in these vesicles. The transport of luciferase into these peroxisomes is competed by molar excess of peptides containing the C-terminal McTS, but not by control peptides. This suggests that the peroxisomal import machinery can be saturated [22].

Another exciting observation is the discovery that the SKL tripeptide does not need to be part of the primary protein sequence in order to function as a PTS. The coupling of a 12 amino acid peptide (H_2N-CRYHLKPLQSKL-COOH) to lysines on human serum albumin (HSA) via the cysteinyl sulfhydryl on the peptide was sufficient to transport the HSA-SKL derivative into peroxisomes of mouse and human cells into which the conjugate was injected [22].

5. Import deficiency in fibroblast cells from Zellweger's syndrome patients

Fibroblasts from normal humans and Zellweger's syndrome patients were tested for their ability to translocate proteins into peroxisomes. While cells from normal humans were quite capable of transporting proteins into peroxisomes, the cells from two different complementation groups of Zellweger patients were incapable of importing proteins containing the SKL signal into peroxisomes [22]. This provides the first convincing demonstration of a genuine protein import defect in Zellweger patients. The microinjection assay can obviously be used to assess import deficiencies in cells from other complementation groups as well.

6. An amino-terminal PTS resides in the cleaved leader peptides of the peroxisomal thiolases

Although the conserved tripeptide McTS is a feature of many peroxisomal proteins, there are several microbody proteins that lack this tripeptide at their C-termini. In the absence of evidence that the tripeptide McTS can function at internal locations in proteins, it would seem likely that other general microbody targeting signals, or peroxisome-, glyoxysome- and glycosome-specific targeting signals exist.

The peroxisomal thiolases A and B contain 36 and 26 amino acid prepieces which are cleaved following import of these precursors into peroxisomes [34,35]. We have shown recently that these prepieces are necessary for the import of the thiolases into peroxisomes of mammalian cells [25]. The prepieces, when fused onto the amino-terminus of CAT, are also sufficient to direct the translocation of the fusion protein into peroxisomes. Deletion analysis of the prepiece of thiolase B showed that the first 11 amino acids are sufficient for peroxisomal targeting. The PTS of thiolase, PTS-2, constitutes a new signal that is distinct from the C-terminal PTS (PTS-1). PTS-1 is a ubiquitous signal whereas PTS-2 is known, at present, to be used only by thiolase. Since PTS-2 is present 11 amino acids away from the amino-terminus in thiolase A, it means that PTS-2 can function at one internal location. This is in contrast with PTS-1 which does not appear to function at amino-terminal or several internal locations of mouse dihydrofolate reductase (Heyman, Gould and Subramani, unpublished data). The PTS-2 does not possess features characteristic of known ER, mitochondrial or chloroplast targeting signals, consistent with its role in targeting proteins only into peroxisomes. Furthermore, the proteolytic cleavage of the thiolase leader can be uncoupled from the translocation of the leader into peroxisomes.

7. Selective import deficiency in Zellweger cells

As stated earlier, Walton et al. [22] have shown a deficiency in the import of proteins containing the PTS-1 signal in Zellweger cells. However, recent work [36] has revealed that cells from Zellweger patients contain aberrant peroxisome ghosts into which the thiolase A and B precursors can be imported. Thus there is no deficiency in the import of proteins containing PTS-2 in cells from Zellweger patients. This result suggests at least two distinct pathways of protein import into peroxisomes, one dependent on PTS-1 and the other on PTS-2. It is tempting to speculate that the lack of processing of thiolase in Zellweger patients is due to the inability of a PTS-1-dependent protease to enter the peroxisome.

Recent genetic evidence from W. Kunau's laboratory shows that certain *S. cerevisiae* mutants are deficient only in the import of thiolase into yeast peroxisomes. Members of this complementation group could not be complemented with a wild-type thiolase gene, ruling out a cis-mutation in the thiolase PTS [37]. Thus both yeast and mammalian cells, and perhaps all eukaryotes, use at least two pathways to import proteins into peroxisomes. The current hypothesis would be that PTS-1 and PTS-2 interact with different receptors. It will be of interest to ascertain whether the same or distinct translocation machineries mediate the subsequent steps in the import of proteins containing these two types of targeting signals.

8. Transport of membrane proteins into peroxisomes

The presence of membrane proteins in peroxisome ghosts of Zellweger patients suggests that the targeting of these proteins is unaffected by the disease. Of the four peroxisomal membrane protein genes that have been cloned [38–41], none appear to have the C-terminal PTS-1 or the N-terminal PTS-2 signal. It is therefore quite likely that peroxisomal membrane proteins are targeted to the organelle using a different signal.

9. Summary

Much progress has been made in the identification of signals that target proteins to the matrix of microbodies. Little is known about the targeting of microbody membrane proteins or the assembly of microbodies.

The current progress in the isolation of mutants [42,43] and genes [40,41,44] involved in peroxisome assembly and import promises to provide further insight into the biogenesis of peroxisomes. It is likely that the general principles will be applicable to other microbodies as well.

Acknowledgements

This work was supported by an NIH grant (DK 41737) to S.S. I thank my collaborators who supplied the data described in this article.

References

1. Rhodin, J. (1954) Ph.D. Thesis, Karolinska Institutet, Aktiebolaget Godvil, Stockholm.
2. De Duve, C. and Baudhuin, P. (1966) Physiol. Rev. 46, 323–357.
3. Cavalier-Smith, T. (1987) Ann. NY Acad. Sci. 503, 55–71.
4. Lazarow, P.B. and Moser, H.W. (1989) in: C.R. Scriver, A.L. Beaudet, W.S. Sly and D. Valle (Eds.), The Metabolic Basis of Inherited Disease, 6th edition, McGraw-Hill, New York, pp. 1479–1509.
5. Breidenbach, R.W. and Beevers, H. (1967) Biochem. Biophys. Res. Commun. 27, 462–469.
6. Cooper, T.G. and Beevers, H. (1969) J. Biol. Chem. 244, 3514–3520.

7. Opperdoes, F.R. and Borst, P. (1977) FEBS Lett. 80, 360–364.
8. Opperdoes, F.R. (1987) Annu. Rev. Microbiol. 41, 127–151.
9. Brul, S., Westerveld, A., Strijland, A., Wanders, R.J.A., Schram, A.W., Heymans, H.S.A., Schutgens, R.B.H., Van Den Bosch, H. and Tager, J.-M. (1988) J. Clin. Invest. 81, 1710–1715.
10. Roscher, A.A., Hoefler, S., Hoefler, G., Paschke, E., Paltauf, F., Moser, A. and Moser, H.W. (1989) Pediatr. Res. 26, 67–72.
11. Zellweger, H. (1987) Dev. Med. Child Neurol. 29, 821–829.
12. Schram, A.W., Strijland, A., Hashimoto, T., Wanders, R.J.A., Schutgens, R.B.H., van den Bosch, H. and Tager, J.M. (1986) Proc. Natl. Acad. Sci. USA 83, 6156–6158.
13. Santos, M.J., Imanaka, T., Shio, H. and Lazarow, P.B. (1988) J. Biol. Chem. 263, 10502–10509.
14. Santos, M.J. Imanaka, T., Shio, H., Small, G.M. and Lazarow, P.B. (1988) Science 239, 1536–1538.
15. Gould, S.J., Keller, G.-A. and Subramani, S. (1987) J. Cell Biol. 105, 2923–2931.
16. Gould, S.J., Keller, G.-A., Hosken, N., Wilkinson, J. and Subramani, S. (1989) J. Cell. Biol. 108, 1657–1664.
17. Distel, B., Gould, S.J., Voorn-Brouwer, T., van der Berg, M., Tabak, H.F. and Subramani, S. (1992) New Biol. 4, 157–165.
18. Gould, S.J., Keller, G.-A. and Subramani, S. (1988) J. Cell Biol. 107, 897–905.
19. Fung, K. and Clayton, C. (1991) Mol. Biochem. Parasitol. 45, 261–264.
20. Gould, S.J., Keller, G.-A., Schneider, M., Howell, S.H., Garrard, L.J., Goodman, J.M., Distel, B., Tabak, H. and Subramani, S. (1990) EMBO J. 9, 85–90.
21. Holt, C.E., Garlick, N. and Cornel, E. (1990) Neuron 4, 203–214.
22. Walton, P.A., Gould, S.J., Feramisco, T. and Subramani, S. (1992) Mol. Cell. Biol. 12, 531–541.
23. Aitchison, J.D., Murray, W.W. and Rachubinski, R.A. (1991) J. Biol. Chem. 266, 23179–23209.
24. Lewin, A.S., Hines, V., Small, G.M. (1990) Mol. Cell. Biol. 10, 1399–1405.
25. Swinkels, B.W., Gould, S.J., Bodnar, A.G., Rachubinski, R.A. and Subramani, S. (1991) EMBO J. 10, 3255–3262.
26. Gould S.J., Krisans, S., Keller, G.-A. and Subramani, S. (1990) J. Cell Biol. 110, 27–34.
27. Keller, G.-A., Krisans, S., Gould, S.J., Sommer, J.M., Wang, C.C., Schliebs, W., Kunau, W., Brody, S. and Subramani, S. (1991) J. Cell Biol. 114, 893–904.
28. Subramani, S. (1991) J. Membr. Biol. 125, 99–106.
29. Imanaka, T., Small, G.M. and Lazarow, P.B. (1987) J. Cell Biol. 105, 2915–2922.
30. Small, G.M., Imanaka, T., Shio, H. and Lazarow, P.B. (1987) Mol. Cell. Biol. 7, 1848–1855.
31. Small, G.M., Szabo, L.J. and Lazarow, P.B. (1988) EMBO J. 7, 1167–1173.
32. Miyazawa, S., Osumi, T., Hashimoto, T., Ohno, K., Miura, S. and Fujiki, Y. (1989) Mol. Cell. Biol. 9, 83–91.
33. Thompson, S.L. and Krisans, S.K. (1990) J. Biol. Chem. 265, 5731–5735.
34. Hijikata, M., Ishii, N., Kagamiyama, H., Osumi, T. and Hashimoto, T. (1990) J. Biol. Chem. 265, 4600–4606.
35. Bodnar, A.G. and Rachubinski, R.A. (1990) Gene 91, 193–199.
36. Balfe, A., Hoefler, G., Chen, W.W. and Watkins, P.A. (1990) Pediatr. Res. 27, 304–310.
37. Marzioch, M., Erdmann, R. and Kunau, W.-H. (1990) 15th Int. Conf. on Yeast Genetics and Molecular Biology, Yeast 6, S469 (abstract).
38. McCammon, M., Dowds, C.A., Orth, K., Moomaw, C.R., Slaughter, C.A. and Goodman, J.M. (1990) J. Biol. Chem. 265, 20098–20105.
39. Kamijo, K., Taketani, S., Yokota, S., Osumi, T. and Hashimoto, T. (1990) J. Biol. Chem. 265, 4534–4540.
40. Tsukamoto, T., Miura, S. and Fujiki, Y. (1991) Nature 350, 77–81.
41. Hohfeld, J., Veenhuis, M. and Kunau, W.-H. (1991) J. Cell Biol. 114, 1167–1178.
42. Erdmann, R., Veenhuis, M., Mertens, D. and Kunau, W.-H. (1989) Proc. Natl. Acad. Sci. USA 86, 5419–5423.

43. Cregg, J.M., Klei, I.J., Sulter, G.J., Veenhuis, M. and Harder, W. (1990) Yeast 6, 87–97.
44. Erdmann, R., Wiebel, F.F., Flessau, A., Rytka, J., Beyer, A., Frohlich, K.U. and Kunau, W.-H. (1991) Cell 64, 499–510.
45. Lazarow, P.B. and De Duve, C. (1976) Proc. Natl. Acad. Sci. USA 73, 2043–2046.
46. Hashimoto, T. (1982) Ann. NY Acad. Sci. 386, 5–12.
47. Bremer, J. and Norum, K.R. (1982) J. Lipid Res. 23, 243–256.
48. Singh, I., Moser, A.E., Goldfischer, S. and Moser, H.W. (1984) Proc. Natl. Acad. Sci. USA 81, 4203–4207.
49. Hajra, A.K., Burke, C.L. and Jones, C.L. (1979) J. Biol. Chem. 251, 5149–5154.
50. Hajra, A.K. and Bishop, J.E. (1982) Ann. NY Acad. Sci. 386, 170–182.
51. Pedersen, J.I., Kase, B.F., Prydz, K. and Bjorkhem, J. (1987) in: H.D. Fahimi and H. Sies (Eds.), Peroxisomes in Biology and Medicine, Springer-Verlag, Berlin, pp. 67–77.
52. Krisans, S.K., Thompson, S.L., Pena, L.A., Kok, E. and Javitt, N.B. (1985) J. Lipid Res. 26, 1324–1332.
53. Keller, G.-A., Pazirandeh, M. and Krisans, S.K. (1986) J. Cell Biol. 103, 875–886.
54. Thompson, S.L. Burrows, R., Laub, R.J. and Krisans, S.K. (1988) J. Biol. Chem. 262, 17420–17425.
55. Tsuneoka, M., Yamamoto, A., Fujiki, Y. and Tashiro, Y. (1988) J. Biochem. 104, 560–564.
56. Keller, G.-A., Scallen, T.J., Clarke, D., Maher, P.A., Krisans, S.K. and Singer, S.J. (1989) J. Cell Biol. 108, 1353–1361.
57. Zaar, K., Angermuller, S., Volkl, A. and Fahimi, H.D. (1986) Exp. Cell Res. 164, 267–271.
58. Wanders, R.J.A., Romeijn, G.J., Van Roermund, C.W.T., Schutgens, R.B.H., Van den Bosch, H. and Tager, J.M. (1988) Biochem. Biophys. Res. Commun. 154, 33–38.
59. Draye, J.-P., Van Hoof, F., de Hoffmann, E. and Vamecq, J. (1987) Eur. J. Biochem. 167, 573–578.
60. Noguchi, T. and Takada, Y. (1979) Arch. Biochem. Biophys. 196, 645–647.
61. Diczfalusy, U., Alexson, S.E.H. and Pedersen, J.I. (1987) Biochem. Biophys. Res. Commun. 144, 1206–1213.
62. Schepers, L., Casteels, M., Vamecq, J., Parmentier, G., Van Veldhoven, P.P. and Mannaerts, G.P. (1988) J. Biol. Chem. 263, 2724–2731.
63. Fujiki, Y., Rachubinski, R.A., Mortensen, R.M. and Lazarow, P.B. (1985) Biochem. J. 266, 697–704.
64. Osumi, T., Hashimoto, T. and Ui, N. (1980) J. Biochem. 87, 1735–1746.
65. Miura, S., Mori, M., Takiguchi, M., Tatibana, M., Furuta, S., Miyazawa, S. and Hashimoto, T. (1984) J. Biol. Chem. 105, 713–722.
66. Fujiki, Y., Tsuneoka, M. and Tashiro, Y. (1989) J. Biochem. 106, 1126–1131.

CHAPTER 18

Lessons for peroxisome biogenesis from fluorescence analyses of Zellweger syndrome fibroblasts

PAUL B. LAZAROW[1], HUGO W. MOSER[2] and MANUEL J. SANTOS[3]

[1]*Department of Cell Biology and Anatomy, Mount Sinai School of Medicine, New York, NY 10029, USA,* [2]*Kennedy Institute and Departments of Neurology and Pediatrics, Johns Hopkins University, Baltimore, MD 21205, USA and* [3]*Department of Cell Biology, Universidad Catolica de Chile, Santiago, Chile*

Abstract

Fibroblasts from seven Zellweger syndrome patients belonging to five complementation groups were examined by immunofluorescence with anti-catalase and an antiserum against peroxisomal membrane proteins. Large peroxisomal membrane ghosts were found in all of the samples. Except for one patient, none of the ghosts contained catalase. Other data indicate that the ghosts must be mostly empty. Thus these mutations interfere with the translocation of proteins through the peroxisome membrane, but not with the assembly of the membrane proteins themselves. These are Peroxisome IMport (PIM) mutations. The results show that peroxisome membrane assembly has fewer requirements, or different requirements, than the translocation of peroxisome matrix proteins into the organelle. These mostly empty peroxisome membranes must be able to grow and divide (as normal peroxisomes do), despite their lack of content. Analysis of the immunofluorescence data revealed three distinct fluorescence patterns; provisionally, these appear to depend on which gene is defective. Fusion of fibroblast samples led to the full recovery of normal peroxisome assembly.

1. Introduction

Peroxisome biogenesis is defective in the human genetic disease, Zellweger's syndrome. As first reported by Goldfischer et al. [1], peroxisomes appear to be missing in the cells and tissues of Zellweger's syndrome patients. Peroxisome functions, including the β-oxidation of very long chain fatty acids and the biosynthesis of phospholipids containing vinyl ether linkages (plasmalogens) are also defective. The most apparent consequence is profound neurological impairment (severe retardation, blindness, lack of neuromuscular control). Death occurs in the neonatal period [2,3].

By the use of immunochemical methods, we discovered that, in fact, peroxisomes are not completely absent; empty membrane ghosts of peroxisomes were detected by immunofluorescence and immunoelectron microscopy in fibroblasts from one patient [4,5]. These ghosts were about two to four times larger than normal peroxisomes (by immunoelectron microscopy) and had a much lower than normal equilibrium density (by cell fractionation) [5]. Because of their lack of contents, they were not recognizable as peroxisomes by routine morphological techniques. The 22-kDa and 70-kDa peroxisomal membrane proteins and some thiolase precursor seem to be associated with these ghosts [6,7].

Complementation analyses have demonstrated that there are at least five complementation groups for peroxisome biogenesis (based on cell fusions and the recovery of peroxisomal functions) [8–11]. We decided to apply our immunofluorescence methods to fibroblasts from each of the five complementation groups identified by Roscher et al. [9], in order to look for peroxisome membranes. We expected that some fibroblasts would probably be entirely lacking in membranes, and that a variety of fluorescent patterns might be observed, depending on the nature of the molecular defect.

2. Results

Fibroblasts from seven Zellweger patients were analyzed by immunofluorescence with anti-catalase and with antibodies against peroxisomal membrane proteins (Table I). Three of the fibroblast samples contained many large peroxisomal membrane ghosts [12]. These were larger in size than normal peroxisomes but fewer in number. None of them contained detectable catalase by immunofluorescence. Some normal-sized peroxisomal membranes were also detected, none of which contained catalase. This was the same pattern as that seen previously [4].

Three other fibroblast samples contained few peroxisomal membrane ghosts. These were also larger than normal peroxisomes and again none contained catalase (Table I) (see [12] for the original photographs).

In the seventh sample, about half the cells contained many ghosts and the other half contained few ghosts. In a few of these cells, a few catalase-containing peroxisomes were also found (Table I).

Most of these fibroblast samples were fused pairwise and tested for the restoration of normal peroxisome biogenesis. Essentially normal peroxisomes were observed by immunofluorescence in many of the fused cells after 48 h. The assignments of the

TABLE I

Immunofluorescence patterns of peroxisomes in Zellweger syndrome

Patient	Complementation group	Immunofluoresence		Catalase-containing peroxisomes
		Peroxisome membranes		
		Normal-sized	Large	
Control	–	Abundant	None	Abundant
Z2	1	Some	Many	None
Z7	4	Some	Many	None
Z9	1	Some	Many	None
Z5	2	Some	Few	None
Z6	3	Some	Few	None
Z8	5	Some	Few	None
Z3	1	Some	Many/few[a]	A few cells have a few[b]

[a] About half the cells have many and the other half have a few.
[b] Most cells have none.

samples into complementation groups on the basis of the recovery of peroxisome formation was identical to the assignments made previously on the same samples by Roscher et al. [9], based on the recovery of plasmalogen biosynthesis.

3. Discussion

3.1. Peroxisome membranes are always present: perhaps they are required for viability.

Peroxisome membranes are present in all of the fibroblast samples from patients with peroxisomal genetic disorders so far examined. This comes as a surprise. One can easily imagine that there could be mutations that would interfere with the assembly of the peroxisome membrane, and this would, of course, result in the absence of the entire peroxisome. Perhaps such defects will turn up in the future. On the other hand, the peroxisomal membrane, and perhaps some residual content, may be essential to life. We recall that mitochondria are believed to be essential. Even in rho° yeast, in which the mitochondrial genome is missing, the small vestigial mitochondrial remnant is believed to play some required role. Thus it could also be the case that the peroxisomal ghost plays some metabolic role that is required for fetal development. In this case, fetuses in which the entire peroxisome, including the membrane, is missing would not develop to term, and thus would be missing from our patient population.

3.2. The peroxisome membranes are nearly empty ghosts: these are Peroxisome IMport (PIM) mutations

The peroxisomal membranes observed by immunofluorescence must be nearly, if not completely, empty. The catalase activity in these fibroblasts was found by cell fractionation to be in the cytosol, not in the peroxisomes. All peroxisomal enzymes tested so far are synthesized in Zellweger's syndrome [13]. A few, including catalase, are stable in the cytosol and become enzymatically active. A cytosolic location for peroxisomal enzymes is to be expected if peroxisome assembly fails, in view of the fact that all peroxisomal proteins assemble by a post-translational mechanism, after synthesis on free polyribosomes in the cytosol [14]. Most newly synthesized peroxisomal proteins are rapidly degraded in Zellweger's syndrome, and thus are absent from cells and tissues (reviewed in [2,3]). Exceptionally, some thiolase precursor appears to be associated with the membranes in Zellweger fibroblasts [7,15].

It is unlikely that the mutations in these patients abolish the targeting information of the peroxisomal proteins. Zellweger's syndrome is an autosomal recessive mutation in those families where the heredity has been tested [3]. Peroxisomal proteins are not modified post-translationally. The targeting information apparently resides in the primary structure of the proteins. In some cases, it consists, at least in part, of a carboxy-terminal SKL [16]. In other cases, different topogenic information is located near the amino-terminus or internally [17–19]. Thus, a single mutation could not affect the primary sequence of all these proteins.

The simplest interpretation is that the mutation affects the translocation of the newly synthesized proteins through the membrane into the matrix space of the peroxisomes. In vitro import studies with rat liver peroxisomes [20] imply at least two pieces of translocation machinery. One is a membrane receptor for the binding of newly made peroxisomal proteins. The second is an ATP-dependent translocator (import ATPase) [21]. The Zellweger mutations could inactivate either of these proteins, and it would explain all of the observed data. The mutations could, of course, also affect as yet unidentified proteins, such as chaperonins that might be required. In any case, these appear to be Peroxisome IMport (PIM) mutations.

3.3. Peroxisome membrane assembly has fewer requirements, or different requirements, from the packaging of peroxisome matrix proteins

The existence of peroxisome membrane structures that are nearly empty means that the membrane proteins, and the membrane phospholipids, are assembling more or less normally, even in the presence of mutations that prevent the import of the matrix proteins. It could be that the targeting of membrane proteins to peroxisomes uses different topogenic features than the targeting of the content proteins. The SKL tripeptide that targets some peroxisome matrix proteins has not yet been found on membrane proteins [22–24]. Post-translational embedding of proteins in the peroxisome membrane may use different machinery than the translocation of proteins through the membrane to the matrix space inside. Alternatively, it is possible that assembly of membrane and matrix proteins shares some machinery, with additional components being required for translocation. It is these additional components which must then be defective in the patients studied so far.

3.4. Empty peroxisome membrane ghosts divide

It is believed that new peroxisomes form by division of pre-existing peroxisomes [14]. The fibroblasts examined in these studies are the progeny of millions of cell divisions in the developing fetus, followed by additional tens or hundreds of divisions in culture. Therefore, these peroxisome ghosts must have divided more or less normally, and the ghosts must have been segregated more or less normally into the daughter cells, at mitosis. The fact that these ghosts are often larger in diameter than the normal peroxisomes suggests that these import mutations are also subtly affecting division of the organelle. Very little is known thus far about the process by which any organelle divides; perhaps these mutants may shed some light on this process.

3.5. Genetic complementation for peroxisome assembly is formally demonstrated

Fibroblasts from patients with genetic disorders of peroxisome assembly have been sorted into complementation groups based on recovery of peroxisome function. Fibroblast samples were fused pairwise, and the binucleate or multinucleate cells were assayed for the recovery of one or another peroxisomal enzyme activity [8–11]. It was assumed in these studies that the regain of peroxisome function was the result of the resumption of normal peroxisome assembly. This seems a reasonable assumption.

The recent experiments formally demonstrate that this is the case [12]. Recovery of peroxisome assembly in fused fibroblasts was assessed by the regain of catalase packaging into particles, observed by immunofluorescence with anti-catalase. Recovery of normal peroxisome biogenesis was also evaluated by the shift in the size distribution of peroxisomal membranes back to normal, observed by immunofluorescence with antiserum against peroxisomal membrane proteins. Recovery of catalase packaging and return to the normal membrane size distribution occurred approximately in parallel. One other instance of the recovery of catalase packaging has been described [25]. The assignment of fibroblasts into complementation groups by the recovery of peroxisome biogenesis agreed perfectly with the assignments into complementation groups based on recovery of peroxisome function. This demonstrates the correctness of the previous assumptions.

3.6. Preliminary partial correlation of genotype and phenotype

At the present time it has been possible to analyze only a single member of most of the complementation groups. Based on these data, it would appear that three of the mutations (groups 2, 3 and 5) give one immunofluorescence pattern (few large ghosts). Two other mutations (groups 1 and 4) give a second pattern (many large ghosts). As more patients become available in these groups, it will be interesting to see whether these apparent correlations persist.

One patient from group 1 also gives a third immunofluorescence pattern in which a few cells have packaged some catalase into a few peroxisomes. Group 1 is the largest complementation group by far, and contains many patients with milder peroxisomal disorders. As will be discussed elsewhere, fibroblasts from these patients show a variety of immunofluorescence phenotypes, with variable amounts of catalase packaging. Perhaps these represent leaky mutations, with variable amounts of residual function in the mutated protein.

3.7. Future directions

It will be of great interest to attempt to identify the defective genes in these patients. Analysis of the gene products may be expected to shed light on the basic mechanisms of peroxisome assembly. Parallel studies are in progress on peroxisome assembly mutants in CHO cells [26,27] and in yeast [28]. These studies have already resulted in the identification of two membrane proteins required for peroxisome assembly: a 35 kDa protein in CHO cells [24] and a 48 kDa protein in yeast [29].

Acknowledgments

We thank Ms. Heather D'Addario for expert assistance. This research was supported by NIH grant DK19384, NSF grant INT 90-02001 and NIH grant RR-00722.

References

1. Goldfischer, S., Moore, C.L., Johnson, A.B., Spiro, A.J., Valsamis, M.P., Wisniewski, H.K., Ritch, R.H., Norton, W.T., Raspin, I. and Gartner, L.M. (1973) Science 182, 62–64.
2. Schutgens, R.B.H., Heymans, H.S.A., Wanders, R.J.A., Bosch, H.V.D. and Tager, J.M. (1986) Eur. J. Pediatr. 144, 430–440.
3. Lazarow, P.B. and Moser, H.W. (1989) in: C.R. Scriver, A.L. Beaudet, W.S. Sly and D. Valle (Eds.), The Metabolic Basis of Inherited Diseases, 6th edition, McGraw-Hill, New York, pp. 1479–1509.
4. Santos, M.J., Imanaka, T., Shio, H., Small, G.M. and Lazarow, P.B. (1988) Science 239, 1536–1538.
5. Santos, M.J., Imanaka, T., Shio, H. and Lazarow, P.B. (1988) J. Biol. Chem. 263, 10502–10509.
6. Weimer, E., Brul, S., Just, W.W., van Driel, R., Brower-Kelder, E., van Derberg, M., Weijers, P.J., Schutgens R.B.H., van den Bosch, H., Schram, A., Wanders, R.J.A. and Tager, J.M. (1989) Eur. J.

Cell Biol. 50, 407–417.
7. Gartner, J., Chen, W.W., Kelley, R.I., Mihalik, S.J. and Moser, H.W. (1991) Pediatr. Res. 29, 141–146.
8. Brul, S., Westerfield, A., Stijland, A., Wanders, A.J.M., Schram, A.W., Heymans, H.S.A., Schutgens, R.B.H., van der Bosch, H. and Tager, J.M. (1988) J. Clin. Invest. 81, 1702–1710.
9. Roscher, A.A., Hoefler, S., Hoefler, G., Paschke, E., Paultauf, F., Moser, A.B. and Moser, H.B. (1989) Pediatr. Res. 26, 67–72.
10. Poll-The, B.T., Skejdal, O.H., Stokke, O., Poulos, A., Demaugre, F. and Saudabray, J.M. (1989) Hum. Genet. 81, 175–181.
11. McGuinness, M.C., Moser, A.B., Moser, H.W. and Watkins, P.B. (1990) Biochem. Biophys. Res. Commun. 172, 364–368.
12. Santos, M.J., Hoefler, S., Moser, A.B., Moser, H.W. and Lazarow, P.B. (1992) J. Cell Physiol. 151, 103–112.
13. Schram, A.W., Strijland, A., Hashimoto, T., Wanders, R.J.A., Schutgens, R.B.H., van den Bosch, H. and Tager, J.M. (1986) Proc. Natl. Acad. Sci. 83, 6156–6158
14. Lazarow, P.B. and Fujiki, Y. (1985) Annu. Rev. Cell Biol. 1, 489–530.
15. Balfe, A., Hoefler, G., Chen, W.W. and Watkins, P. (1990) Pediatr. Res. 27, 304–310.
16. Gould, S.J., Keller, G.A., Hosken, J., Wilkinson, J. and Subramani, S. (1989) J. Cell Biol. 108, 1657–1664.
17. Small, G.M., Szabo, L.S. and Lazarow, P.B. (1988) EMBO J. 7, 1167–1173.
18. Osumi, T., Tsukamoto, T., Hata, S., Yokota, S., Miura, S., Fujika, Y., Hijikata, M., Miyazawa, S. and Hashimoto, T. (1991) Biochem. Biophys. Res. Commun. 181, 947–954.
19. Swinkels, B.W., Gould, S.J., Bodnar, A.G., Rachubinski, R.A. and Subramani, S. (1991) EMBO J. 10, 3255–3262.
20. Imanaka, T., Small, G.M. and Lazarow, P.B. (1987) J. Cell Biol. 105, 2915–2922.
21. Lazarow, P.B. (1989) Curr. Opin. Cell Biol. 1, 630–634.
22. Gould, S.J., Krisans, S., Keller, G.A. and Subramani, S. (1990) J. Cell Biol. 110, 27–34.
23. Kamijo, K., Taketani, S., Yokota, S., Ozumi, T. and Hashimoto, T. (1990) J. Biol. Chem. 265, 4534–4540.
24. Tsukamoto, T., Miura, S. and Fujiki, Y. (1991) Nature 350, 77–81.
25. Brul, S., Wiemer, E.A.C., Oosthuizen, M., Westerveld, A., van der Driel, R., Strijland, A., Schutgens, R.B.H., van der Bosch, H., Wanders, R.J.A. and Tager, J.M. (1989) in: A. Azzi, Z. Drakota, and S. Papa (Eds.), Molecular Basis of Membrane Associated Disease, Springer-Verlag, Heidelberg, pp. 420–428.
26. Zoeller, R.A., Allen, L.A.H., Santos, M.J., Lazarow, P.B., Hashimoto, T., Tartakoff, A.M. and Raetz, C.R.H. (1989) J. Biol. Chem. 264, 21872–21878.
27. Tsukamoto, T., Yokota, S. and Fujiki, Y. (1990) J. Cell Biol. 110 651–660.
28. Erdmann, R., Wiebel, F.F., Flessau, A., Rytka, J., Beyer, A., Frohlich, K.-U. and Kunau, W.-H. (1991) Cell 64, 499–510.
29. Hohfeld, J., Veenhuis, M. and Kunau, W.-H. (1991) J. Cell Biol. 114, 1167–1178.

Part E

Mitochondria

CHAPTER 19

The mitochondrial protein import machinery of *Saccharomyces cerevisiae*

VICTORIA HINES

Chiron Corporation, 4560 Horton Street, Emeryville, California, USA

1. Introduction

Mitochondria contain an outer and an inner membrane which enclose two aqueous compartments, the intermembrane space and the matrix. Each of these four compartments has a unique polypeptide composition and the vast majority of these proteins are encoded by nuclear genes and synthesized on cytoplasmic ribosomes. Most mitochondrial proteins are synthesized as larger precursors with an amino-terminal presequence. While these sequences lack a specific consensus sequence, they usually contain positively charged, hydrophobic and hydroxyl amino acids. Precursor proteins bind specifically to the mitochondria by interacting with receptors on the mitochondrial outer membrane. Translocation across the inner mitochondrial membrane requires both energy in the form of ATP and an electrochemical potential. After translocation, the presequence is removed by specific proteases located in the matrix and on the outer face of the inner membrane. During the import process, proteins must be correctly sorted and targeted to one of four submitochondrial compartments. The information for sorting to the matrix and the intermembrane space is usually contained within the amino-terminal presequence. The sorting signals for the outer membrane and the inner membrane are still poorly defined.

Recent progress in this field has led to the identification of many of the components involved in this process as well as of the mechanism by which the import machinery facilitates this complex process. Many of these advances have come from studies of mitochondrial biogenesis in the yeast *Saccharomyces cerevisiae* and the fungus *Neurospora crassa*. While both systems are amenable to biochemical characterization in vitro, the ease with which *S. cerevisiae* can be genetically manipulated greatly facilitates the extension of in vitro findings to studies with intact cells. This is particularly important because the rate-limiting steps in protein import into isolated mitochondria may not accurately reflect in vivo import kinetics. Such concerns may

TABLE I

Components of the yeast mitochondrial import system

Component	Gene	Essential?	References
Import receptors			
Mas70p	*MAS70*	No	[14]
p32	*MIR1*	No	[15,16]
Protein translocation channel			
ISP42	*ISP42*	Yes	{21,24]
Refolding and processing proteins			
hsp60	*MIF4*	Yes	[31]
mhsp70	*SSC1*	Yes	[25,26]
Mas1p	*MAS1*	Yes	[35]
Mas2p	*MAS2*	Yes	[36]
Inner membrane protease I	*IMP1*	No	[37,38]

be addressed in *S. cerevisiae* by deleting the gene corresponding to the protein under study and analyzing the resulting null mutant. Since contributions from *N. crassa* studies are reviewed in Chapter 21, this chapter concentrates on recent studies with *S. cerevisiae* which have identified components of the mitochondrial import machinery. The different cytosolic factors implicated in protein import, including presequence binding factors [1] and cytosolic chaperones are not considered here (but see [2]). Table I summarizes the currently identified yeast components. Figure 1 depicts the interaction of a matrix-targeted precursor with various components during the import process which are discussed below.

2. Components of the import machinery

2.1. Import receptors

Outer membrane proteins involved in the early steps of precursor binding to the mitochondrial outer membrane were postulated several years ago based upon two findings: (1) pretreatment of mitochondria with proteases resulted in inhibition of import [3–6]; (2) addition of antibodies against mitochondrial outer membrane proteins resulted in inhibition of import [7]. However, neither of these treatments caused a complete block of protein import for all precursors, suggesting that several functionally overlapping import receptors might exist. To date, two such receptors have been identified in yeast.

One of these receptors is an abundant protein of the yeast mitochondrial outer membrane, termed Mas70p, with a molecular weight of 70 000. The protein is

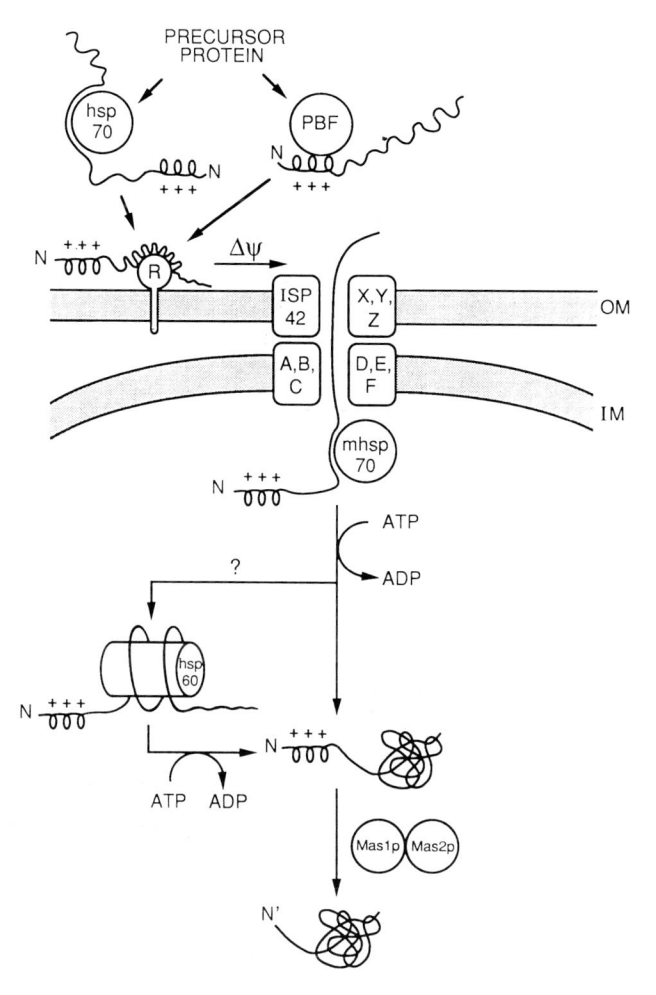

Fig. 1. Transport of a mitochondrial precursor protein with a cleavable amino-terminal, matrix targeting signal into the matrix. See text for details. The relationship between mitochondrial hsp70 and hsp60 is not yet clear, nor is it known precisely when the targeting sequence is removed by the matrix protease. PBF, presequence binding factor. R, one of several import receptors such as MOM19, Mas70p or p32 of the mitochondrial outer membrane. X, Y, Z and A–F, hypothetical additional subunits of the import channels across the outer and inner membrane, respectively. Mas1p and Mas2p, the two subunits of the matrix processing protease. Reproduced, with permission, from the Annual Review of Genetics, Vol. 25, 1991 by Annual Reviews, Inc.

anchored in the outer membrane by an N-terminal hydrophobic domain which also serves to target the protein to the mitochondria [8,9]. The remaining 60 kDa domain faces the cytoplasm. This cytoplasmic domain contains several 34-residue repeats with partial sequence identity to a protein family whose members are involved in cell cycle control and cytoskeletal functions [10,11]. A homologous protein, termed MOM72, has also been identified in *N. crassa* [12].

Deletion of the *MAS70* gene from yeast is not lethal, nor is mitochondrial import blocked in vivo [13]. However, pulse-chase studies with intact cells from a *mas70* null mutant demonstrated that the kinetics of import for all precursors tested was substantially reduced compared to wild-type cells [14]. Thus, while Mas70p function is not essential for protein import, it appears to accelerate the import of all precursors tested so far.

Initial in vitro import studies implicated Mas70p in the import of several, but not all, mitochondrial precursors [14]. Pretreatment of mitochondria with antibodies against Mas70p inhibited the import of the F_1-ATPase β-subunit or the ADP/ATP translocator up to 75%. Additional studies with the ADP/ATP translocator indicated that Mas70p facilitated an early binding step.

Recent in vitro studies help explain why the general import function observed for Mas70p in vivo is not apparent in all in vitro import experiments. Import of the precursor to alcohol dehydrogenase isozyme III from a rabbit reticulocyte lysate shows no dependence upon the presence of Mas70p. However, when this precursor is presented to yeast mitochondria after denaturation by urea, removal of Mas70p from the mitochondrial surface dramatically inhibits the import rate (V. Hines, unpublished results). Under these conditions, import into isolated mitochondria is very rapid and may more accurately reflect in vivo import conditions than the usual in vitro import conditions. In fact, pulse-chase experiments show that in vivo import is very rapid. It therefore appears that Mas70p facilitates the import of several, if not most, authentic precursors, but with isolated mitochondria, this interaction is only apparent when the step facilitated by Mas70p becomes rate-limiting for import.

Since protein import still occurs in the absence of Mas70p, additional import receptors must be present on the mitochondrial surface. Using an anti-idiotypic antibody to a mitochondrial signal peptide, Pain et al. [15] identified an integral membrane protein of 32 kDa (p32) which appears to function as a signal sequence receptor. A stable complex between p32 and mitochondrial precursors could be isolated and antibodies against p32 inhibited the import of certain precursors in vitro. This protein is not essential, since disruption of the gene encoding p32 (*MIR1*) produced viable haploid cells [16]. However, the steady state levels of several mitochondrial proteins were markedly reduced in a *mir1* null mutant. The deduced amino acid sequence of the *MIR1* gene product is identical to the sequence reported for the mitochondrial phosphate translocator, an inner membrane protein [17]. The reason for the apparent dual location and function of this protein is not yet clear.

A third import receptor, termed MOM19, has been identified in *N. crassa* [18]. This protein facilitates the binding and import of many precursors in in vitro import studies including the import of the MOM72 receptor [12]. Pretreatment of mitochondria with antibodies against MOM19 strongly blocks import of several precursors, but not that of the ADP/ATP translocator. The residual import activity observed under these conditions indicates the presence of other functional import receptors. Thus, while all three receptor proteins facilitate binding of various precursors, none appears to be essential for mitochondrial import, indicating that functionally overlapping receptor proteins facilitate this first stage of protein import.

2.2. The protein translocation channel

The second stage of protein import involves translocation of precursor proteins across the outer and inner mitochondrial membranes. Import can be arrested at this stage by lower temperatures [19] or other tools which block complete translocation of precursors [20]. Such a translocation-arrested precursor spans both membranes: the amino-terminus is exposed to the mitochondrial matrix and the carboxy-terminus remains outside the mitochondrion. This intermediate was used in conjunction with a photochemical crosslinking reagent to identify a 42 kDa import site protein (ISP42) [21]. This outer membrane protein must be in close contact with the precursor and probably forms part of a hetero-oligomeric translocation channel through which proteins pass. Studies with the *N. crassa* homologue, MOM38, have identified several other proteins which co-purify as a complex with MOM38 and might be additional subunits of this channel [22]. The size of this protein channel has been examined by chemically linking molecules of various sizes to a mitochondrial precursor protein. Even a double-stranded DNA molecule can be imported into mitochondria if it is linked to a precursor protein [23]. Since the diameter of a double-stranded DNA helix is about 20 Å, these translocation channels can be rather large and flexible.

The central role of ISP42 in protein import has been demonstrated in vitro and in vivo. In vitro, antibodies to ISP42 block import of several precursor proteins [21]. When expression of ISP42 is blocked in vivo, the cells accumulate mitochondrial precursor proteins outside the mitochondrial inner membrane and cease growing [24]. Disruption of the ISP42 gene is lethal. This represents the first mitochondrial membrane protein known to be essential for life.

2.3. Refolding and processing proteins

The third stage of protein import involves folding, assembly and processing of newly translocated proteins in the matrix of the mitochondrion. Partially translocated proteins interact initially with a mitochondrial member of the 70 kDa heat-shock protein family (mhsp70) [25,26]. Precursor proteins which still contain the cleavable matrix targeting presequence associate with mhsp70, as demonstrated by chemical cross-linking studies and co-immune precipitations. This binding could represent the earliest interaction of the imported precursor with a matrix protein. Hydrolysis of ATP by mhsp70 results in the release of the protein into the mitochondrial matrix and may provide the driving-force for translocation of precursors across the inner membrane [27]. Release from mitochondrial hsp70 may also trigger the refolding of the mature protein in the matrix, although release is much more rapid than refolding. Other proteins that are assembled into larger complexes, interact transiently with mhsp70 and then bind to a second chaperonin termed mitochondrial hsp60 [27,28]. This protein resembles the well-characterized groEL of bacteria and is composed of 14 identical 60 kDa subunits [29]. ATP-dependent release of precursors from mitochondrial hsp60 results in correct folding and oligomer assembly [28].

Both of these mitochondrial chaperonins are essential not only for mitochondrial protein import, but also for viability of yeast cells. Deletion of the gene coding for mhsp70 (*SSC1*) is lethal [30] and a yeast mutant containing a temperature sensitive *ssc1* gene accumulates precursors which are stuck across both mitochondrial membranes at the non-permissive temperature [26]. The gene for mitochondrial hsp60 is also essential for yeast viability; yeast mutants containing a temperature-sensitive allele of mitochondrial hsp60 die at the non-permissive temperature [31].

Removal of the amino-terminal targeting signal is catalyzed by a matrix-localized metalloprotease. The precise timing of the processing step is not clear, but the processing can occur while the precursor is still bound to a chaperonin [27]. The matrix protease is a soluble enzyme in the yeast matrix, consisting of two non-identical subunits [32]. The genes encoding the two subunits of the protease (*MAS1* and *MAS2*) were identified during a screen of a bank of yeast mutants temperature sensitive for growth and mitochondrial import [33–36]. The amino acid sequences of the two subunits have extensive homology over their entire length and both are synthesized with a typical matrix-targeting presequence. Disruption of either of the two *MAS* genes is lethal; inactivation of either subunit results in an in vivo block of protein translocation across the inner membrane and cell inviability [33]. Additional processing of the precursor protein by additional proteases occurs for a small number of matrix and intermembrane space proteins. One of these additional proteases has been localized to the outer surface of the inner mitochondrial membrane and has been partially purified [37,38].

3. The mechanism of protein import

3.1. Energy requirements

Protein translocation across the inner membrane requires a membrane potential and hydrolysis of nucleoside triphosphates [39]. Dissipation of the mitochondrial membrane potential in vivo causes accumulation of precursor proteins in the yeast cytoplasm [40]. Dissipation of the membrane potential of isolated mitochondria blocks import, but still allows binding of precursor proteins to the mitochondrial surface. The membrane potential appears to facilitate an early interaction of the matrix-targeting sequence with the inner membrane [19,41]. Since most matrix-targeting sequences form amphiphilic α-helices [42], the potential could serve to electrophorese the targeting sequence across the inner membrane [43,44].

The role of NTP hydrolysis in protein import has been extensively studied. In particular, it has been investigated whether NTP is required outside the inner membrane, inside the matrix, or both. External ATP is required to maintain an import competent conformation of precursor proteins. If a precursor is stabilized in a tightly folded conformation, protein import is blocked [45,46]. Conversely, the in vitro import rate is significantly increased when the precursor conformation is destabilized by either urea [41] or mutagenesis [47]. Maintenance of an import competent conformation may be facilitated by cytosolic chaperone proteins. These proteins hydrolyze ATP and prevent tight folding of precursor proteins in the cytoplasm [48]. Members of the hsp70 class of heat shock proteins have been implicated in the import of some yeast proteins both in vivo [49] and in vitro [50]. In addition, external ATP can stimulate binding of some precursors to the mitochondrial surface [51].

However, external ATP is not essential for membrane translocation. In the absence of ATP, both artificial precursor proteins and urea-denatured authentic precursor proteins are able to translocate the mitochondrial outer membrane. The translocation intermediate generated in the absence of ATP is located between the two mitochondrial membranes, with the presequence integrated into the inner membrane. Addition of ATP within the matrix is sufficient to chase this intermediate into the mitochondrial matrix [52].

The requirement for internal ATP also seems to be associated with chaperone proteins. Hwang and Schatz [53] demonstrated that ATP was required in the mitochondrial matrix for complete translocation across the inner membrane. Precursor association with the mitochondrial hsp70 protein may be the first interaction of precursors within the matrix and dissociation of this complex requires NTP hydrolysis [27]. Those proteins that subsequently associate with mitochondrial hsp60 might require additional NTP hydrolysis prior to complete folding and assembly.

3.2. Contact sites

The precise location of protein import sites is of considerable interest. Morphological examination of mitochondria consistently show regions of close contact between the outer and inner membrane [54]. When yeast cells are treated with cycloheximide, cytoplasmic polysomes are found attached to isolated mitochondria at these contact sites [55], indicating that these sites are the region at which precursor proteins are imported into mitochondria. If these contact sites do contain the protein import sites, they should be enriched in the protein components of the import machinery. In fact, both Mas70p/MOM72 [12,14] and the p32 import receptor [15] are found to be enriched at contact sites. In addition, mitochondria can be subfractionated into three vesicle populations, one of which is enriched in contact sites [56]. These vesicles contain both outer and inner membrane proteins. When precursor import is arrested at a stage wherein the precursor spans both inner and outer membranes, the resulting translocation intermediate co-fractionates with the contact site vesicles.

While this evidence suggests that protein translocation may occur at fixed contact sites, several lines of evidence indicate that the import channels of the two membranes appear to be dynamic structures, which can diffuse into, and out of, membrane contact sites [57].

The first indication that contact sites were not the only site of protein import came from studies of osmotically shocked mitochondria [7]. If the mitochondrial outer membrane is rendered import incompetent by trypsin treatment, exposure of the mitochondrial inner membrane by osmotic shock restores efficient protein import. This inner membrane import has the same energy and presequence requirements as import into whole mitochondria. Isolated inner membrane vesicles can also efficiently import mitochondrial precursors [58].

A recently described translocation intermediate also favors a dynamic model for contact sites [52]. In the absence of ATP and the presence of a membrane potential, several authentic and artificial precursor proteins become arrested between the two membranes, with their amino-terminus stuck across the inner membrane. Addition of ATP chases these intermediates into the mitochondrial matrix. Therefore, under ATP-depletion conditions, the outer membrane becomes disengaged from the inner membrane and translocation across the two membranes becomes uncoupled.

Finally, several outer membrane proteins required for protein import are not found enriched in contact sites. This includes the *N. crassa* import receptor MOM19 [18] and ISP42 [24].

3.3. Protein sorting

The protein sequences necessary for targeting to the correct mitochondrial compartment have been under study for several years. Although there is no strong sequence homology for any targeting sequence, some general structural patterns have emerged. Proteins destined for different compartments contain different targeting information and may also use different import mechanisms. In addition, several proteins appear to employ unique sorting mechanisms.

Protein import to the mitochondrial outer membrane diverges in many respects from import into other mitochondrial compartments. Assembly into the outer membrane requires ATP outside the mitochondria but no electrochemical potential [51,53]. Experimental evidence suggests that at least some outer membrane proteins utilize the import receptors used by other proteins (MOM19) [18], but at some point, the import process must branch off from the import pathway which directs proteins to the matrix space. The outer membrane proteins studied to date do not contain a cleavable presequence. The targeting information for the import receptor Mas70p has been localized within the first 29 residues, which can be functionally divided into two domains [9,59]. The first 12 amino acids, when fused to a passenger protein, are sufficient for mitochondrial targeting, but are not sufficient for outer membrane anchoring [60]. Such fusion proteins are targeted to the mitochondrial matrix. Residues 10–19 are uncharged and hydrophobic and function in membrane anchoring. Shortening this domain results in redirecting fusion proteins to the matrix [59]. These results suggest that Mas70p is targeted to the mitochondria by an amino-terminal matrix-targeting signal and the hydrophobic domains function as a stop-transfer signal. Other outer membrane proteins, like IPS42, lack identifiable targeting signals. ISP42 contains neither typical matrix-targeting signals nor transmembrane stretches [24].

The sorting signals for many intermembrane space proteins are also composed of two functional domains: an amino-terminal matrix-targeting signal followed by a hydrophobic stretch which is required for intermembrane space sorting. Unlike outer membrane signals, these sequences are cleaved off during import. A few intermembrane space proteins, such as cytochrome c, do not contain a cleavable presequence and are imported to the intermembrane space in mechanisms which appear to be unique for the individual proteins [61].

Two models have been proposed for how the bipartite signal sequence might function in sorting. The stop-transfer model suggests that the amino-terminal domain is completely translocated to the matrix, but when the hydrophobic domain interacts with the inner membrane, further precursor translocation across the inner membrane is blocked [2,62]. The matrix protease cleaves the matrix-targeting signal and a second intermembrane space protease cleaves the protein on the outer face of the inner membrane, releasing the mature protein into the intermembrane space. Recent evidence by Glick et al. [62a] provides further support for the stop-transfer model. The conservative sorting model proposes that the precursor is initially completely translocated into the matrix [63–65]. Processing by the matrix protease exposes a re-export signal (represented by the hydrophobic domain) which directs export of the protein from the matrix in a process analogous to bacterial secretion.

Two pathways for inner membrane sorting have been described. In one case, exemplified by the ADP/ATP carrier, the targeting information is found within internal domains of the protein rather than in a cleavable presequence [66]. While complete import requires ATP outside the mitochondria and an electrochemical potential, the protein does not pass through the matrix, but rather integrates into the inner membrane directly from the translocation channel [51]. Other proteins, such as the subunit IV of cytochrome oxidase, are sorted to the matrix, processed and then assembled into the inner membrane [67].

Finally, matrix-targeting signals are generally amino-terminal extensions which can fold into an amphipathic helix and are usually removed by the matrix protease [42,44,68]. The signal can be as short as 7–10 residues and still retain matrix-targeting function [69,70]. Additional redundant targeting information has also been found in the mature domains of some matrix proteins [71].

4. Outlook

The combination of genetic and biochemical studies of mitochondrial import has resulted in the identification of several components involved in protein import in yeast. Several steps of the import process can now be explained by the function of individual protein components. The coming years should see the identification of additional import receptors and other outer and inner membrane proteins which together form the two translocation channels. It will be interesting to see which of these proteins is essential for protein import. The answer to several mechanistic questions should also be forthcoming. Why is ATP required both externally and internally? How dynamic are import sites? What is the mechanism for protein sorting to the outer membrane?

References

1. Murakami, K. and Mori, M. (1990) EMBO J. 9, 3201–3208.
2. Glick, B. and Schatz, G. (1991) Annu. Rev. Genet. 25, 21–44.
3. Riezman, H., Hay, R., Witte, C., Nelson, N. and Schatz, G. (1983) EMBO J. 2, 1113–1118.
4. Zwizinski, C., Schleyer, M. and Neupert, W. (1983) J. Biol. Chem. 258, 4071–4074.
5. Zwizinski, C., Schleyer, M. and Neupert, W. (1984) J. Biol. Chem. 259, 7850–7856.
6. Ohba, M. and Schatz, G. (1987) EMBO J. 6, 2109–2115.
7. Ohba, M. and Schatz, G. (1987) EMBO J. 6, 2117–2122.
8. Hase, T., Riezman, H., Suda, K. and Schatz, G. (1983) EMBO J. 2, 2169–2172.
9. Hase, T., Müller, U., Riezman, H. and Schatz, G. (1984) EMBO J. 3, 3157–3164.
10. Boguski, M.S., Sikorski, R.S., Hieter, P. and Goebl, M. (1990) Nature (London) 346, 114.
11. Sikorski, R.S., Boguski, M.S., Goebl, M. and Hieter, P. (1990) Cell 60, 307–317.
12. Söllner, T., Pfaller, R., Griffiths, G., Pfanner, N. and Neupert, W. (1990) Cell 62, 107–115.
13. Riezman, H., Hase, T., van Loon, A.P.G.M., Grivell, L.A., Suda, K. and Schatz, G. (1983) EMBO J. 2, 2161–2168.
14. Hines, V., Brandt, A., Griffiths, G., Horstmann, H., Brütsch, H. and Schatz, G. (1990) EMBO J. 9, 3191–3200.
15. Pain, D., Murakami, H. and Blobel, G. (1990) Nature (London) 347, 444–449.
16. Murakami, H., Blobel, G. and Pain, D. (1990) Nature (London) 347, 488–491.
17. Meyer, D.I. (1990) Nature (London) 347, 424–425.
18. Söllner, T., Griffiths, G., Pfaller, R., Pfanner, N. and Neupert, W. (1989) Cell 59, 1061–1070.
19. Schleyer, M. and Neupert, W. (1985) Cell 43, 339–350.
20. Vestweber, D. and Schatz, G. (1988) J. Cell Biol. 107, 2037–2043.
21. Vestweber, D., Brunner, J., Baker, A. and Schatz, G. (1989) Nature (London) 341, 205–209.
22. Kiebler, M., Pfaller, R., Söllner, T., Griffiths, G., Horstmann, H., Pfanner, N. and Neupert, W. (1990) Nature (London) 348, 610–616.
23. Vestweber, D. and Schatz, G. (1989) Nature (London) 338, 170–172.
24. Baker, K.P., Schaniel, A., Vestweber, D., and Schatz, G. (1990) Nature (London) 348, 605–609.
25. Scherer, P.E., Krieg, U.C., Hwang, S.T., Vestweber, D. and Schatz, G. (1990) EMBO J. 9, 4315–4322.
26. Kang, P.-J., Ostermann, J., Shilling, J., Neupert, W., Craig, E.A. and Pfanner, N. (1990) Nature (London) 348, 137–143.
27. Manning-Krieg, U.C., Scherer, P.E. and Schatz, G. (1991) EMBO J. 10, 3273–3280.

28. Ostermann, J., Horwich, A.L., Neupert, W. and Hartl, F.-U. (1989) Nature (London) 341, 125–130.
29. McMullin, T.W. and Hallberg, R.L. (1988) Mol. Cell. Biol. 8, 371–380.
30. Craig, A.E., Kramer, J. and Kosic-Smithers, J. (1987) Proc. Natl. Acad. Sci. USA 84, 4156–4160.
31. Cheng, M.Y., Hartl, F.-U., Martin, J., Pollock, R.A., Kalousek, F., Neupert, W., Hallberg, E.M., Hallberg, R.L. and Horwich, A.L. (1989) Nature (London) 337, 620–625.
32. Yang, M., Jensen, R.E., Yaffe, M.P. and Schatz, G. (1988) EMBO J. 7, 3857–3862.
33. Yaffe, M.P. and Schatz, G. (1984) Proc. Natl. Acad. Sci. USA 81, 4819–4823.
34. Yaffe, M.P., Ohta, S. and Schatz, G. (1985) EMBO J. 4, 2069–2074.
35. Witte, C., Jensen, R.E., Yaffe, M.P. and Schatz, G. (1988) EMBO J. 7, 1439–1447.
36. Jensen, R.E. and Yaffe, M.P. (1988) EMBO J. 7, 3863–3871.
37. Schneider, A., Behrens, M., Scherer, P., Pratje, E., Michaelis, G. and Schatz, G. (1991) EMBO J. 10, 247–254.
38. Behrens, M., Michaelis, G. and Pratje, E. (1991) Mol. Gen. Genet. 228, 167–176.
39. Eilers, M. and Schatz, G. (1988) Cell 52, 481–483.
40. Reid, G. A. and Schatz, G. (1982) J. Biol. Chem. 257, 13056–13061.
41. Eilers, M., Hwang, S. and Schatz, G. (1988) EMBO J. 7, 1139–1145.
42. von Heijne, G. (1986) EMBO J. 5, 1335–1342.
43. Pfanner, N. and Neupert, W. (1985) EMBO J. 4, 2819–2825
44. Roise, D. and Schatz, G. (1988) J. Biol. Chem. 263, 4509–4511.
45. Eilers, M. and Schatz, G., (1986) Nature (London) 322, 228–232.
46. Chen, W.-J. and Douglas, M.G. (1987) J. Biol. Chem. 262, 15605–15609.
47. Vestweber, D. and Schatz, G. (1988b) EMBO J. 7, 1147–1151.
48. Ellis, J. (1987) Nature (London) 328, 378–379.
49. Deshaies, R.J., Koch, B.D., Werner-Washburne, M. and Craig, E.A. (1988) Nature (London) 332, 800–805.
50. Murakami, H., Pain, D. and Blobel, G. (1988) J. Cell Biol. 107, 2051–2057.
51. Pfanner, N., Tropschug, M. and Neupert, W. (1987) Cell 49, 815–823.
52. Hwang, S.T., Wachter, C. and Schatz, G. (1991) J. Biol. Chem. 266, 21083–21089.
53. Hwang S.T. and Schatz, G. (1989) Proc. Natl. Acad. Sci. USA 86, 8432–8436.
54. Hackenbrock, C.R. (1968) Proc. Natl. Acad. Sci. USA 61, 598–605.
55. Kellems, R.E., Allison, V. F. and Butow, R.A., (1975) J. Cell Biol. 65, 1–14.
56. Pon, L., Moll, T., Vestweber, D., Marshallsay, B. and Schatz, G. (1989) J. Cell Biol. 109, 2603–2616.
57. Glick, B., Wachter, C. and Schatz, G. (1991) Trends Cell Biol. 1, 99–103
58. Hwang, S.T., Jascur, T., Vestweber, D., Pon, L. and Schatz, G. (1989) J. Cell Biol. 109, 487–493.
59. Nakai, M., Hase, T. and Matsubara, T. (1989) J. Biochem. 105, 513–519.
60. Hurt, E.C., Müller, U. and Schatz, G. (1985) EMBO J. 4, 3509–3518.
61. Stuart, R. and Neupert, W. (1990) Biochimie 72, 115–121.
62. van Loon, A.P., Brändli, A.W., Pesold-Hurt, B., Blank, D. and Schatz, G. (1987) EMBO J. 6, 2433–2439.
62a. Glick, B.S., Brandt, A., Cunningham, K., Müller, S., Hallberg, R.L. and Schatz, G. (1992) Cell 69, 809–822.
63. Hartl, F.U., Schmidt, B., Wachter, E., Weiss, H. and Neupert, W. (1986) Cell 47, 939–951.
64. Hartl, F.U., Ostermann, J., Guiard, B. and Neupert, W. (1987) Cell 51, 1027–1037.
65. Hartl, F.U. and Neupert, W. (1990) Science 247, 930–938.
66. Smagula, C. S. and Douglas, M. G. (1988) J. Cell. Biochem. 36, 323–327.
67. Hurt, E.C., Pesold–Hurt, B. and Schatz, G. (1984) EMBO J. 3, 3149–3156.
68. Lemire, B.D., Fankhauer, C., Baker, A. and Schatz, G. (1989) J. Biol. Chem. 264, 20206–20215.
69. Hurt, E.C., Pesold-Hurt B., Suda, K., Oppliger, W. and Schatz, G. (1985) EMBO J. 4, 2061–2069.
70. Verner, K. and Lemire, B.D. (1989) EMBO J. 88, 1491–1496.
71. Bedwell, D.M., Klionsky, D.J. and Emr, S.D. (1987) Mol. Cell. Biol. 7, 4038–4047.

CHAPTER 20

Protein insertion into mitochondrial outer and inner membranes via the stop-transfer sorting pathway*

GORDON C. SHORE, DOUGLAS G. MILLAR and JIAN-MING LI

Department of Biochemistry, McIntyre Medical Sciences Building, McGill University, Montreal, H3G 1Y6, Canada

Abstract

Because mitochondria have two membranes, the problem of protein insertion into these membranes is also a problem of protein sorting. Here, we discuss possible mechanisms for insertion into the mitochondrial outer or inner membrane for those proteins that are imported unidirectionally into the organelle, via a stop-transfer (non-conservative) pathway. For polytopic proteins, sorting presumably pertains only to the first insertion event since this predetermines membrane selection for subsequent insertion of the rest of the molecule. Models are proposed in which selection of the outer membrane is determined by an OMM signal-anchor sequence, in which the translocation and membrane anchor signals either overlap or are in close juxtaposition, and selection of the inner membrane is by a combination of a matrix-targeting signal and a distal stop-transfer sequence. We have employed manipulations that allow a protein containing an OMM signal-anchor sequence to by-pass the outer membrane and efficiently insert into the inner membrane, raising the possibility that potential regulation of the outer membrane protein translocation machinery (e.g., if applied by cytosolic factors or receptors) might influence sorting. Our discussion focuses on proteins that are anchored in the membrane by hydrophobic helical segments. Integral proteins that lack such structures (e.g., porin and ISP42/MOM38) may intercalate into the bilayer by a different mechanism.

---

*This study was financed by operating grants from the Medical Research Council and National Cancer Institute of Canada.

1. Introduction

In comparison to other membrane systems, in particular those of the ER of mammalian cells and the cytoplasmic membrane of *Escherichia coli*, an understanding of how proteins are inserted into mitochondrial membranes, and how they assume their final topology, is far less advanced. In the former systems, targeting and insertion are generally accomplished by a potential combination of four topogenic domains: signal sequences, signal-anchor sequences, stop-transfer sequences, and insertion domains [1,2]. The orientation of individual transmembrane helical segments in the majority of these proteins is determined by the first insertion event and, in the case where this is a signal-anchor sequence or an insertion domain, is determined in turn by the charged nature of the sequences flanking the hydrophobic core [3,4].

The extent to which similar topogenic information extends to mitochondria remains to be determined. Moreover, the situation for mitochondria is complicated by a number of issues. First, mitochondria contain two translocation-competent membranes. The problem of protein insertion into these membranes, therefore, is also a problem of protein sorting. Second, the structure of many mitochondrial membrane proteins is complex and, to date, remains poorly defined. In addition, some of the best studied membrane proteins in terms of biogenesis exhibit such complex structures, and this may influence their mechanism of insertion. The outer membrane protein, porin, for example, is devoid of hydrophobic stretches [5] and the kind of helical transmembrane segments that are typical of some of the best studied examples of proteins assembled into the ER membrane. Rather, porin is intercalated into the lipid-bilayer via β-strands [6]. For reasons outlined elsewhere [7], insertion of such structures (although mediated by mitochondrial proteins [8,9]) will have different requirements than hydrophobic helical structures. Another example is the carrier protein family of the inner membrane (e.g., ADP/ATP carrier and uncoupling protein), whose biogenesis has been extensively investigated [10]. AAC and UCP contain helical transmembrane segments, but these are amphiphilic in nature and are probably stabilized in the membrane as paired structures [11,12]. We have suggested that each of the three transmembrane pairs of amphiphilic helices in these proteins is inserted into the membrane as a hairpin unit, led by a matrix-targeting signal located in the matrix ectodomain connecting the two helical strands [13] (discussed later). Finally, the problem is complicated by the fact that two pathways appear to have evolved for delivery of proteins to mitochondrial membranes: the conservative sorting pathway [14,15] and the stop-transfer (non-conservative or direct insertion) sorting pathway [14,16–20]. The two are described diagrammatically in Fig. 1.

The conservative sorting pathway describes proteins that are routed first to the matrix, where their N-terminal matrix-targeting signal is usually removed, and then are re-directed for export by a second sorting sequence within the molecule [15]. Although this pathway has been best documented for proteins destined for the

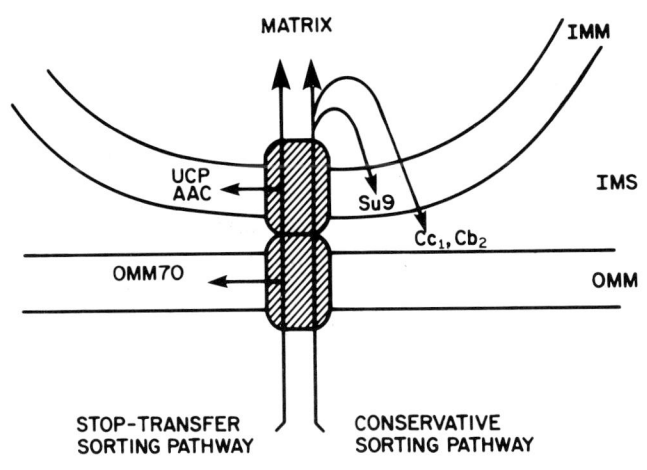

Fig. 1. Two sorting pathways in mitochondria. Import via the stop transfer and conservative sorting pathways is shown to occur at contact sites [57,60,61]. See the text for an explanation. OMM, outer mitochondrial membrane; IMS, intermembrane space; IMM, inner mitochondrial membrane; OMM70 (MAS70), a 70 kDa integral OMM protein in yeast [16] (the *Neurospora* homolog is MOM72 [38]); UCP, uncoupling protein; AAC, ADP/ATP carrier; Su9, subunit 9 of *Neurospora crassa* F_o-ATPase; Cc_1, cytochrome c_1; Cb_2, cytochrome b_2.

intermembrane space [15,21,22], it may extend as well to certain inner membrane proteins. An example of the latter is subunit 9 of *Neurospora crassa* F_o-ATPase, a small, highly hydrophobic protein [14]. A common feature of all proteins identified to date that follow the conservative pathway is their ancestral origins in bacteria. It has been proposed, therefore, that a bacterial export mechanism for delivery of proteins from the matrix to the inner membrane or intermembrane space has been conserved during the endosymbiotic evolution of mitochondria in eukaryotic cells [15].

The stop-transfer sorting pathway describes those proteins that are integrated into either the outer or inner membrane via a direct insertion mechanism [14,16–20] rather than passing first to the matrix and then being redirected for export. Presumably most outer membrane proteins follow this pathway because these proteins do not require $\Delta\psi$ for insertion [23–25] and are not proteolytically processed, i.e., they do not pass into or across the inner membrane. Additionally, two proteins of the inner membrane, UCP and AAC, neither of which have homologs in bacteria, are probably inserted via a stop-transfer mechanism [14,20]. In common with models developed for the ER [1,7,26], the mechanism of insertion of these proteins into the lipid bilayer is likely to be intimately linked to the process of protein translocation that has evolved for the unidirectional import of nuclear-encoded precursors.

Finally, there is evidence that import to the intermembrane space may also occur during unidirectional import, in which the sorting domain arrests translocation at the inner membrane (see [66]).

2. Results and discussion

2.1. Mitochondrial topogenic sequences and stop-transfer sorting

The cleavable N-terminal signal sequence that characterizes most proteins that are destined for the mitochondrial matrix has been well characterized (reviewed in [27]). Typically, they are positively charged amphiphilic structures that form a helix upon interaction with a membrane surface, yielding a hydrophobic moment perpendicular to the helical axis [28,29]. Their function is to target proteins to the mitochondrial protein translocation apparatus and, in most cases, to trigger and commit translocation of the precursor to the matrix [30]. Although such sequences have membrane surface-seeking properties, they are unlikely to significantly penetrate the lipid bilayer [31] and, certainly, they cannot form stable transmembrane segments. On their own, therefore, they lack a critical feature of ER and bacterial signal sequences, the potential to function as a signal-anchor domain. Nevertheless, the functional equivalent of a signal-anchor sequence can apparently be created when a hydrophobic membrane-anchor segment is located in close juxtaposition to a targeting signal.

A naturally occurring example of such a domain is the signal-anchor sequence of the bitopic protein of yeast outer membrane, OMM70 (MAS70), which is targeted to and anchored in the outer membrane via its N-terminal 29 amino acids [32–34], leaving a large (>60 kDa) fragment exposed to the cytosol [32]. In yeast, amino acids 1–12 can function as a matrix-targeting signal, albeit the extent of import to the matrix of fusion proteins containing this sequence was very low [33]. Recently, we fused amino acids 1–29 of OMM70 to dihydrofolate reductase (DHFR) and examined its import into rat heart mitochondria. The hybrid protein, designated pOMD29, was efficiently inserted into the outer membrane in the expected orientation by a process dependent on ATP and protease-sensitive mitochondrial surface components [35]. However, deletion of a large portion of the membrane anchor segment (amino acids 16–29) created a protein that neither entered, nor stably bound to, mitochondria [35,36]. We suggest that the hydrophilic, positively charged N-terminus (amino acids 1–10) and the hydrophobic membrane-anchor segment (amino acids 11–29) cooperate to form the requisite topogenic domain: the OMM signal-anchor sequence. pOMD29, like OMM70 [32], has an N_{in}-C_{cyto} topology [36,37].

2.2. The OMM signal-anchor sequence

We propose that integral proteins of the outer membrane are targeted and inserted via a signal-anchor sequence and that, for polytopic proteins, a signal-anchor must be the first topogenic domain to interact with the translocation machinery of the outer membrane. A priori, the signal-anchor sequence could be located anywhere in the molecule as part of bitopic or polytopic [1] proteins and could be inserted in either the N_{in} or N_{cyto} orientation. Signal-anchor sequences may have overlapping targeting and anchor sequences (as in the case of pOMD29/OMM70) or the two may be in close physical juxtaposition. An example of the latter is the artificial protein, pOCT-G35, which is the precursor to the matrix protein, ornithine carbamyl transferase, containing the 19 amino acid stop-transfer sequence of VSV G located adjacent to the amino-terminal matrix-targeting signal of pOCT [18]. pOCT-G35 was found to insert into the outer membrane of rat heart mitochondria in the N_{cyto}-C_{in} orientation [18], i.e., an orientation opposite to that of pOMD29 [35]. When amino acids 1–11 of pOMD29 (OMM70) were replaced with amino acids 1–36 of pOCT, the protein was inserted into the outer membrane, again in the reverse orientation compared to pOMD29 (i.e., in the N_{cyto}-C_{in} orientation of pOCT-G35 [37]), indicating that orientation can be determined by sequences in front of the membrane-anchor segment [37].

Assuming that vectorial translocation of the signal-anchor sequence by a protein-conducting apparatus is a necessary prerequisite for insertion into the membrane [1,7,26], the signal-anchor sequence presumably contains information to trigger the translocation process, in addition to providing information for targeting and membrane-anchoring. A priori, this could be accomplished by a matrix-targeting domain (as in the cases described above), but it remains to be determined if other types of sequences provide this function in native outer membrane proteins. The positively charged N-terminus of OMM70, for example, is a very weak, if not ineffectual, matrix-targeting domain [33,36] and, interestingly, this region in the *N. crassa* homolog of OMM70 (MOM72), is quite different and does not carry a net positive charge [38]. As discussed elsewhere [7], such mechanisms may not extend to β-barrel proteins like porin [7], and perhaps ISP42/MOM38 [39,40].

2.3. Stop-transfer sorting to the inner membrane

If proteins are targeted to the mitochondrial outer membrane by OMM signal-anchor sequences, what specifies targeting to the inner membrane? We suggest that this is accomplished, initially, by a combination of a matrix-targeting domain and a distal stop-transfer sequence. At least two of the probable three internal targeting sequences of UCP, for example, have been shown to specify translocation to the matrix [13; K.B. Freeman, pers. commun.]; each matrix-targeting domain is proposed to connect a pair of amphiphilic transmembrane helices which, together, could provide the stop-transfer function [13] (discussed later). Another example is the matrix-destined protein, pOCT, which was modified to insert into the inner membrane by engineering the 19 amino acid stop-transfer sequence of VSV G toward the C-terminus of the polypeptide [18]. Conversely, attaching the pOCT matrix-targeting signal to a large C-terminal fragment of VSV G (containing the single G stop-transfer sequence) also resulted in efficient insertion into the inner membrane [17]. Because the VSV G stop-transfer sequence does not correspond to the tripartite structure typical of a bacterial export signal [41,42], and notwithstanding its location in the protein, there is little reason to assume that either of these artificial proteins followed the conservative sorting pathway. Finally, the small, native protein corresponding to subunit V_a of the yeast inner membrane cytochrome c oxidase complex contains an N-terminal matrix-targeting domain and a membrane-anchor sequence located toward the C-terminus of the protein [43]. Deletion of the membrane-anchor sequence resulted in translocation of the protein across both mitochondrial membranes to the soluble matrix compartment [43]. It remains to be determined, however, whether the membrane-anchor domain of V_a functions as a stop-transfer sequence or as a bacterial-like signal-anchor operating on the matrix export pathway [44].

The distinction between OMM signal-anchor sequences and stop-transfer sequences is operational. In the former, the membrane-anchor segment is part of, or closely juxtaposed to, that part of the topogenic sequence that triggers translocation across the outer membrane, i.e., the membrane anchor gains access to the translocation apparatus as the first step of polypeptide translocation. Stop-transfer sequences, on the other hand, rely on a matrix-targeting signal located at a critically minimum distance away, i.e., the membrane-spanning (stop-transfer) segment gains access to the translocation machinery of the outer membrane only after transmembrane translocation has commenced [18,45].

It remains to be explained, however, why the outer membrane translocator can integrate an OMM signal-anchor sequence but allows the stop-transfer sequence of an inner membrane protein to pass. We [18] and others [45,46] have suggested that the outer membrane translocator is regulated in this regard. If translocation occurs at contact sites, physical interaction with the inner membrane translocator might prevent the outer membrane translocator from allowing a hydrophobic segment to be released laterally into the bilayer [18,45]. One possibility is that such prevention is itself triggered when a matrix-targeting domain engages the inner membrane translocator. Perhaps it could be regulated by the type of cytosolic factor or import receptor with which a particular precursor protein interacts. Alternatively, the inability of the OMM translocator to release transmembrane segments (including signal-anchor sequences) laterally into the bilayer may be a permanent feature of contact sites. If this is the case, it may be that OMM signal-anchor sequences are integrated into the bilayer by ISP42/MOM38 complexes, the majority of which are located outside contact sites [39,40], before the protein has had a chance to be recruited to a contact site. It is interesting in this regard that isolated OMM vesicles are capable of importing and integrating outer membrane proteins like porin [47] and monoamine oxidase B [48]. Ono and Tuboi, on the other hand, have suggested that porin is preferentially integrated into the OMM at contact sites [50]. Again, however, proteins like porin and monoamine oxidase B may follow a different pathway than proteins like OMM70.

2.4. Polytopic proteins

Presumably, the problem of sorting of integral proteins to the outer and inner mitochondrial membranes pertains only to the first insertion event because, once established, the respective membrane is selected and subsequent transmembrane insertion events are predetermined. For the ER, evidence has accumulated to show that integration of multiple transmembrane segments is accomplished by sequential action of signal/signal-anchor and stop-transfer sequences [1,2,51,52]. Thermodynamic arguments [7,45] suggest that this mechanism, or some variation, should extend as well to the insertion of α-helical transmembrane segments into other types of membranes, including those of the mitochondrion.

A hypothetical model, based on the ER prototype, for the insertion of UCP (and AAC) into the mitochondrial inner membrane is shown in Fig. 2. Earlier studies provided evidence that import of AAC occurs at contact points [53,54]. An interesting question, however, is whether the same inner membrane translocator, i.e., at the contact point, is responsible for the complete insertion of UCP/AAC, or just the first insertion event (steps 1 and 2 in Fig. 2), with the translocation intermediate then diffusing laterally in the bilayer, and subsequent insertion events (steps 3–5, Fig. 2) being mediated by inner membrane translocators located outside the contact site. It has been shown, for example, that the inner membrane of yeast mitochondria contains abundant, and efficient, translocators at regions that do not appear to be part of stable contact sites [55–57]. Moreover, these translocators can apparently function independently of the outer membrane [56].

Finally, whereas we suggest that the first (and membrane-selective) insertion event of a polytopic protein into the inner membrane is mediated by the combination of a matrix-targeting signal and a distal stop-transfer sequence, subsequent insertion events could be mediated either by a repetition of this mode (as suggested for UCP and AAC in Fig. 2) or by a signal-anchor sequence, or by the kind of insertion domain that characterizes the *sec*-independent, but ATP- and $\Delta\psi$-dependent, insertion of the H_1 and H_2 segments of leader peptidase into the cytoplasmic membrane of *E. coli* (see [58,59]).

2.5. Default sorting

According to the stop-transfer model [1,18,45], the OMM signal-anchor sequence results in protein insertion into the outer membrane because translocation is abrogated by the lateral release of the transmembrane segment into the lipid bilayer. To determine if the OMM signal-anchor sequence can also be recognized by the translocation machinery of the inner membrane, we have employed a system developed in yeast for bypassing the outer membrane [55]. This can be achieved by selectively rupturing the outer membrane by osmotic shock, a procedure that leaves the inner membrane intact and does not affect $\Delta\psi$ [35,55]. These techniques, as applied to rat

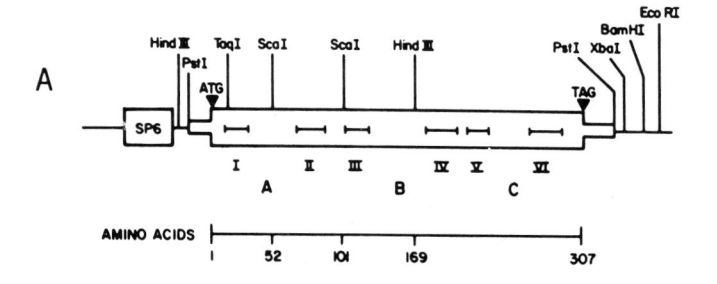

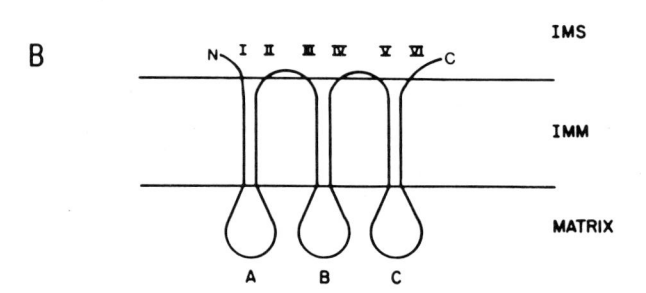

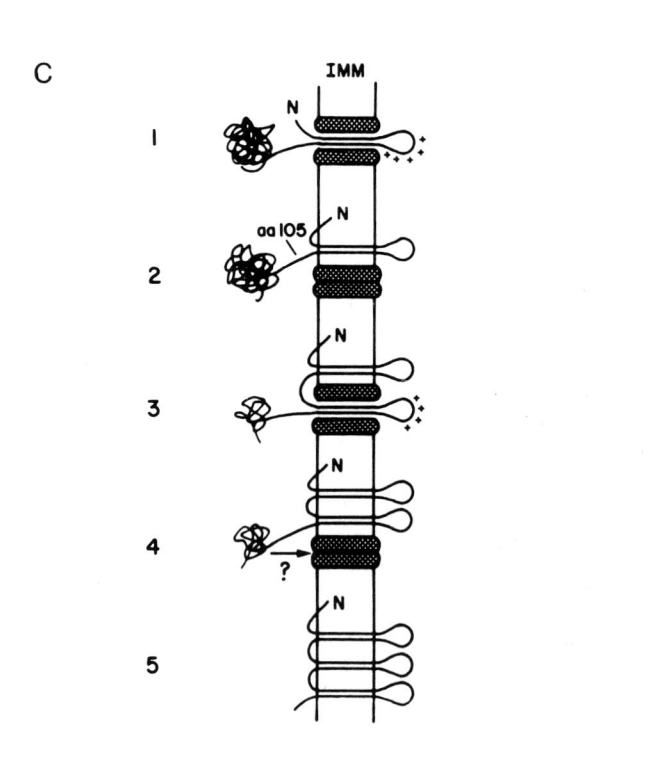

heart mitochondria, resulted in a system that can efficiently import the precursor of the matrix enzyme, ornithine carbamyl transferase, and the inner membrane protein, UCP, by a mechanism that is independent of the protease-sensitive OMM binding site for these precursors [35; and unpublished results]. Likewise, pOMD29 was inserted into the inner membrane in a manner very similar to insertion into the outer membrane, except that, unlike insertion into the outer membrane, insertion into the inner membrane was dependent on $\Delta\psi$ [35].

The results indicate that the protein translocation machineries of the outer and inner membrane interpret topogenic information and insert membrane proteins in a remarkably similar way. Further, these findings suggest that, if the outer membrane translocator is negatively regulated with respect to responding to the anchor domain, the OMM signal-anchor can, by default, be recognized by the inner membrane.

3. Conclusions

Our discussion has focused on proteins that follow the stop-transfer pathway for delivery to the mitochondrial outer and inner membrane, and which are anchored in the membrane by hydrophobic helical segments. We suggest that selective targeting of such transmembrane integral proteins to the mitochondrial outer membrane is specified by an OMM signal-anchor sequence, in which the targeting/translocation domain and the hydrophobic membrane anchor either functionally overlap (e.g., pOMD29) or are in close juxtaposition (e.g., pOCT-G35). Although the tar-

---

←

Fig. 2. Hypothetical model for insertion of UCP into the mitochondrial inner membrane. (A) Expression plasmid containing UCP cDNA in which positions of the six transmembrane segments (I–VI) and the connecting matrix ectodomains (A–C) are shown. For details, see [13]. (B) The predicted disposition of UCP in the inner membrane (after [11,12]). Hybrid proteins, constructed by fusing a portion of UCP ectodomain A to dihydrofolate reductase and ornithine carbamyl transferase, were imported in vitro to the soluble mitochondrial matrix (K.B. Freeman, pers. commun.). Deletion of UCP amino acids 2–101 (i.e., the first loop domain I-A-II) caused the rest of the molecule (i.e., amino acids 102–307 containing the second and third loop domains III-B-IV and V-C-VI) to be imported across both membranes, and deposited in the matrix [13]. Fusion of UCP amino acids 1–105 to a carrier protein resulted in import and insertion into the inner membrane [13]. The results suggest that insertion of domain I-A-II is required to allow insertion of domains III-B-IV and V-C-VI, perhaps by seeding insertion into the membrane and allowing the remaining transmembrane segments to arrest translocation of ectodomains B and C to the matrix. Ectodomains A and B contain clusters of positively charged amino acids, whereas this is less pronounced in ectodomain C [11,12,62]. (C) Insertion of UCP into the inner membrane. Like AAC [53,54], UCP is presumably imported via contact sites. For simplicity, only the inner membrane is shown. By analogy with the proposed mechanisms for sequential insertion of transmembrane segments of polytopic proteins into the ER [1,63] via a pore complex [26], we suggest that UCP is sequentially threaded into the inner membrane led by matrix-targeting signals in ectodomains A, B, and C, and arrested by pairs of amphiphilic transmembrane helices (I and II, III and IV, V and VI, respectively). For additional details, see the text.

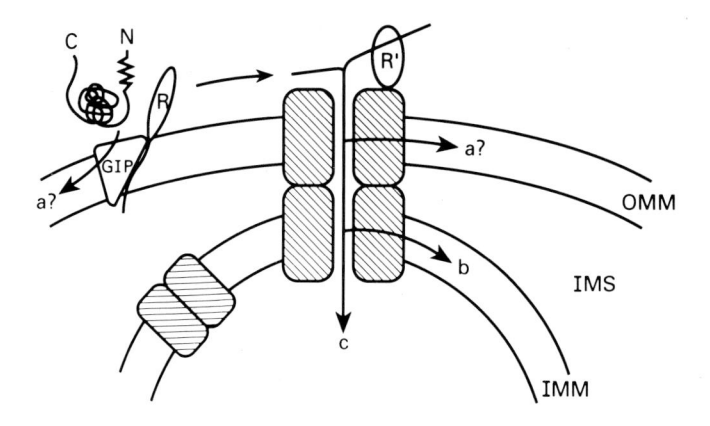

a. PROTEINS WITH
OMM SIGNAL-ANCHOR SEQUENCES

b. PROTEINS WITH MATRIX-TARGETING
SIGNALS AND DISTAL STOP-TRANSFER
SEQUENCES

c. PROTEINS WITH MATRIX-TARGETING
SIGNALS

Fig. 3. Protein sorting via the stop-transfer pathway. R,R′, import receptors [9,10,64]; GIP, general insertion site [65] (ISP42/MOM38 [39,40]). See text for discussion.

geting/translocation function can be achieved using a matrix-targeting signal, it remains to be determined if this motif is employed by other native outer membrane proteins. Furthermore, it is not yet known if outer membrane proteins are inserted into the outer membrane at a step that precedes delivery of precursor proteins to translocation contact sites (Fig. 3). Whichever the case, however, it is clear that the outer membrane translocation machinery at contact sites is organized or regulated in such a fashion that it does not arrest translocation of proteins destined for the inner membrane (Fig. 3). We suggest that initial targeting of proteins into the inner membrane is mediated by a combination of a matrix-targeting signal and a distal stop-transfer sequence. Although very little is known concerning the mechanism of insertion of proteins into the inner membrane bilayer, we suggest that models developed to explain stop-transfer insertion of proteins into ER and bacterial membranes are useful starting points, even for proteins with relatively complex structures like UCP and AAC (Fig. 2).

References

1. Blobel, G. (1980) Proc. Natl. Acad. Sci. USA 77, 1496–1499.
2. Wickner, W. and Lodish, H.F. (1985) Science 230, 400–407.
3. Von Heijne, G. (1986) EMBO J. 5, 3021–3027.
4. Hartmann, E., Rapoport, T.A. and Lodish, H.F. (1989) Proc. Natl. Acad. Sci. USA 86, 5786–5790.

5. Kleene, R. (1987) EMBO J. 6, 2627–2633.
6. Jap, B.K. (1989) J. Mol. Biol. 205, 407–419.
7. Singer, S.J. (1990) Annu. Rev. Cell Biol. 6, 247–296.
8. Pfaller, R. and Neupert, W. (1987) EMBO J. 6, 2635–2642.
9. Pfanner, N., Söllner, T. and Neupert, W. (1991) Trends Biochem. Sci. 16, 63–67.
10. Pfanner, N. and Neupert, W. (1990) Annu. Rev. Biochem. 59, 331–353.
11. Aquila, H., Link, T.A. and Klingenberg, M. (1985) EMBO J. 4, 2369–2376.
12. Runswick, M.J., Powell, J.T., Nyren, P. and Walker, J.E. (1987) EMBO J. 6, 1367–1373.
13. Liu, X, Bell, A.W., Freeman, K.B. and Shore, G.C. (1988) J. Cell Biol. 107, 503–509.
14. Mahlke, K., Pfanner, N., Martin, J., Horwich, A.L., Hartl, F.-U. and Neupert, W. (1990) Eur. J. Biochem. 192, 551–555.
15. Hartl, F.-U. and Neupert, W. (1990) Science 247, 930–938.
16. Hase, T., Müller, U., Reizman, H. and Schatz, G. (1984) EMBO J. 3, 3157–3164.
17. Nguyen, M. and Shore, G.C. (1987) J. Biol. Chem. 262, 3929–3931.
18. Nguyen, M., Bell, A.W. and Shore, G.C. (1988) J. Cell Biol. 106, 1499–1505.
19. Nakai, M., Hase, T. and Matsubara, H. (1989) J. Biochem. 105, 513–519.
20. Liu, X., Freeman, K.B. and Shore, G.C. (1990) J. Biol. Chem. 265, 9–12.
21. Hartl, F.-U., Schmidt, B., Wachter, E., Weiss, H. and Neupert, W. (1986) Cell 47, 939–951.
22. Hartl, F.-U., Osterman, J., Guird, B. and Neupert, W. (1987) Cell 51, 1027–1037.
23. Freitag, H., Janes, M. and Neupert, W. (1982) Eur. J. Biochem. 126, 197–202.
24. Mihara, K., Blobel, G. and Sato, R. (1982) Proc. Natl. Acad. Sci. USA 79, 7102–7106.
25. Noel, C., Nicalaou, V., Rachubinski, R.A., Argan, C. and Shore, G.C. (1985) Biochim. Biophys. Acta 814, 35–42.
26. Simon, S.M. and Blobel, G. (1991) Cell 65, 371–380.
27. Roise, D. and Schatz, G. (1988) J. Biol. Chem. 263, 4509–4511.
28. Roise, D., Horvath, S.J., Tomich, J.M., Richards, J.H. and Schatz, G. (1986) EMBO J. 5, 1327–1334.
29. Epand, R.M., Hui, S.-W, Argan, C., Gillespie, L.L. and Shore, G.C. (1986) J. Biol. Chem. 261, 10017–10020.
30. Attardi, G. and Schatz, G. (1988) Annu. Rev. Cell Biol. 4, 289–333.
31. Skerjanc, I.S., Sheffield, W.P., Silvius, J.R. and Shore, G.C. (1988) J. Biol. Chem. 263, 17233–17236.
32. Hase, T., Müller, U., Reizman, H. and Schatz, G. (1984) EMBO J. 3, 3157–3164.
33. Hurt, E.C., Müller, U. and Schatz, G. (1985) EMBO J. 4, 3509–3518.
34. Nakai, M., Hase, T. and Matsubara, H. (1989) J. Biochem. 105, 513–519.
35. Li, J.-M. and Shore, G.C. (1992) Biochim. Biophys. Acta 1106, 233–241.
36. McBride, H.M., Millar, D.G., Li, J.-M. and Shore, G.C., submitted.
37. Li, J.-M. and Shore, G.C. (1992) Science 256, 1815–1817.
38. Steger, H.F., Sölner T., Kiebler, M., Dietmeier, K.A., Pfaller, R., Trülzsch, K.S., Tropschug, M., Neupert, W. and Pfanner, N. (1990) J. Cell Biol. 111, 2353–2363.
39. Baker, K.P., Schaniel, A., Vestweber, D. and Schatz, G. (1990) Nature 348, 605–609.
40. Kiebler, M., Pfaller, R., Söllner, T., Griffiths, G., Horstmann, H., Pfanner, N. and Neupert, N. (1990) Nature 348, 610–616.
41. von Heijne, G. (1986) J. Mol. Biol. 192, 287–290.
42. Boyd, D. and Beckwith, J. (1990) Cell 62, 1031–1033.
43. Glaser, S.M., Miller, B.R. and Cumskey, M.G. (1990) Mol. Cell Biol. 10, 1873–1881.
44. Miller, B.R. and Cumsky, M.G. (1991) J. Cell Biol. 112, 833–841.
45. Singer, S.J. and Yaffe, M.P. (1990) Trends Biochem. Sci. 15, 369–373.
46. Adrian, G.S., McCammon, M.T., Montgomery, D.L. and Douglas, M.G. (1986) Mol. Cell Biol. 6, 626–634.
47. Gasser, S. and Schatz, G. (1983) J. Biol. Chem. 258, 3427–3430.
48. Zhuang, Z. and McCauley, R. (1989) J. Biol. Chem. 264, 14594–14596.

49. Vestweber, D., Brunner, J., Baker, A. and Schatz, G. (1989) Nature 341, 205–209.
50. Ono, H. and Tuboi, S. (1987) Eur. J. Biochem. 168, 509–514.
51. Audigier, Y., Friedlander, M. and Blobel, G. (1987) Proc. Natl. Acad. Sci. USA 84, 5783–5787.
52. Wessels, H.P. and Spiess, M. (1988) Cell 55, 61–70.
53. Pfanner, N. and Neupert, W. (1987) J. Biol. Chem. 262, 7528–7531.
54. Pfanner, N. Tropschug, M. and Neupert, W. (1987) Cell 49, 815–823.
55. Ohba, M. and Schatz, G. (1989) EMBO J. 6, 2117–2122.
56. Hwang, S., Jascur, T., Vestweber, D., Pon, L. and Schatz, G. (1989) J. Cell Biol. 109, 487–493.
57. Pon, L., Moll, T., Vestweber, D., Marshallsay, B. and Schatz, G. (1989) J. Cell Biol. 109 2603–2616.
58. Nilsson, I. and von Heijne, G. (1990) Cell 62, 1135–1141.
59. Dalbey, R.E. (1990) Trends Biochem. Sci. 15, 253–257.
60. Schleyer, M. and Neupert, W. (1985) Cell 43, 339–350.
61. Schwaiger, M., Herzog, V. and Neupert, W. (1987) J. Cell Biol. 105, 235–246.
62. Ridley, R.G., Patel, H.V., Gerber, G.E., Morton, R.C. and Freeman, K.B. (1986) Nucleic Acids Res. 14, 4025–4035.
63. Singer, S.J., Maher, P.A. and Yaffe, M.P. (1987) Proc. Natl. Acad. Sci. USA 84, 1960–1964.
64. Baker, K.P. and Schatz, G. (1991) Nature 349, 205–208.
65. Pfaller, R., Steger, H.F., Rassow, J., Pfanner, N. and Neupert, W. (1988) J. Cell Biol. 107, 2483–2490.
66. Schatz, G. (1991) The Harvey Lectures, Series 85, 109–126.

CHAPTER 21

General and exceptional pathways of protein import into sub-mitochondrial compartments

ROLAND LILL, CHRISTOPH HERGERSBERG, HELMUT SCHNEIDER, THOMAS SÖLLNER, ROSEMARY STUART and WALTER NEUPERT

Institut für Physiologische Chemie, Physikalische Biochemie und Zellbiologie der Universität München, Goethestraße 33, 8000 München 2, Germany

1. The general pathways for protein import into sub-mitochondrial compartments

In the past few years, considerable progress has been made to elucidate the mechanisms of how nuclear-encoded proteins are imported into and sorted within mitochondria (for detailed reviews, see [1–5]). The consensus pathways for protein sorting into the different mitochondrial compartments are depicted in Fig. 1. Nuclear encoded mitochondrial proteins are synthesized on cytoplasmic ribosomes as precursors which usually contain N-terminal targeting (signal) sequences. These signal peptides are positively charged and have the potential to form amphipathic α-helices [6]. Some precursor proteins, however, lack such cleavable pre-sequences, such as those destined for the mitochondrial outer membrane (OM). To avoid premature folding or aggregation in the cytosol, the precursors are transiently bound to cytosolic chaperone proteins which have been shown to belong to the heat shock protein 70 (hsp70) class [7]. Dissociation from hsp70 requires the hydrolysis of ATP which, in many cases, is considered to be a rate-limiting step for translocation.

Specific interaction of most precursor proteins with the OM is mediated by protease-sensitive surface receptors. *Neurospora crassa* MOM19 acts as the receptor for the majority of proteins analyzed so far [8], whereas the participation of *N. crassa* MOM72 has been demonstrated only in connection with the import of the ADP/ATP carrier [9]. Its yeast counterpart, MAS70, appears to have a broader substrate specificity [10,11]. Translocation through or insertion into the OM is catalyzed by a number of proteins which are associated in a complex in the OM (Fig. 1). This complex (termed the receptor/GIP complex) has been identified by co-immunoprecipitation with the surface receptors MOM19 or MOM72 which are part of the

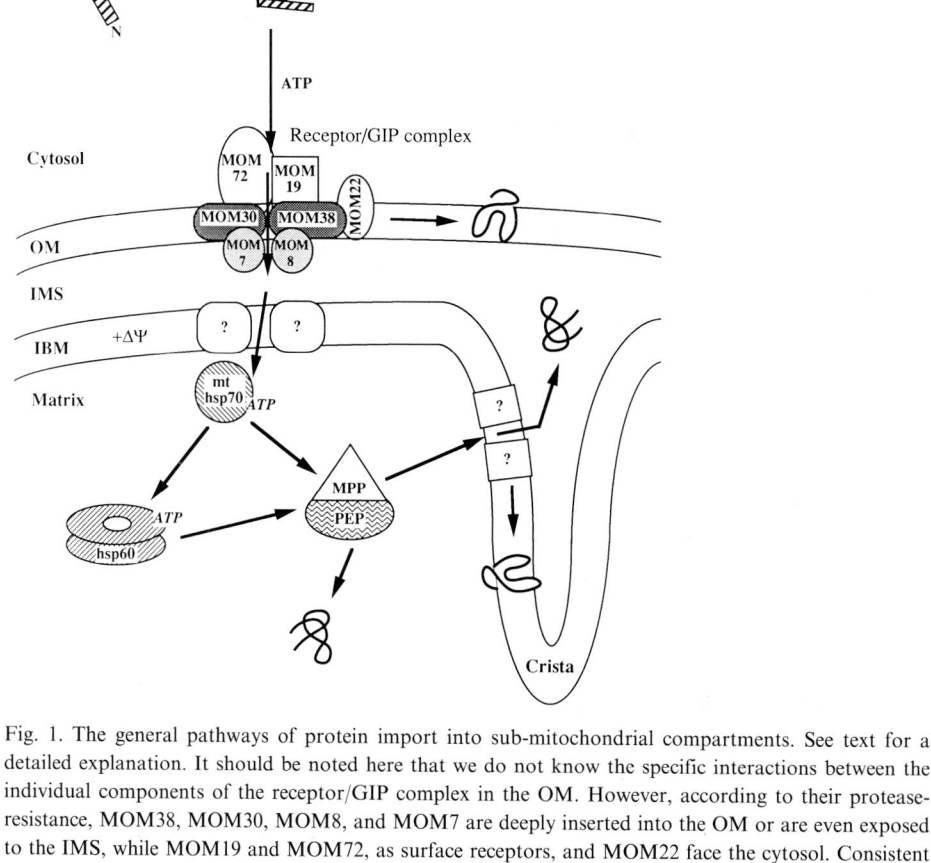

Fig. 1. The general pathways of protein import into sub-mitochondrial compartments. See text for a detailed explanation. It should be noted here that we do not know the specific interactions between the individual components of the receptor/GIP complex in the OM. However, according to their protease-resistance, MOM38, MOM30, MOM8, and MOM7 are deeply inserted into the OM or are even exposed to the IMS, while MOM19 and MOM72, as surface receptors, and MOM22 face the cytosol. Consistent with this observation it has been demonstrated that the former set of proteins forms (at least part of) GIP, a site where precursor proteins like AAC are already deeply inserted into the OM and are protected from externally added protease. Aside from its co-immunoprecipitation with MOM19, the role of MOM22 is unknown at present. For the sake of clarity, the second proteolytic step occurring in the IMS after export of proteins from the matrix has been omitted. Proteins inserted into the OM normally do not contain cleavable signal sequences (see text). IBM, inner boundary membrane; MOMx, mitochondrial outer membrane protein with a molecular mass of x kDa. MPP, matrix processing peptidase; PEP, processing enhancing protein; Cyt and mt hsp70, cytosolic and mitochondrial isoforms of heat shock protein of 70 kDa; Hsp60, heat shock protein of 60 kDa; ?, unknown components of the protein import and export apparatus of the IM.

complex [12]. One of its components, *N. crassa* MOM38, is part of a translocation site in which the precursor proteins are already deeply inserted in the OM and thus are resistant to externally added protease. Since this site is used by most precursor proteins, it has been termed general insertion pore (GIP) [13]. The yeast counterpart of MOM38, ISP42, is essential for cell growth [14]. Both functional and structural analysis of the receptor/GIP complex has been achieved by crosslinking the precursor of the ADP/ATP carrier (AAC) to various components of the complex [15]. Crosslinking of AAC arrested at different stages of its passage through the OM [16] allowed the components acting early and late in the OM translocation reaction to be defined. Receptor-bound AAC could be crosslinked to MOM19 and MOM72, whereas AAC arrested at the level of GIP could be shown to react efficiently with MOM7, MOM8, and with MOM30 proteins thus identifying new members of the complex.

Proteins destined for the intermembrane space (IMS), the inner membrane (IM), and the matrix become further translocated through the IM at sites of close contact between OM and IM [17]. Under in vivo conditions, more than 90% of the OM surface is in close contact with parts of the IM, the inner boundary membrane, which is competent for protein import (Fig. 1). Only a fraction of this area (10%) is made up of the morphological contact sites which comprise stable attachments of OM and IM even after shrinking of the IM. The initial interaction of the precursor proteins with the IM is mediated through an N-terminal signal sequence, and translocation is dependent on a membrane potential across the IM. Nothing is known so far about the nature of the components of the IM import machinery. The action of mitochondrial hsp70 (located in the matrix) and ATP hydrolysis are thought to be needed to pull the proteins into the matrix and are required for subsequent folding of matrix proteins [18]. N-terminal signal sequences are cleaved off by the matrix processing peptidase consisting of two components (MPP and PEP in *N. crassa*, MAS1 and MAS2 in yeast). Some of the incoming proteins also interact with hsp60 in the matrix which assists in folding and assembly of the incoming proteins [19–21].

Proteins of the IM or the IMS are then retargeted into the IM. A second signal sequence which shows a close similarity to bacterial export signals mediates this interaction. IM proteins become inserted into the IM, while proteins destined for the IMS are completely translocated through the IM. Finally, the second signal sequence becomes cleaved, e.g., by the inner membrane protease [22], whereafter IM proteins are functionally assembled, and IMS proteins are released as soluble proteins into the IMS. Since this membrane transit step mechanistically resembles the path taken by proteins in the bacterial ancestors of mitochondria, it has been termed the 'conservative sorting pathway' [2]. Nothing is known so far about the re-translocation (export) apparatus in the IM. One may speculate though that the components of such a translocation complex may bear some similarity to the members of the protein translocase identified in the plasma membrane of bacteria [23].

2. Exceptional pathways of protein import

A few exceptions to the general protein import pathways into the sub-mitochondrial compartments have been described so far. In these cases, only some steps or components of the general pathways are used while others are bypassed or alternative routes are taken. For example, import of yeast cytochrome c oxidase subunit Va, a protein of the IM (Cox Va) [24], is independent of protease-sensitive surface receptors in the OM. The protein enters the OM obviously at a later stage, most likely at the GIP stage, thereby circumventing the participation of surface receptors. The import of Cox Va is independent of surface receptors, and is similar to the situation found for MOM19 (see below) and for a low-efficiency bypass import of several proteins that was observed even after the surface receptors had been completely degraded [25]. For delivery of the Cox Va precursor protein to the OM, hsp70 seems not to be involved, since comparatively low levels of ATP are needed, and import occurs efficiently even at low temperatures. On the other hand, import of Cox Va is reported to share the features of the general pathway in that (i) translocation occurs via translocation contact sites, (ii) it needs a membrane potential for translocation, and (iii) its pre-sequence is cleaved after import.

A non-conservative import pathway into the IM without a transient passage through the matrix has been reported for the *N. crassa* ADP/ATP carrier protein (AAC)[26]. As mentioned above, AAC is so far the only known precursor that uses MOM72 instead of MOM19, as an import receptor. From MOM72, AAC is transferred into the protease-resistant location of GIP. Dependent on a membrane-potential across the IM, but independent of ATP and hsp60, as found for proteins of the general pathway, AAC is directly inserted into the IM and functionally assembled [26]. In vivo experiments with subunit VI of yeast QH_2 cytochrome c reductase (SubVI) suggest a direct transfer of this protein to the outer face of the IM [27]. The import into mitochondria is thought to be governed by its unusual N-terminal pre-sequence which is 25 amino acids long, but, in contrast to normal signal sequences, is highly negatively charged. Fusion proteins of the pre-sequence with the mature part of superoxide dismutase (SOD) as a reporter protein (pre-SubVI-SOD) were transported into the IMS as analyzed by sub-mitochondrial fractionation after in vivo import into mitochondria [27]. Furthermore, pre-SubVI-SOD failed to complement a mutant in the matrix-localized SOD also suggesting no transport of the fusion protein to the matrix. Thus, a negatively charged pre-sequence is also able to direct proteins into mitochondria, and seems to contain the information that precludes transfer of the attached protein across the IM.

Import of AAC and a few other mitochondrial proteins is unusual with respect to the absence of a cleavable signal sequence at the N-terminus of these proteins. In some cases, the sorting signals have been located somewhere in the mature region of the proteins, usually close to the N-terminus. Examples with non-cleavable signal sequences include the OM proteins MAS70 of yeast, *N. crassa* MOM72, and porin (see above), as well as AAC, 3-oxoacyl-CoA thiolase (reviewed in [1]), and some ribosomal proteins [28]. Many of these ribosomal proteins might lack a separate signal, since this class of proteins is usually highly positively charged, and thus already contains sequences that are similar to standard signal sequences.

In contrast to these proteins which use at least parts of the consensus pathway, cytochrome c, a soluble protein of the IMS, has developed quite an exceptional pathway that has no connection to the general routes. Cytochrome c is imported through the OM without the need for protease-sensitive surface receptors or the OM import receptor complex described above (for a review see [29]). Rather, cytochrome c utilizes a high-affinity interaction with cytochrome c heme lyase (CCHL) for its specific import into mitochondria. CCHL covalently attaches the heme group to cytochrome c and is located in the IMS at the outer face of the IM. Heme attachment may already occur when cytochrome c is still bound to the OM in a trans-membrane fashion [30]. Folding of the protein chain around the hydrophobic heme group may then lead to the release of functional holocytochrome c into the IMS. Thus, CCHL, during the biogenesis of cytochrome c, provides a dual function: (i) as an import receptor specific for cytochrome c, and (ii) as an enzyme which covalently adds the prosthetic group to the incoming cytochrome c thus rendering the import reaction irreversible.

The study of exceptional import pathways into mitochondria is rewarding in many aspects. First, it provides information about new components involved in protein transport across membranes. Second, it helps to elucidate the function of certain components of the general pathway, e.g., the determinants of recognition by these components. Third, it may serve to understand how import into mitochondria is regulated in order to achieve a balance between imported material and import catalysts. We have studied the import pathways of two proteins, the OM protein MOM19 and CCHL of the IMS, in some detail. Both proteins are imported along interesting, novel pathways into mitochondria.

3. MOM19 is imported into the OM without the aid of surface receptors

MOM 19 is the major surface receptor for protein import into *N. crassa* mitochondria which is used by most precursor proteins analyzed so far [8]. It was interesting to see how MOM19 itself is delivered to mitochondria. To study its import pathway, we first cloned the cDNA and determined the protein sequence [31]. As expected for an OM protein, MOM19 does not contain a cleavable N-terminal signal sequence. Residues 42–59 may form a highly positively charged, amphipathic α-helix and may serve as an internal, uncleaved import signal. The first 26 amino acids are hydrophobic and may represent the membrane anchor for the protein. The rest of the protein is hydrophilic and most likely protrudes into the cytosol.

For the analysis of the in vitro import of MOM19 into isolated *N. crassa* mitochondria, we made use of a property of endogenous MOM19. Treatment of mitochondria with elastase produces characteristic fragments of MOM19. When MOM19 precursor was imported in vitro from reticulocyte lysate, the same proteolytic pattern as that with endogenous MOM19 was produced after elastase treatment. MOM19 in reticulocyte lysate or MOM19 bound to the surface of mitochondria did not give rise to the specific proteolytic pattern, but was further degraded under the conditions used. These results indicate that imported MOM19 was functionally inserted into the OM [31].

Which components are involved in the insertion of MOM19 into the OM? When the mitochondria were pre-treated with protease to degrade the endogenous surface receptors, namely MOM19 and MOM72, the import of MOM19 was unaffected. This result shows that MOM19 does not utilize one of the protease-sensitive surface receptors in the OM for its own import. How, then, does MOM19 become specifically imported into mitochondria and does not associate with other cellular membranes like microsomes [31]? One possibility is that MOM19 enters the mitochondria via MOM38 with which it forms a tight complex [12]. Since *N. crassa* MOM38 is inaccessible to protease and its function cannot be blocked by antibodies, we made use of the yeast homolog of MOM38, namely ISP42, which in contrast is accessible to antibodies [32]. *N. crassa* MOM19 was imported into yeast mitochondria and functionally assembled into the receptor complex as shown by co-immunoprecipitation of MOM19 with anti ISP42 antibodies. This heterologous import and assembly of MOM19 also suggests that the receptor complexes of yeast and *N. crassa* are functionally very similar. Mitochondria which were pre-blocked with anti-ISP42 antibody failed to import MOM19 [31], demonstrating that ISP42 is directly involved in the import of MOM19. Since the *N. crassa* homolog of ISP42, MOM38, is a component of GIP, these results suggest that MOM19 enters the mitochondria directly at the level of the GIP site thereby bypassing the surface receptors. The same conclusion can be drawn from import experiments with yeast mitochondria pre-treated with protease at concentrations which degraded the receptors but not ISP42. Import of MOM19 into such mitochondria remained unaffected, when import of other, receptor-dependent proteins had already vanished.

For other components of the mitochondrial protein import machinery it has been shown that they participate in their own biogenesis. Examples include mitochondrial hsp60 [20], hsp70 [33], and the matrix processing peptidase [34]. Why does MOM19 circumvent the use of protease-sensitive surface receptors like MOM19 itself? The receptor-independent pathway entering the OM at MOM38/GIP might represent an evolutionary remnant form of mitochondrial protein import allowing specific import without the need of surface receptors. The invention of surface receptors might then serve a dual function: (i) it increases the specificity for import by the introduction of another checkpoint in addition to MOM38/GIP; (ii) it increases the efficiency and the rate of import considerably. In line with these assumptions, a low efficiency import of several proteins into mitochondria was found even after the surface receptors had been removed by protease digestion [25]. Most likely, these proteins enter the mitochondria at a later step.

4. Cytochrome c heme lyase is imported directly through the OM via a non-conservative sorting pathway

N. crassa cytochrome *c* heme lyase (CCHL) is localized in the IMS, where it is found associated with the outer face of the IM [30]. The enzyme catalyzes the covalent attachment of a reduced heme group to cytochrome *c*, and it serves as a high-affinity receptor for cytochrome *c* during its import across the OM [29]. *N. crassa* CCHL is about 60% homologous (40% identity on amino acid level) to both yeast CCHL and to yeast cytochrome c_1 heme lyase (CC$_1$HL [35,36; A. Haid, pers. commun.]). All three heme lyases lack cleavable N-terminal signal sequences. We have analyzed the import pathway of *N. crassa* CCHL in vitro and found an interesting import mechanism which, on the one hand, shares characteristics of the general pathway (see Fig. 1), and on the other hand, uses novel ways to reach the IMS.

CCHL synthesized in an in vitro transcription-translation system efficiently and rapidly associates with mitochondria [37]. During further incubation, CCHL becomes translocated into the organelles, as assayed by protection of imported CCHL against digestion by externally added protease. As revealed by digitonin fractionation of mitochondria, imported CCHL is correctly localized to the IMS, its functional environment. Here, it is bound to membranes, and thus behaves indistinguishably from the endogenous protein [30] and its enzyme activity [38].

CCHL does not require free ATP for import into mitochondria, an observation which is in sharp contrast to what has been reported for other mitochondrial precursor proteins. Except for a requirement of ATP in the matrix (see above), ATP hydrolysis is needed for the dissociation of the precursor proteins from cytosolic hsp70 proteins. Hsp70s bind to precursor proteins in order to keep them competent for translocation by stabilizing an unfolded conformation [7]. Dissociation from hsp70 is assumed to be the rate-limiting step for translocation. For the import of CCHL, however, several observations argue against a participation of cytosolic hsp70 proteins [37]. (i) CCHL is rapidly imported into the IMS. Even at 0°C import occurs with a half-time of 10 min, and thus is much faster than observed for other precursors. (ii) The time course and the efficiency of the import are unchanged when CCHL is denatured in 8 M urea and imported into mitochondria after rapid dilution of the denaturant. Usually, unfolding of the precursor proteins and disruption of the interaction with hsp70 accelerates the import kinetics markedly. (iii) CCHL import occurs independently of the presence of free ATP. When ATP is hydrolyzed by apyrase (an ATP- and ADP-hydrolyzing enzyme activity from potato), import of CCHL is unaffected, while the import of control proteins like the β-subunit of the F_1-ATPase is completely inhibited. CCHL imported in the absence of

ATP is still correctly localized in the IMS indicating that CCHL is not simply trapped in an intermediate stage during import. The observation that CCHL import does not involve both hsp70 and the hydrolysis of free ATP may be related to the unusually high content of proline residues (11.6%) which might keep the protein in an extended, import-competent structure without the aid of chaperone proteins.

For translocation through the OM, CCHL uses the receptor/GIP complex. Import of CCHL was completely abolished when mitochondria were pre-treated with protease indicating that protease-sensitive receptors mediate the entry of CCHL into the OM [37]. To find out which specific receptor is used, mitochondria were pre-incubated with immunoglobulin G (IgG) against MOM19, MOM72, and as a control, porin. Subsequent import of CCHL was specifically reduced by 80% in mitochondria that had been blocked with anti-MOM19 antibodies as compared to unblocked mitochondria, while the other antibodies did not reduce the import efficiency. These results demonstrate that CCHL, as most other mitochondrial precursor proteins, uses the main receptor of *N. crassa* mitochondria, MOM19, for initial interaction with mitochondria. Furthermore, competition experiments with chemical amounts of import-competent porin demonstrate that CCHL uses the same OM import machinery, namely the receptor/GIP complex which is effective for most other mitochondrial precursor proteins [37].

How does CCHL reach the IMS after transit through the OM? Two possibilities can be envisaged. Either, the protein is further translocated via contact sites through the IM and the matrix along the conservative sorting pathway (Fig. 1), or it enters the IMS directly after its passage through the OM (Fig. 2A). The first possibility seems unlikely from the fact that CCHL lacks the usual bipartite, cleavable pre-sequence (as found in, e.g., cytochrome c_1- and cytochrome b_2-precursors). The positively charged signal sequence is thought to mediate the interaction with the IM. For insertion and translocation through the IM, a membrane potential has been shown to be essential [3,39]. Consistent with the view that CCHL translocates directly into the IMS, its import into mitochondria occurs independently of the existence of a membrane potential [37]. Depletion of the membrane potential by uncouplers like CCCP did not influence the import of CCHL into mitochondria. The same result was obtained when either ionophores (valinomycin) or inhibitors of the electron transport chain and of the F_1F_0-ATPase (antimycin A and oligomycin, respectively) were used to prevent the formation of a membrane potential. These results strongly suggest that CCHL does not pass through the IM during its import into mitochondria. Rather, it reaches the IMS directly after translocation through the OM import channel via a non-conservative sorting pathway.

The fact that CCHL requires neither ATP hydrolysis nor a membrane potential as energy sources for import, raises the important question of what energetically drives the import reaction. Three possibilities may be anticipated. (i) The driving force for import may come from folding of CCHL within the IMS. It will be interesting to see whether such folding reaction is catalyzed by yet to be discovered factors. (ii)

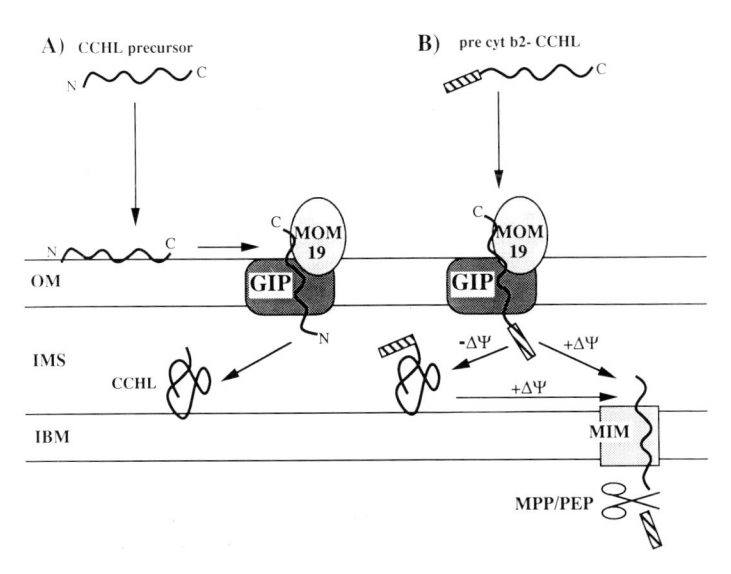

Fig. 2. (A) The non-conservative import pathway of CCHL across the OM and (B) the two-step import mechanism of CCHL carrying an N-terminal matrix targeting sequence. See text for a detailed explanation. Pre cyt b2-CCHL represents CCHL with the positively charged pre-sequence of cytochrome b_2 (shown by the hatched rectangle) attached to its N-terminus; MIM, mitochondrial inner membrane import apparatus; GIP, general insertion pore. All other abbreviations are as in Fig. 1.

Alternatively, CCHL may be pulled into the IMS by specific interaction with a binding partner. Since CCHL avidly binds to liposomes, such an interaction may include binding to lipid molecules in the IM and OM. (iii) Finally, it cannot be excluded on the basis of the previous experiments (see above) that the hydrolysis of miniscule amounts of factor-bound ATP, inaccessible to digestion by apyrase, may drive the import. It should be mentioned here that MOM38 contains an ATP binding motif [12], and thus may be a candidate for such an ATP binding factor. These questions are now open to direct experimental approach. Recently, an in vitro import system with purified OM vesicles has been developed. CCHL as well as OM proteins are imported into these vesicles in a receptor-dependent fashion (A. Mayer, R. Lill and W. Neupert, unpublished results). The ability of OM vesicles to import CCHL nicely confirms the conclusions drawn above that the IM is not involved in the passage of CCHL into the IMS.

Which signals guide CCHL along its path to the IMS and why is further translocation through the IM precluded? The import signal in CCHL is unknown. Experiments with antibodies raised against the N- and C-terminus of CCHL show that the N-terminal half of CCHL inserts into the OM first and thus must contain the information for OM interaction (R. Lill and W. Neupert, unpublished results). Both antibodies efficiently block translocation of CCHL. When import is inhibited by the C-terminal antibodies, specific N-terminal fragments of CCHL are produced after protease digestion. Occurrence of these fragments is dependent on functional MOM19 protein and is not observed with either control or N-terminal antibodies. These experiments show that CCHL with an attached C-terminal antibody may form a membrane-spanning translocation intermediate in the OM. From the size of the observed fragments of 34, 28, and 22 kDa, it can be concluded that the N-terminal half of the protein (i) enters the OM first, and (ii) contains the information for both interaction with and translocation through the OM. Preliminary experiments with truncated CCHL proteins corroborate this result. The putative import signal must be contained within 150 central amino acids, since 80 N-terminal and 130 C-terminal amino acids can be deleted from CCHL without any influence on the import efficiency (G. Kispal, R. Lill and W. Neupert, unpublished results). Further experiments are necessary to define the precise location and the chemical character of the import signal in CCHL.

Aside from describing a novel import pathway into the IMS, our studies on CCHL import bear relevance to the general protein import pathway into mitochondria. Import of proteins across the OM is not necessarily coupled to translocation across the IM, i.e. the two translocation machineries in the two membranes do not necessarily form a continuous channel. Indication for this has come already from studies on translocation intermediates in the IMS en route to the matrix. Segments of the polypeptide chains in transit are accessible to added protease after rupture of the OM indicating that the intermediates are exposed to the IMS [40,41]. Sequential transfer across the OM and IM in intact mitochondria was observed for CCHL fusion proteins carrying a matrix targeting sequence at the N-terminus (Fig. 2B). Without a membrane potential, the fusion proteins were imported into the IMS like CCHL itself. Upon establishing a membrane potential, at least the N-terminus of the proteins was transferred from the IMS to the matrix as indicated by the cleavage of the signal sequence (G. Kispal, R. Lill and W. Neupert, unpublished results). Taken together, these data suggest that transport across the OM and IM can occur sequentially by using two principally independent translocation machineries [17,42]. Contact sites between the OM and IM may then be considered as dynamic structures that kinetically accelerate import across both membranes and thus may represent the structure that is predominantly used in vivo for the import of proteins into the matrix.

5. Perspectives

Future work on mitochondrial protein import has to concentrate on questions in two major directions. First, the as yet unknown components of the mitochondrial protein sorting apparatus have to be identified. Most importantly, these are the constituents of the import and export machineries in the IM. Furthermore, transit through and folding within the IMS may be mediated by still unidentified factors.

Drawing the analogy to the bacterial system, the participation of other chaperones in protein folding and assembly into oligomeric structures may be anticipated, e.g., activities like that of GroES or DnaJ homologs (see Chapters 25 and 26). Finally, other catalysts of post-translational modifications like new proteolytic processing activities may be discovered.

The second question is how the known components function and cooperate to finally achieve the correct localization and assembly of incoming proteins. This is a major challenge, but the discovery that translocation through the two membranes may be considered as two independent steps, will eventually allow to individually study the transit through the two membranes. It is thus easy to predict that future research on mitochondrial protein targeting and sorting will be as exciting as it has been in the past.

Acknowledgements

We thank U. Hanemann, C. Kraus and K. Aschenbrenner for expert technical assistance. The work was supported by a grant of Sonderforschungsbereich 184 (Teilprojekte B1 and B4) and by the Fonds der Chemischen Industrie.

References

1. Hartl, F.-U., Pfanner, N., Nicholson, D.W. and Neupert, W. (1989) Biochim. Biophys. Acta 988, 1–45.
2. Hartl, F.-U. and Neupert, W. (1990) Science 247, 930–938.
3. Pfanner, N. and Neupert, W. (1990) Annu. Rev. Biochem. 59, 331–353.
4. Pfanner, N., Söllner, T. and Neupert, W. (1991) Trends Biochem. Sci. 16, 63–67.
5. Baker, K.P. and Schatz, G. (1991) Nature 349, 205–208.
6. von Heijne, G. (1986) EMBO J. 5, 1335–1342.
7. Deshaies, R.J., Koch, B.D. and Schekman, R. (1988) Trends Biochem. Sci. 13, 384–388.
8. Söllner, T., Griffiths, G., Pfaller, R., Pfanner, N. and Neupert, W. (1989) Cell 59, 1061–1070.
9. Söllner, T., Pfaller, R., Griffiths, G., Pfanner, N. and Neupert, W. (1990) Cell 62, 107–115.
10. Steger, H., Söllner, T., Kiebler, M., Dietmeier, K., Pfaller, R., Trülzsch, K.S., Tropschug, M., Neupert, W. and Pfanner, N. (1990) J. Cell Biol. 111, 2353–2363.
11. Hines, V., Brandt, A., Griffiths, G., Horstmann, H., Brütsch, H. and Schatz, G. (1990) EMBO J. 9, 3191–3200.
12. Kiebler, M., Pfaller, R., Söllner, T., Griffiths, G., Horstmann, H., Pfanner, N. and Neupert, W. (1990) Nature 348, 610–616.
13. Pfaller, R., Steger, H.F., Rassow, J., Pfanner, N. and Neupert, W. (1988) J. Cell Biol. 107, 2483–2490.
14. Baker, K.P., Schaniel, A., Vestweber, D. and Schatz, G. (1990) Nature 348, 605–609.
15. Söllner,T., Rassow, J., Wiedmann, M., Schlossmann, J., Keil, P., Neupert, W. and Pfanner, N. (1992) Nature 355, 84–87.
16. Pfanner, N. and Neupert, W. (1987) J. Biol. Chem. 262, 7528–7536.
17. Pfanner, N., Rassow, J., vander Klei, I. and Neupert, W. (1992) Cell 68, 999–1002.
18. Neupert, W., Hartl, F.-U., Craig, E.A. and Pfanner, N. (1990) Cell 63, 447–450.

19. Ostermann, J., Horwich, A.L., Neupert, W. and Hartl, F.-U. (1989) Nature 341, 125–130.
20. Cheng, M.Y., Hartl, F.-U., Martin, J., Pollock, R.A., Kalousek, F., Neupert, W., Hallberg, E.M., Hallberg, R.L. and Horwich, A.L. (1989) Nature 337, 620–625.
21. Manning-Krieg, U., Scherer, P.E. and Schatz, G. (1991) EMBO J. 10, 3273–3280.
22. Schneider, A., Behrens, M., Scherer, P., Pratje, E., Michaelis, G. and Schatz, G. (1991) EMBO J. 10, 247–254.
23. Wickner, W., Driessen, A.J.M. and Hartl, F.-U. (1991) Annu. Rev. Biochem. 60, 101–124.
24. Miller, B.R. and Cumsky, M.G. (1991) J. Cell. Biol. 112, 833–841.
25. Pfaller, R., Pfanner, N. and Neupert, W. (1989) J. Cell. Biol. 264, 34–39.
26. Mahlke, K., Pfanner, N., Horwich, A.L., Hartl, F.-U. and Neupert, W. (1990) Eur. J. Biochem. 192, 551–555.
27. Oudshoorn, P. (1991) Dissertation, University of Amsterdam.
28. Kang, W., Matsushita, Y. and Isono, K. (1990b) Mol. Gen. Genet. 225, 474–482.
29. Stuart, R.A. and Neupert, W. (1990) Biochimie, 72, 115–121.
30. Hergersberg, C., Griffiths, G., Lill, R. and Neupert, W. (1992) submitted.
31. Schneider, H., Söllner, T., Dietmeier, K., Eckerskorn, C., Lottspeich, F., Trülzsch, B., Neupert, W., Pfanner, N. (1991) Science 254, 1659–1662.
32. Vestweber, D., Brunner, J., Baker, A. and Schatz, G. (1989) Nature 341, 205–209.
33. Kang, P.J., Ostermann, J., Shilling, J., Neupert, W., Craig, E.A. and Pfanner, W. (1990a) Nature 348, 137–143.
34. Witte, C., Jensen, R.E., Yaffe, M.P. and Schatz, G. (1988) EMBO J. 7, 1439–1447.
35. Drygas, M.E., Lambowitz, A.M. and Nargang, F.E. (1989) J. Biol. Chem. 264, 17897–17906.
36. Dumont, M.E., Ernst, J.F. and Sherman, F. (1988) J. Biol. Chem. 263, 15928–15937.
37. Lill, R., Stuart, R., Drygas, M.E., Nargang, F.E. and Neupert, W. (1992) EMBO J. 11, 449–456.
38. Nicholson, D.W., Hergersberg, C. and Neupert, W. (1988) J Biol. Chem. 263, 19034–19042.
39. Martin, J., Mahlke, K. and Pfanner, N. (1991) J. Biol. Chem. 266, 18051–18057.
40. Rassow, J. and Pfanner, N. (1991) FEBS Lett. 293, 85–88.
41. Hwang, S.T., Wachter, C. and Schatz, G. (1991) J. Biol. Chem. 266, 21083–21089.
42. Glick, B., Wachter, C. and Schatz, G. (1991) Trends Cell Biol. 1, 99–103.

Part F

Chloroplasts

CHAPTER 22

Targeting of proteins into and across the chloroplastic envelope

H.-M. LI, S.E. PERRY and K. KEEGSTRA

Department of Botany, University of Wisconsin, Madison, Wisconsin 53706, USA

Abstract

The transport of precursor proteins into chloroplasts is composed of at least two steps: a specific binding of precursor proteins to the surface of the chloroplastic envelope followed by translocation of precursor proteins across the envelope. Precursor binding to the envelope can only be detected when the translocation step is blocked and is greatly stimulated by low concentrations of ATP. When envelope membranes were isolated, bound precursors were found in a fraction previously shown to contain contact sites connecting the two membranes of the envelope. This observation supports the hypothesis that protein transport into chloroplasts occurs at contact sites. Translocation of precursor proteins across the envelope requires ATP in the stroma as its sole energy source. Synthetic peptide analogues, corresponding to different regions of the transit peptide of the small subunit of ribulose-1,5-bisphosphate carboxylase precursor, can inhibit either the binding or the translocation step of protein transport into chloroplasts. These peptides can also inhibit import of proteins destined for various compartments of the chloroplast, except a protein destined for the outer membrane. This indicates that precursor proteins of most chloroplastic compartments share the same import apparatus for their transport across the envelope. Proteins destined for the outer membrane of the envelope follow a unique import pathway. They are not synthesized as higher molecular weight precursors. Neither ATP nor a thermolysin-sensitive receptor is required for their insertion. The targeting mechanisms of proteins destined for the inner envelope membrane and for the intermembrane space of the envelope are still unknown. Some possible sorting pathways are discussed.

1. Introduction

Chloroplasts are functionally complex organelles that perform a diverse array of metabolic processes in addition to their well known role in photosynthesis. For example, they are the site of starch synthesis and degradation, fatty acid biosynthesis, nitrate reduction and nitrogen assimilation as well as several other important metabolic functions. Consistent with their functional complexity, chloroplasts are structurally complex organelles, possessing three different lipid bilayer membranes enclosing three different aqueous compartments (Fig. 1). Because of the limited

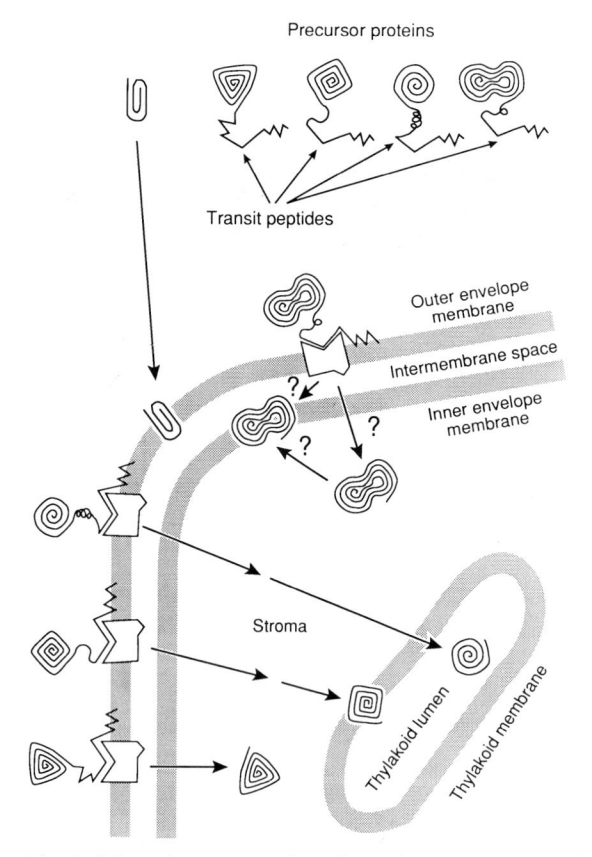

Fig. 1. Schematic representation of protein transport into different compartments of chloroplasts. All nuclear-encoded chloroplastic proteins, except proteins of the outer membrane, are synthesized as higher molecular weight precursors with amino terminal extensions called transit peptides. Precursor proteins targeted to the stroma, the thylakoid membrane and the thylakoid lumen are all first translocated into the stroma. This translocation is then followed by processing and/or further sorting of the proteins to their proper compartments. Proteins of the outer membrane insert into the outer membrane directly from the cytosol. The targeting pathway of the inner membrane proteins is not clear. Two possible pathways are depicted in this model (see text for discussion).

coding capacity of the plastid genome, most of the proteins in each compartment are encoded by nuclear genes and synthesized as precursors in the cytoplasm before they are transported into chloroplasts and targeted to their proper locations (Fig. 1). Understanding how these precursor proteins are targeted from the cytoplasm to their proper locations within chloroplasts is an intriguing problem that has received considerable attention in the last decade. In this chapter, we briefly summarize recent results from our laboratory on transport of precursor proteins across and into the chloroplastic envelope, and discuss these results in terms of the many important contributions from other laboratories.

2. Transport across the envelope membranes

Transport of precursor proteins across the envelope consists of (1) binding of the precursor proteins to the chloroplastic surface, and (2) translocation of the precursor proteins across the envelope membranes into the stroma [1]. These steps are shared by precursor proteins destined for the stroma, for the thylakoid membrane and for the thylakoid lumen, i.e., despite the difference of their final locations, all of these precursor proteins are transported first into the stroma. Evidence that transport to the stroma is shared by all precursors derives in part from studies using chimeric constructs in which the transit peptides of the precursor to ribulose-1,5-bisphosphate carboxylase (prSS) and of the precursor to the light-harvesting chlorophyll a/b protein (prLHCP) are switched. These experiments show that the transit peptide of prLHCP functions as a stromal targeting sequence [2,3]. The transit peptides of precursors to thylakoid lumen proteins have two domains [4]: a domain that targets the protein to the stroma [5], followed by a region responsible for sorting to the thylakoid lumen.

2.1. Binding of precursors to the chloroplastic surface

Specific binding must be studied isolated from the translocation step. If translocation is not blocked, any precursor that is specifically bound will be translocated across the envelope, leaving only nonspecifically bound precursor on the chloroplastic surface. One method to isolate the binding step is to keep the temperature below 10°C. Binding will occur readily on ice, but little or no translocation occurs below 10°C [6]. Alternately, it is possible to separate binding from translocation based on the different energy requirements for these two steps. A higher level of exogenously added ATP (1–2 mM) is needed to support translocation than to support binding (50–100 μM ATP), so binding can be separated from translocation by limiting the ATP concentration [7]. The energy requirement for binding has a broad specificity in that hydrolyzable nucleotides other than ATP can stimulate binding, but only ATP supports translocation across the envelope [7]. Finally, the ATP involved in translocation is needed in the stroma [8], whereas the nucleoside triphosphate (NTP) requirement for binding is in the intermembrane space between the two envelope membranes (Olsen and Keegstra, unpublished results).

Precursor binding to the chloroplastic surface involves one or more protein components of the envelope membranes. Evidence for involvement of a receptor protein includes the observation that precursor binding requires a protease-sensitive component on the chloroplastic surface, and is saturable at about 1500–3500 binding sites per chloroplast [9]. As discussed above, binding requires low levels of NTP [7]. This energy requirement provides an additional piece of evidence for the existence of a proteinaceous receptor in that it is difficult to assign a role to the NTP requirement without invoking a protein. The role of the NTP remains unidentified, but protein kinases [10,11] and ATPases [12] are present in the chloroplastic envelope and could be involved in the protein transport process.

Binding of the precursor proteins to chloroplasts is not readily reversible. Bound precursors are stably associated with the envelope membranes [7]. Chloroplasts with bound precursor were lysed and the envelope membranes were isolated by sucrose density gradient centrifugation as described by Cline et al. [13] to localize the bound precursor (Ostrom and Keegstra, unpublished results). Bound precursors were not found in the fractions containing the outer envelope membrane as anticipated. Rather, precursors were found in a region of the gradient containing mixed envelope membranes. Membranes in this region of the gradient had previously been demonstrated to contain putative contact sites [13]. These observations provide support for the popular hypothesis that protein transport occurs at special contact sites, where the two membranes of the chloroplastic envelope are tightly appressed. In view of these results, it may be more appropriate to consider bound precursors as an arrested early transport intermediate.

Another feature of precursor binding to the envelope is a productive and specific interaction between receptors and ligands. Binding is productive in that 60–85% of bound prSS can subsequently be imported [7,9]. Binding is also ligand specific in that a transit peptide is necessary for binding; the mature protein lacking a transit peptide will not bind to the surface of chloroplasts [9]. However, transit peptides of different precursors have no primary sequence homology [14] raising an important question concerning the specificity of the putative receptors. There could be many different receptors, each with some primary sequence specificity. Or there could be one to a few types of receptors that recognize some general characteristics, such as a higher order of polypeptide structure, common to many precursors. We have found that a 20-residue synthetic peptide corresponding to the middle region of the transit peptide from prSS inhibits binding not only of prSS (a stromal protein), but also inhibits binding of precursor to ferredoxin (prFD, a second stromal protein), of prLHCP (a thylakoid membrane protein), and of precursor to plastocyanin (prPC, a thylakoid lumen protein) [15]. Schnell et al. [16] found that synthetic peptides corresponding to the carboxyl-terminal 30 amino acids of the transit peptides from prFD and from prSS were both able to block prSS and prFD import into chloroplasts at the binding step of transport. These results indicate that these precursors share a common receptor protein even though they are sorted to different compartments within the chloroplast.

A second important question involves the identity of the putative receptor protein. One approach taken by Cornwell and Keegstra [17] was to use a heterobifunctional, photoactivatable cross-linking reagent to crosslink bound radiolabelled prSS to its receptor protein. A 86 kDa conjugate was identified. Because prSS is a 20 kDa protein, the putative receptor is predicted to have a molecular weight of approximately 66 kDa. A different approach was taken by Pain et al. [18]. They raised antiidiotypic antibodies against a synthetic peptide corresponding to the carboxylterminal 30 residues of the prSS transit peptide. These antibodies should mimic the transit peptide and bind to the same site on a receptor protein. When used in immunoblotting, these antibodies react with a 30 kDa protein localized to the contact sites of the chloroplastic envelope. The cDNA clone encoding this protein was obtained [19] and was found to encode the same protein previously identified as the phosphate translocator protein, which should be in the inner membrane of the envelope [20,21]. Further work will be necessary to localize the 30 kDa protein and to determine its function.

2.2. Translocation of precursor across the envelope membranes

With adequate temperatures and sufficient levels of ATP, precursor proteins are translocated across the envelope into the stroma. Although the molecular mechanism of translocation is unknown, it is clear that translocation requires energy in the form of ATP [8,22,23]. The ATP can be provided either by photophosphorylation or by adding exogenous ATP at a level of 1–2 mM in the external solution. However, the ATP needed for translocation is utilized inside chloroplasts [8,23], so the concentration at the site of use is unknown. Protein transport across the envelope does not require a membrane potential, as does mitochondrial protein transport [24].

Currently, nothing is known about the identity of proteins of the chloroplastic envelope, if any, that are involved in precursor translocation across the envelope. We have found that prSS, prFD, prPC, and prLHCP share a component in the import apparatus past the binding step. Synthetic peptides corresponding to the ends of the transit peptide of prSS have little or no effect on binding of these proteins to the surface of chloroplasts. However, these peptides do have an inhibitory effect on the import of these proteins across the envelope [15]. We concluded that these peptides are inhibiting translocation of the precursors by competing for a translocation protein, perhaps similar to the general insertion protein proposed in mitochondria [25,26]. In the mitochondrial model, precursors first bind to protease-sensitive receptors [25]. They are then transferred to a protease-resistant general insertion protein, from which outer membrane proteins are released to assemble into functional complexes, while precursors destined for other compartments undergo membrane potential-dependent translocation, perhaps mediated by yet other membrane proteins.

In mitochondria, there is evidence that members of the hsp70 protein family are involved in the protein import process [24,27]. These hsp70s are involved both in the cytoplasm and inside mitochondria. They are proposed to unfold precursors in the cytoplasm so that precursor proteins can be translocated through a proteinaceous pore at contact sites, and then to refold the precursor protein inside the mitochondrion. We have recently identified at least two, and possibly three hsp70 homologues in chloroplasts [28]. A cDNA clone encoding an hsp70 homologue located in the stroma has been obtained (Marshall and Keegstra, unpublished results). This clone was sequenced and found to have many similarities to mitochondrial hsp70 proteins, but it is even more similar to an hsp70 from the cyanobacteria *Synecocystis* [29]. It is not yet known whether any of these chloroplastic hsp70 homologues are involved in chloroplastic protein import.

3. Targeting of proteins into the envelope membranes

In addition to the three compartments located inside chloroplasts (Fig. 1), there are three additional compartments that are part of the chloroplastic envelope. Although it is clear that most, if not all, envelope proteins are synthesized in the cytoplasm, only a limited amount of information is available on the mechanisms that direct them to the two membranes of the envelope. Nothing is known about protein targeting to the intermembrane space of the envelope.

3.1. Targeting to the outer envelope membrane

The chloroplastic outer envelope membrane is a unique membrane in that it has features of both prokaryotic and eukaryotic membranes. It contains large amounts of galactolipids like other chloroplastic membranes and the membranes of cyanobacteria [30], which agrees well with the putative endosymbiotic origin of chloroplasts. However, the outer membrane is also rich in phosphatidylcholine, which is almost absent from other chloroplastic membranes and is a major component of all extraplastidial membranes [30]. Thus it is possible that the outer membrane may have a different evolutionary origin from other chloroplastic compartments and the biogenesis of proteins in the outer membrane may have some unique features not found with proteins in other chloroplastic compartments. Interestingly, recent work with two outer membrane proteins, OM14 of pea chloroplasts and E6.7 of spinach chloroplasts, provide evidence that targeting of proteins to the outer membrane is distinctly different from transport of proteins into the interior of chloroplasts [31,32]. A summary of some of the characteristics that distinguish the two

TABLE I

Comparison of the import of internal proteins with the insertion of outer membrane proteins

Feature	Internal proteins	Outer membrane proteins
Larger precursors	+	−
ATP required	+	−
Thermolysin-sensitive receptors involved	+	−

processes is presented in Table I. One of the major differences is that all known examples of the proteins transported across the envelope membranes are synthesized as higher molecular weight precursors, but the two outer membrane proteins studied so far are synthesized without cleavable transit peptides. Another distinguishing feature is that transport across the envelope membranes requires ATP, whereas insertion of proteins into the outer envelope membrane does not. Also, transport of proteins across the envelope membranes is diminished or abolished by mild thermolysin pretreatment of chloroplasts, whereas insertion of proteins into the outer membrane is not.

The insertion of OM14 into the outer envelope membrane of pea chloroplasts was not inhibited by synthetic peptide analogues of the transit peptide of prSS [32]. These peptides were shown to inhibit the binding or translocation of several precursor proteins destined for various internal chloroplastic compartments [15]. Combined with the observation that neither ATP nor a thermolysin-sensitive receptor is required for the insertion of the outer membrane proteins (Table I), it seems that chloroplastic outer membrane proteins use a different transport apparatus from the one used by the majority of chloroplastic proteins. Outer membrane proteins may use a receptor system that is not susceptible to thermolysin digestion. Alternatively, they may interact directly with the lipid components of the outer membrane so that a proteinaceous receptor is not required.

Altogether, current data suggest that chloroplastic outer membrane proteins have adopted a unique import pathway during evolution. This could be a result of the possible unique origin of the outer membrane. It could also be a reflection of different mechanisms evolved either to translocate proteins across two membranes (as in the case protein translocation into the inside of chloroplasts), or to insert proteins into only one membrane (as in the case of protein import to the outer membrane). A direct insertion from the cytosol, without need for an extra peptide extension and its subsequent removal via processing, may be the most viable and efficient import mechanism that could be evolved for the outer membrane proteins under evolutionary pressure. As for the ATP and a thermolysin-sensitive receptor independent aspect of the insertion, it may also be a general feature of all chloroplastic outer membrane proteins. However, one noteworthy point is that both of the chloroplastic outer membrane proteins studied so far are relatively small in size. Thus, this spontaneous insertion mechanism may only be used by small proteins. Future work is required to determine whether the pathway used by OM14 and E6.7 represents a general feature of all chloroplastic outer membrane proteins. It also remains to be elucidated where the targeting information resides in the polypeptides of outer membrane proteins since they do not possess cleavable transit peptides, and also what mechanisms provide the specificity for these proteins to be targeted to the chloroplastic outer membrane.

3.2. Targeting to the inner envelope membrane

The transport of cytoplasmically synthesized proteins to the inner envelope membrane shares more similarities with transport of proteins across the envelope than with insertion of proteins into the outer envelope membrane. The transport of several inner membrane proteins has been studied [20,33; Li and Keegstra, unpublished results], and in each case, the proteins are synthesized as higher molecular weight precursors. The import of these precursors requires energy and a thermolysin-sensitive component on the surface of chloroplasts.

Although it has not yet been shown that inner membrane proteins interact with the same transport apparatus as other internal proteins, this conclusion seems likely judging from their similar requirements for import into chloroplasts. The question then arises as to how these proteins are directed to the inner envelope membrane. Two hypotheses can be considered, based on analogy with the targeting of proteins to the inner mitochondrial membrane. The first is the stop-transfer hypothesis [34]. In this situation, the inner membrane proteins contain a stop-transfer signal that halts their translocation across the inner envelope membrane and functions to anchor the protein in the membrane. This hypothesis has been suggested for certain proteins of the inner mitochondrial membrane [35].

The second possibility for targeting of inner membrane proteins is the conservative sorting hypothesis [35]. In this case, the inner membrane proteins are transported across the envelope into the organelle, where the targeting signal is removed. Then the processed protein is inserted into the inner envelope membrane from the interior of the organelle. This second step resembles the export process in prokaryotes. Thus, although this pathway seems convoluted, it makes sense in terms of the postulated endosymbiotic origin of mitochondria and chloroplasts. However, if this is the pathway used by the chloroplastic inner membrane proteins, then the question arises as to how the inner membrane proteins distinguish between the thylakoid membrane and the inner envelope membrane once they are transported into the stroma. The same question also applies to targeting of proteins to the intermembrane space of the envelope. Some mitochondrial intermembrane space proteins have been shown to use the conservative sorting pathway for their targeting to the intermembrane space. Complications again arise for chloroplastic intermembrane space proteins since they will need an additional mechanism to distinguish between the intermembrane space and the thylakoid lumen if they are transported first into the stroma. An amphiphilic α-helix structure in the transit peptide of some chloroplastic inner membrane proteins has been suggested to be the envelope-targeting domain of chloroplastic inner membrane proteins [33]. However, it is not clear why the amphiphilic α-helix is in the N-terminus of some transit peptides and is in the extreme C-terminus of some other transit peptides. Also, this kind of amphiphilic α-helix is not found in the transit peptides of yet other inner membrane proteins (Li and Keegstra, unpublished results). Further work will be required to determine which of these hypotheses applies to proteins of the chloroplastic inner envelope membrane, and what the topogenic sequences are that direct a certain targeting pathway.

4. Summary and future prospects

The past decade has seen many advances in our understanding of how cytoplasmically synthesized proteins are targeted into chloroplasts. For example, several groups have demonstrated that the transit peptide of a precursor is both necessary and sufficient for targeting a protein into chloroplasts [36]. The amino acid sequences and general features of many transit peptides have been described [37]. Many features of the transport process are now known [14]. For example, several groups have demonstrated that ATP, but not a proton motive force, is needed for transport across the envelope membranes [8,22,23]. Transport into or across the thylakoid membrane is a separate event that can be distinguished from envelope transport by several criteria [5,38]. More recently, some progress has been made in identifying putative components of the transport apparatus [16]. The major challenge now is to identify all the components of the transport apparatus and to determine their functions during the translocation of precursor proteins. Only when this has been accomplished will it be possible to provide a molecular level description of the transport events.

References

1. Keegstra, K. (1989) Cell 56, 247–253.
2. Hand, J.M., Szabo, L.J., Vasconcelos, A.C. and Cashmore, A.R. (1989) EMBO J. 8, 3195–3206.
3. Lamppa, G.K. (1988) J. Biol. Chem. 263, 14996–14999.
4. Smeekens, S., Bauerle, C., Hageman, J., Keegstra, K. and Weisbeek, P. (1986) Cell 46, 365–375.
5. Hageman, J., Baecke, C., Ebskamp, M., Pilon, R., Smeekens, S. and Weisbeek, P. (1990) Plant Cell 2, 479–494.
6. Bauerle, C., Dorl, J. and Keegstra, K. (1991) J. Biol. Chem. 266, 5884–5890.
7. Olsen, L.J., Theg, S.M., Selman, B.R. and Keegstra, K. (1989) J. Biol. Chem. 264, 6724–6729.
8. Theg, S.M., Bauerle, C., Olsen, L.J., Selman, B.R. and Keegstra, K. (1989) J. Biol. Chem. 264, 6730–6736.

9. Friedman, A.L. and Keegstra, K. (1989) Plant Physiol. 89, 993–999.
10. Soll, J. (1988) Plant Physiol. 87, 898–903.
11. Soll, J., Fischer, I. and Keegstra, K. (1988) Planta 176, 488–496.
12. McCarty, D.R., Keegstra, K. and Selman, B.R. (1984) Plant Physiol. 76, 584–588.
13. Cline, K., Keegstra, K. and Staehelin, L.A. (1985) Protoplasma 125, 111–123.
14. Keegstra, K., Olsen, L.J. and Theg, S.M. (1989) Annu. Rev. Plant Physiol. Plant Mol. Biol. 40, 471–501.
15. Perry, S.E., Buvinger, W.E., Bennett, J. and Keegstra, K. (1991) J. Biol. Chem. 266, 11882–11889.
16. Schnell, D.J., Blobel, G. and Pain, D. (1991) J. Biol. Chem. 266, 3335–3342.
17. Cornwell, K.L. and Keegstra, K. (1987) Plant Physiol. 85, 780–785.
18. Pain, D., Kanwar, Y.S. and Blobel, G. (1988) Nature 331, 232–237.
19. Schnell, D.J., Blobel, G. and Pain, D. (1990) J. Cell Biol. 111, 1825–1838.
20. Flügge, U.I., Fischer, K., Gross, A., Sebald, W., Lottspeich, F. and Eckerskorn, C. (1989) EMBO J. 8, 39–46.
21. Willey, D.L., Fischer, K., Wachter, E., Link, T.A. and Flügge, U.-I. (1991) Planta 183, 451–461.
22. Flügge, U.I. and Hinz, G. (1986) Eur. J. Biochem. 160, 563–570.
23. Pain, D. and Blobel, G. (1987) Proc. Natl. Acad. Sci. USA 84, 3288–3292.
24. Neupert, W., Hartl, F.-U., Craig, E.A. and Pfanner, N. (1990) Cell 63, 447–450.
25. Pfaller, R., Steger, H.F., Rassow, J., Pfanner, N. and Neupert, W. (1988) J. Cell Biol. 107, 2483–2490.
26. Vestweber, D., Brunner, J. Baker, A. and Schatz, G. (1989) Nature 341, 205–209.
27. Kang, P.-J., Ostermann, J., Shilling, J., Neupert, W., Craig, E.A. and Pfanner, N. (1990) Nature 348, 137–143.
28. Marshall, J.S., DeRocher, A.E., Keegstra, K. and Vierling, E. (1990) 87, 374–378.
29. Chitnis, P.R. and Nelson, N. (1991) J. Biol. Chem. 266, 58–65.
30. Douce, R. and Joyard, J. (1990) Annu. Rev. Cell Biol. 6, 173–216.
31. Salomon, M., Fischer, K., Flügge, U.-I. and Soll, J. (1990) Proc. Natl. Acad. Sci. USA 87, 5778–5782.
32. Li, H.-m., Moore, T. and Keegstra, K. (1991) Plant Cell 3, 709–717.
33. Dreses-Werringloer, U., Fischer, K., Wachter, E., Link, T.A. and Flügge, U.-I. (1991) Eur. J. Biochem. 195, 361–368.
34. van Loon, A.P.G.M. and Schatz, G. (1987) EMBO J. 6, 2441–2448.
35. Mahlke, K., Pfanner, N., Martin, J., Horwich, A.L., Hartl, F.-U. and Neupert, W. (1990) Eur. J. Biochem. 192, 551–555.
36. Lubben, T.H., Theg, S.M. and Keegstra, K. (1988) Photosynthesis Res. 17, 173–194.
37. von Heijne, G., Hirai, T., Klösgen, R.-B., Steppuhn, J., Bruce, B., Keegstra, K. and Herrmann, R. (1991) Plant Mol. Biol. Rep. 9, 104–126.
38. Reed, J.E., Cline, K., Stephens, L.C., Bacot, K.O. and Viitanen, P.V. (1990) Eur. J. Biochem. 194, 33–42.

CHAPTER 23

Transport of proteins into the thylakoids of higher plant chloroplasts

COLIN ROBINSON

Department of Biological Sciences, University of Warwick, Coventry, CV4 7AL, UK

Abstract

Nuclear-encoded thylakoid lumen proteins, such as plastocyanin and the three subunits of the oxygen-evolving complex (OEC) are imported by a complex process that involves the operation of two distinct protein translocation systems and two processing peptidases. In order to analyse the later stages of this pathway in detail, we have developed an assay for the in vitro import of lumenal proteins by isolated thylakoids. Using this assay, it is found that the translocation of the three OEC proteins (but not plastocyanin) is driven by the transthylakoid proton gradient. Other differences in import requirements are also observed between different OEC proteins; the import of the 33 kDa subunit into thylakoids requires the activity of at least one stromal factor and the presence of ATP. In contrast, the import of the 23 kDa and 16 kDa subunits requires only the ΔpH. The combined results suggest that either multiple protein translocation systems operate in the thylakoid membrane, or that a single translocation system is particularly adaptable in its mode of action.

1. Introduction

The assembly of the photosynthetic machinery in higher plant chloroplasts is a complex process, in part because the component proteins are synthesized by two distinct genetic systems. Approximately half of the known thylakoidal proteins are synthesized within the organelle, whereas the remainder are imported after synthesis in the cytosol. The biogenesis of the latter group of proteins has attracted considerable interest in recent years, since these proteins must cross both of the envelope membranes and the soluble stromal phase to reach the thylakoid membrane. We are particularly interested in the import of hydrophilic thylakoid lumen proteins, which must in addition traverse the thylakoid membrane to reach their site of function.

Studies on the import of thylakoid lumen proteins have focused primarily on four prominent photosynthetic proteins: plastocyanin, a mobile electron carrier, and the 33 kDa, 23 kDa and 16 kDa proteins (33K, 23K, 16K) of the photosystem II oxygen-evolving complex. It is now believed that these proteins are imported by a two-step mechanism after synthesis in the cytosol. The proteins are initially synthesized with bipartite pre-sequences which consist of two targeting signals in tandem, specifying envelope transit and thylakoid transfer. The transit sequence first directs transport into the stroma, after which it is removed by a stromal processing peptidase (SPP). The transfer signal then mediates transport of the intermediate form across the thylakoid membrane, after which processing to the mature size is carried out by a thylakoidal processing peptidase, TPP [1–3]. This two-stage import model, which is depicted in Fig. 1, was prompted by three key observations.

(1) The presequences of lumenal proteins contain two structurally distinct domains. The amino-terminal domains clearly resemble the pre-sequences of stromal proteins, being hydrophilic, basic and rich in hydroxylated residues. In contrast, the carboxy-terminal thylakoid transfer domains contain a hydrophobic core region reminiscent of so-called signal sequences [4,5].

(2) Transient, intermediate-size stromal forms of several lumenal proteins have been observed during the import of in vitro synthesized precursors into isolated chloroplasts [2,6].

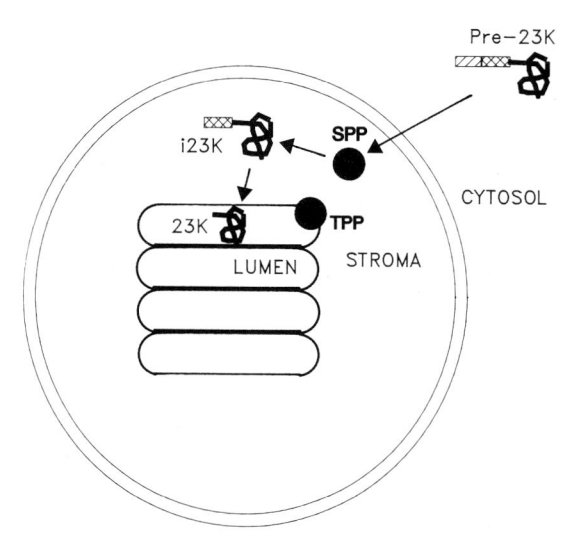

Fig. 1. Two-step model for the import of thylakoid lumen proteins. Lumenal proteins such as the 23 kDa oxygen-evolving complex protein (23K) are synthesized with a bipartite pre-sequence consisting of an envelope transit and a thylakoid transfer signal in tandem. After synthesis in the cytosol, pre-23K is imported into the chloroplast stroma and cleaved to an intermediate form (i23K) by a stromal processing peptidase (SPP). i23K is subsequently transported across the thylakoid membrane and processed to the mature size by a thylakoidal processing peptidase, TPP.

(3) The two-step proteolytic maturation sequence can be reconstructed in vitro by the addition of the appropriate processing peptidases. Both SPP and TPP have been highly purified [7,8] and shown to process lumenal protein precursors to intermediate and mature sizes, respectively [1,2].

The above observations strongly suggested that the import of thylakoid lumen proteins involves the operation of two distinct protein translocation systems, one of which is located in the chloroplast envelope and the other in the thylakoid membrane. In order to understand the mechanism of the thylakoid system in greater detail, we have sought to develop an efficient assay for the import of proteins into isolated thylakoids. The purpose of this chapter is, firstly, to describe the development and refinement of such an import assay, and, secondly, to describe how this assay has been used to analyse the mechanisms by which proteins are transported across the thylakoid membrane.

2. Results

2.1. Development of an in vitro assay for the import of proteins by isolated thylakoids

A prerequisite for the detailed study of any protein transport system is the availability of an efficient, reliable, in vitro transport assay. It has been possible since the late 1970s to reconstitute protein import into isolated, intact chloroplasts and mitochondria, and numerous studies on the chloroplast and mitochondrial protein import mechanisms have relied on the use of these in vitro assays (reviewed in Chapters F1 and F3). In the case of the chloroplast, it is now clear that protein transport across the envelope membranes takes place post-translationally, requires ATP, and proceeds via surface-exposed receptor molecules [9–13]. The intact chloroplast import assay is, however, of limited use for the study of thylakoidal protein transport, because the latter system cannot be studied separately from the envelope-based translocation system. In view of this problem, we considered it essential to develop an assay for the in vitro import of lumenal proteins by isolated thylakoids. Our approach was to incubate, under various conditions, washed pea thylakoids, crude stromal extract and precursors of two lumenal proteins, 33K and 23K. The precursors were synthesized by transcription of full-length cDNAs followed by translation in a wheat-germ lysate [2].

Initially, it was found that the import of 33K could be driven by adding ATP to the reaction mixture. In the presence of stromal extract, thylakoids and ATP, it was found that pre-33K was converted to i33K (by SPP in the stroma) and also to mature-size 33K. The mature-size 33K was resistant to added protease, indicating that import into the lumenal space had occurred [14].

However, two problems emerged during studies to improve and characterize this import assay. Firstly, it was found that the efficiency of import of 33K could not be improved beyond a figure of about 20% of available precursor. More seriously, it was found that virtually no import of 23K took place under these conditions. We therefore attempted to improve the efficiency of import of both proteins, and found that a critical requirement in each case was light. In the presence of light of moderate intensity (300 μeinsteins m^{-2} s^{-1}), isolated thylakoids are capable of importing up to 40% of available 33K and up to 70% of available 23K [15,16]. Recent work has shown that 16K can be imported with even higher efficiency [17,18].

2.2. Energy requirements for the import of proteins into isolated thylakoids

Our initial observation of light-driven import of 33K and 23K into thylakoids raised the obvious possibility that translocation is driven by the thylakoidal proton motive force. To test this possibility, we included inhibitors in the import assay which would either block the formation of the proton motive force, or collapse the proton gradient (ΔpH) or electrical potential ($\Delta\psi$) components selectively. We found that import does indeed require an energized membrane, and that the dominant component in this respect is ΔpH, and not $\Delta\psi$ [15].

In the case of 23K and 16K, the ΔpH appears to be the only energy requirement for import into thylakoids; Cline et al. [17] have found that removing nucleotide triphosphates from the incubation mixture has no effect on the rate of transport. In contrast, we have found that the import of 33K does require ATP. Import is inhibited if the pre-33K translation mixture is spun through a Sephadex G-50 column to remove small molecules, but can be restored by the addition of 1 mM Mg-ATP to the import incubation.

In view of the critical role of the ΔpH in the import of the three oxygen-evolving complex proteins, it might be expected that the ΔpH would be involved, at least to some extent, in the import of all other lumenal proteins. Surprisingly, this is not the case. Theg et al. [12] showed that the transport of plastocyanin into the thylakoid lumen of intact chloroplasts was not affected by the presence of ionophores or electron transport inhibitors. In collaborative experiments with R. Klösgen and R.G. Herrmann, we have similarly found that a ΔpH is not required for the import of plastocyanin into isolated thylakoids. This collaboration has indicated that a second protein, CF$_0$ subunit II, is also imported in the absence of a ΔpH. At present it is difficult to explain these differences in requirements, since all of these proteins are synthesized with apparently similar thylakoid transfer signals. Clearly, however, these findings raise the possibility that the thylakoid membrane contains more than one type of translocation system.

2.3. Events in the stroma

On the basis of the available evidence, there seems to be little doubt that two distinct translocation systems operate in the targeting of proteins from the cytosol to the thylakoid lumen. Clearly, it is of interest to investigate the fate of these proteins while in the stromal phase between the translocation systems, since a number of fundamentally different possibilities can be envisaged. For example, after translocation into the stroma, do the proteins refold, traverse the stroma, and then become unfolded by the thylakoid import machinery? Or are they maintained in an unfolded, translocation-competent state by stromal elements of the import machinery?

In the early stages of this work, we assumed that one essential soluble factor would be the stromal processing peptidase. To our surprise, we have found precisely the opposite: there is as yet no evidence to indicate that the activity of this enzyme is required for the targeting of any protein into the thylakoid lumen. It is clear that isolated thylakoids are capable of importing the full precursor forms of 23K, 16K and plastocyanin [16–19]. In the case of 23K, we were able to chemically modify the precursor and completely block cleavage by the stromal peptidase; the protein is nevertheless targeted into the lumen of intact chloroplasts with kinetics that are indistinguishable from those of unmodified precursor [16]. In at least some cases, therefore, the cleavage signal for the stromal peptidase appears to be essentially redundant, and it is not clear why this signal has been retained in the pre-sequence.

Both pre-23K and pre-16K can be efficiently imported by isolated thylakoids in the complete absence of added stromal extract [16–18]. These findings tend to suggest that, at least in vitro, no soluble stromal factors operate in the translocation of these proteins across the thylakoid membrane. It should, however, be stressed that we cannot rule out the possibility that the wheat-germ lysate contains soluble proteins which assist in the transport of 23K and 16K into thylakoids.

The requirements for 33K are notably different in that the presence of stromal extract is essential for efficient import into thylakoids [16]. In initial studies using the full precursor as substrate, we suggested that this may reflect a requirement for stromal processing peptidase; the efficiency of import appeared to correlate well with the extent to which pre-33K was processed to the intermediate size. More recently, however, we have conducted studies using an artificial, intermediate-size 33K polypeptide which no longer contains a cleavage site for the stromal peptidase. Figure 2 shows that this polypeptide is imported by isolated thylakoids only in the presence of stromal extract, indicating that at least one stromal factor is required for efficient import. Preliminary results indicate that the factor is heat-sensitive and non-dialysable, strongly suggesting that it is proteinaceous (although we have not yet ruled out the possible involvement of RNA). Current work is aimed at investigating the structure, mode of action and substrate specificity of this factor. Results so far obtained indicate that stromal extract is also required for the import of plastocyanin into thylakoids, but not for 16K import.

Why is a stromal factor required for the import of 33K (and probably plastocyanin) but not 23K or 16K? At present we are unable to answer this question, since we know nothing about the mode of action of this factor. However, one possibility under consideration is that the factor has an unfolding or antifolding activity which is

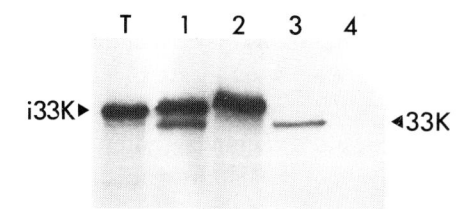

Fig. 2. Import of an artificial i33K into isolated thylakoids. Percoll purified pea chloroplasts were lysed in hypotonic buffer and the thylakoids were washed twice in the same buffer before resuspension in either stromal extract (lanes 1 and 3) or lysis buffer (lanes 2 and 4) and incubation with i33K which had been synthesized by in vitro transcription-translation of a truncated cDNA clone. After incubation for 30 min under illumination, samples were analysed by SDS-polyacrylamide gel electrophoresis and fluorography either directly (lanes 1 and 2) or after protease K treatment of the thylakoids (lanes 3 and 4). Lane T, i33K translation product; 33K, mobility of authentic mature 33K.

required to maintain 33K in an import-competent conformation. If this is the case, it is possible that 23K and 16K are rather loosely folded proteins which can be unfolded and translocated in the absence of this factor. Whatever the answer, these findings represent another significant example of how different lumenal proteins can have widely differing requirements for translocation across the thylakoid membrane. These different requirements, which are listed in Table I, are particularly striking when it is considered that only a few lumenal proteins have been studied to date. Table I also includes some recent findings on the import of two other thylakoid proteins, the Rieske FeS protein and CF_0 subunit II.

TABLE I

Requirements for the import of proteins into isolated thylakoids

Protein	Import requirements				Reference
	ΔpH	SPP cleavage	Stromal extract	ATP	
23K, 16K	Yes	No	No	No	Mould and Robinson [15], Mould et al. [16], Cline et al. [17], Klösgen et al. [18]
33K	Yes	ND[a]	Yes	Yes	Mould and Robinson [15], Mould and Robinson (unpublished),
Plastocyanin	No	No	Yes	ND	Bauerle and Keegstra [19], Klösgen, Robinson and Herrmann (unpublished)
CF_0II	No	No	ND	ND	Klösgen, Robinson and Herrman (unpublished)
Rieske FeS	Yes	No	ND	ND	J.C. Gray (pers. commun.)

[a] ND, not determined.

2.4. Maturation of imported thylakoid lumen proteins

During or shortly after translocation across the thylakoid membrane, lumenal proteins are processed to the mature size by TPP. Early studies showed that TPP is highly specific for imported lumenal proteins, and that this enzyme is tightly bound to the thylakoid membrane with the active site on the lumenal face [2,5,8,20]. However, Halpin et al. [5] made a particularly significant observation which was prompted by the emergence of a number of primary sequences for lumenal protein precursors. A feature common to all thylakoid transfer signals is the presence of a hydrophobic core domain, and short-chain residues (usually alanine) at the -3 and -1 positions, relative to the TPP cleavage site [4]. These are also some of the primary characteristics of signal sequences which direct protein export in bacteria, and Halpin et al. [5] indeed found that the reaction specificities of TPP and *Escherichia coli* signal peptidase are extremely similar.

These findings raised the interesting possibility that the thylakoidal import/maturation system had evolved from a prokaryotic export system, which is by no means improbable considering that chloroplasts are widely believed to have originated from endosymbiotic cyanobacteria. Furthermore, a striking example of just such a scenario has emerged in studies on mitochondrial protein transport. The import of cytochrome b_2 into the intermembrane space involves two proteolytic processing steps, one of which is carried out by the matrix processing peptidase, and the other by inner membrane protease 1. Cloning and sequencing of the latter enzyme has revealed a region of considerable homology with bacterial signal peptidase [21,22]. One of the top priorities in this laboratory is to isolate clones encoding TPP in order to assess whether this enzyme may also be a direct descendant from a prokaryotic ancestor.

3. Discussion

The field of thylakoidal protein transport is still in its infancy, and many questions concerning the structure and mechanism of this system remain to be answered. Nevertheless, recent studies have already produced significant progress on some points, and have also produced some surprises. Two particularly noteworthy points have emerged, both of which require urgent further study.

(1) The thylakoidal ΔpH clearly plays a critical role in the import of some of the lumenal proteins. Indeed, with 23K and 16K, it may be that this is the only requirement for translocation, since these proteins can be imported by thylakoids in the absence of ATP or added stromal extract. At present, however, we have no idea how translocation is driven by the ΔpH. It will be particularly important to determine the precise driving force for translocation, since two radically different mechanisms can be envisaged: it may be the proton flux into the lumen which is important (perhaps in a type of symport mechanism) or, alternatively, it may be the pH

difference which is critical, i.e. an acidic lumen may be a dominant requirement. Resolving this issue will be essential for a detailed understanding of the translocation system.

(2) Despite the fact that only a handful of lumenal proteins have been studied, we are already faced with a bewildering variety of import requirements. Although a ΔpH is vital for the translocation of 33K, 23K, 16K and the Rieske FeS protein, it is unnecessary for the transport of plastocyanin and CF_0II. On the other hand, import of 33K and plastocyanin requires the activity of at least one stromal factor, whereas the import of 23K and 16K does not. At present it is very difficult to rationalize these observations; most of these proteins have rather similar targeting signals which would tend to suggest that they are imported by a single translocation system. The results raise the interesting possibility that a single translocation system is particularly adaptable in its mode of action, depending on the characteristics of the protein to be transported.

Obviously, there are other critical issues which also need to be examined. For example, we have found no evidence so far for the existence of import receptors on the thylakoid surface. To date, these experiments have been technically difficult, because protease treatment of thylakoids would destroy the electron transport system. This would completely block formation of the proton motive force and inhibit import of the oxygen-evolving complex subunits, thereby masking any effects due to the digestion of putative import receptors. However, these experiments should be possible if plastocyanin or CF_0II is used as a substrate, since these proteins can now be imported in the absence of a proton motive force.

References

1. Hageman, J., Robinson, C., Smeekens, S. and Weisbeek, P. (1986) Nature 324, 567–569.
2. James, H.E., Bartling, D., Musgrove, J.E., Kirwin, P.M., Herrmann, R.G. and Robinson, C. (1989) J. Biol. Chem. 264, 19573–19576.
3. Ko, K. and Cashmore, A.R. (1989) EMBO J. 8, 3187–3192.
4. von Heijne, G., Steppuhn, J. and Herrmann, R.G. (1989) Eur. J. Biochem. 180, 535–541.
5. Halpin, C., Elderfield, P.D., James, H.E., Zimmermann, R., Dunbar, B. and Robinson, C. (1989) EMBO J. 8, 3917–3921.
6. Smeekens, S., Bauerle, C., Hageman, J., Keegstra, K. and Weisbeek, P. (1986) Cell 46, 365–375.
7. Robinson, C. and Ellis, R.J. (1985) Eur. J. Biochem. 152, 67–73.
8. Kirwin, P.M., Elderfield, P.D. and Robinson, C. (1987) J. Biol. Chem. 262, 16386–16390.
9. Highfield, P.E. and Ellis, R.J. (1978) Nature 271, 420.
10. Chua, N.-H. and Schmidt, G.W. (1978) Proc. Natl. Acad. Sci. 75, 6110–6117.
11. Grossman, A.R., Bartlett, S.G. and Chua, N.-H. (1980) Nature 285, 625–628.
12. Theg, S.M., Bauerle, C., Olsen, L.J., Selman, B.R. and Keegstra, K. (1989) J. Biol. Chem. 264, 6730.
13. Cline, K., Werner-Washburne, M., Lubben, T. and Keegstra, K. (1985) J. Biol. Chem. 260, 3691–3696.
14. Kirwin, P.M., Meadows, J.W., Shackleton, J.B., Musgrove, J.E., Elderfield, P.D., Mould, R., Hay, N.A. and Robinson, C. (1989) EMBO J. 8, 2251–2255.
15. Mould, R.M. and Robinson, C. (1991) J. Biol. Chem. 266, 12189–12193.

16. Mould, R.M., Shackleton, J.B. and Robinson, C. (1991) J. Biol. Chem. 266, 17286–17290.
17. Cline, K., Ettinger, W. and Theg, S.M. (1992) J. Biol. Chem. 267, 2688–2696.
18. Klösgen, R., Brock, I.W., Herrmann, R.G. and Robinson, C. (1992) Plant Mol. Biol. 18, 1031–1034.
19. Bauerle, C. and Keegstra, K. (1991) J. Biol. Chem. 266, 5876–5883.
20. Kirwin, P.M., Elderfield, P.D., Williams, R.S. and Robinson, C. (1988) J. Biol. Chem. 263, 18128–18132.
21. Schneider, A., Behrens, M., Scherer, P., Pratje, E., Michaelis, G. and Schatz, G. (1991) EMBO J. 10, 247–254.
22. Behrens, M., Michaelis, G. and Pratje, E. (1991) Mol. Gen. Genet 228, 167–176.

CHAPTER 24

Comparison of two different protein translocation mechanisms into chloroplasts

JÜRGEN SOLL[1], HEIKE ALEFSEN[2], BIRGIT BÖCKLER[1], BIRGIT KERBER[3], MICHAEL SALOMON[2] and KARIN WAEGEMANN[1]

[1]Botanisches Institut, Universität Kiel, Olshausenstraße 40 W-2300 Kiel 1, Germany,
[2]Botanisches Institut, Universität München, Menzinger Straße 67 W-8000 München 19, Germany and [3]Fachrichtung Botanik, Universität des Saarlandes, W-6600 Saarbrücken, Germany

1. Introduction

Chloroplasts are highly structured plant specific organelles. They possess three discrete membrane systems which differ in composition and function, i.e. the outer/inner envelope and the thylakoid membranes. In addition three solute spaces can be distinguished, i.e. the space between the envelope membranes, the stroma and the thylakoid lumen [1]. While most of the chloroplastic proteins, which are synthesized as precursors in the cytosol, seem to follow a common route of translocation into the organelle, proteins of the outer envelope, which is in direct contact with the cytosol, are inserted (imported) by a very different and distinct mechanism [2,3].

A typical polypeptide destined for the inside of the organelle, possesses a cleavable target sequence, retains a loosely folded conformation with the help of molecular chaperones, is recognized by proteinaceous receptors on the organellar surface, requires low concentration (μM) ATP for binding but high concentrations (mM) for complete translocation through the membranes [4,5]. Outer envelope polypeptides (OEP) studied so far, do not possess a cleavable target sequence, do not require protease sensitive receptors on the organellar surface and do not require ATP for either binding or insertion into the outer envelope [3,6].

2. Results and discussion

2.1. Import characteristics of pSSU and OEP 7

A typical import experiment for a plastidic precursor protein destined for the inside

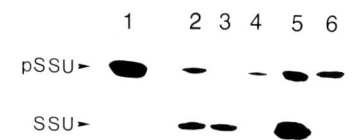

Fig. 1. Characteristics of pSSU import into chloroplasts. pSSU translation product (lane 1) is imported into chloroplasts and processed to its mature form (lane 2). SSU appears protease protected inside the organelle after thermolysin treatment (lane 3). Chloroplasts pretreated with thermolysin bind pSSU only to a very small extent in a non-productive way (lane 4). The import assay was either not depleted (lane 5) or depleted of ATP (lane 6) through the action of apyrase. Methods are described by Waegemann and Soll [15].

of the organelle is shown in Fig. 1. The precursor protein (pSSU) binds to the chloroplast outer envelope, it is subsequently translocated inside the organelle, processed to its mature form and protected against externally added protease. Chloroplasts pretreated by thermolysin, a protease which only digests surface exposed polypeptides [7,8], bind pSSU only to a very small extent in a non-productive manner [2]. Similar results are obtained if ATP is removed from the import incubation mixture by the ATP hydrolyzing enzyme apyrase. Binding is greatly reduced and import not observed, demonstrating the ATP dependence of binding as well as translocation [5].

The insertion (import) of OEP 7 seems to follow a quite distinct pathway. No shift in molecular weight can be observed between the translation product and the inserted form, demonstrating the absence of a cleavable transit sequence [3]. Translocation experiments carried out in the light, i.e. in the presence of ATP, show no greater OEP 7 translocation efficiency than those carried out in the dark, i.e. in the absence of ATP (Fig. 2). Neither apyrase treatment nor the simultaneous inclusion of a non-hydrolysable ATP analog, adenylylimidodiphosphate, influenced the yield of OEP 7

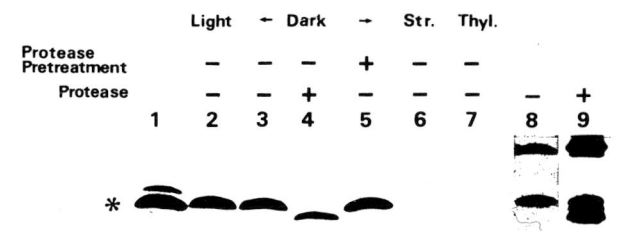

Fig. 2. Characteristics of OEP 7 import (insertion) into chloroplasts. OEP 7 translation product (lane 1) was incubated with intact chloroplasts either in the light (lane 2), i.e. presence of ATP, or in the dark (lane 3), i.e. absence of ATP (translation product was apyrase treated). Protease treatment after import yields a protease protected breakdown product (lane 4). Pretreatment of intact chloroplasts by thermolysin does not influence the efficiency of OEP 7 import (lane 5). OEP 7 is localized in the envelope membranes (lanes 6, 7). Lanes 8 and 9 show a silver stained gel of envelope membranes either not treated (lane 8) or treated (lane 9) with thermolysin. Protease treatment of imported OEP 7 and OEP 7 in situ gives identical proteolytic breakdown products (compare lanes 4 and 9). Methods are described by Salomon et al. [3].

insertion into the outer envelope membrane of chloroplasts. Translocation of OEP 7 followed by thermolysin treatment resulted in a lower molecular weight breakdown product identical to that found if envelope membranes were treated with protease (Fig. 2). OEP 7 contains only the N-terminal methionine, which is not removed by either maturation or thermolysin treatment. This clearly indicates that the N-terminus is exposed to the intermembrane space and therefore protease protected, while the C-terminus is on the cytosolic leaflet of the membrane and susceptible to external protease. Analysis of the amino acid sequence of OEP 7 corroborates these results and predicts only one membrane span [3]. The outline of this translocation route is supported by findings described in [6] where an identical insertion mechanism is described for an outer envelope protein of pea chloroplasts.

2.2. Specificity and mechanism of OEP 7 insertion

The import route outlined above for OEP 7 is very distinct and also differs from that described for proteins localized in the outer mitochondrial membrane, which require protease sensitive receptors and ATP for correct routing and efficient translocation [9]. Experiments were carried out to address the problems of specificity and mechanism of OEP 7 import. In an initial experiment, chloroplasts and mitochondria, both isolated from pea leaves, were incubated in the same import assay with OEP 7 translation product. After completion of the import reaction, chloroplasts were separated from mitochondria by differential centrifugation and each organelle type analyzed, respectively. The results (Fig. 3) demonstrate that OEP 7 binds to the surface of chloroplasts as well as mitochondria. Treatment of the organelles with thermolysin, however, clearly demonstrates that OEP 7 integrates only into the outer envelope of chloroplasts in the proper way but not into the outer membrane of mitochondria as judged from the protease protected breakdown product.

It has been shown that precursor proteins have to retain a loosely folded (transport competent) conformation in order to be translocated through the import apparatus of either mitochondria [9,10] or chloroplasts [4]. They do this with the help of

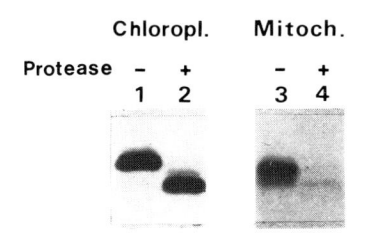

Fig. 3. OEP 7 specifically inserts into chloroplast but not mitochondrial membranes. Chloroplasts and mitochondria, isolated from pea leaves, were incubated in the same incubation assay with OEP 7 translation product. After completion of the reaction, organelles were separated by differential centrifugation and either not treated or treated by thermolysin prior to SDS-PAGE and fluorography.

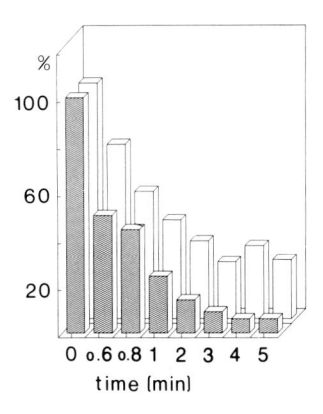

Fig. 4. Efficient insertion of OEP 7 into chloroplasts requires a special conformation. OEP 7 was synthesized either at 26°C (▨) or 37°C (□) in a reticulocyte lysate system. Only OEP 7 synthesized at 26°C inserts efficiently into the outer envelope of chloroplasts (not shown). OEP 7 synthesized at 26 °C (□) is more susceptible to trypsin treatment (units/ml) than OEP 7 synthesized at 37°C(□).

molecular chaperones, e.g. hsc70 [11,12]. We therefore addressed the question whether it was possible to distinguish between import competent and import incompetent OEP 7.

This was indeed found to be the case. OEP 7 synthesized in a reticulocyte lysate system at 37°C under improper cofactor conditions and prolonged reaction time did, no longer insert into the outer envelope of chloroplasts (not shown). Trypsin treatment of import incompetent OEP 7 demonstrates that it is much less sensitive to protease than the import competent form of OEP 7 (Fig. 4). The data strongly indicate that OEP 7 like other precursor polypeptides needs to retain a special conformation until it has been inserted into the outer envelope membrane. The outer chloroplast envelope has a unique lipid and protein composition in comparison not only to other organellar membranes exposed to the cytosol but also to the other membrane systems of the chloroplast. Phosphatidylcholine for example is not present in thylakoids. Monogalactosyldiglyceride and digalactosyldiglyceride which are exclusively found in plastidal membrane systems are also major lipid constituents of the outer envelope [1]. The specificity of OEP 7 insertion into the outer envelope could therefore be, at least in part, due to the specific lipid composition.

Most likely the interaction of OEP 7 with other outer envelope proteins aids the insertion specificity. This can be deduced from experiments presented in Fig. 5. Purified chloroplast membranes, i.e. outer envelope, inner envelope and thylakoids, were incubated with OEP 7 translation product. Only the interaction of OEP 7 with outer envelope membrane vesicles, either pretreated with or without protease, resulted in the correct insertion of the protein into the membrane bilayer. OEP 7 also bound to the other chloroplast membranes but we could not detect the typical proteolytic breakdown product, indicating that OEP 7 was either surface exposed and thus protease sensitive or inserted incorrectly into the membranes.

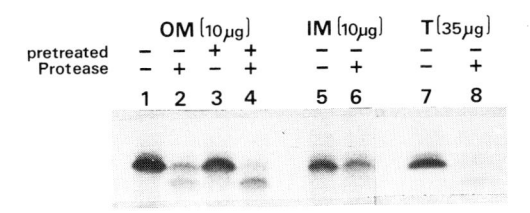

	OM (10μg)	IM (10μg)	T (35μg)
pretreated	− − + +	− −	− −
Protease	− + − +	− +	− +
	1 2 3 4	5 6	7 8

Fig. 5. OEP 7 integrates correctly only into isolated outer chloroplast envelopes. Chloroplasts were separated into outer/inner envelopes and thylakoids prior to incubation with OEP 7. All other manipulations are indicated on the top of each fluorogram. Methods as in Salomon et al. [3].

It has been shown that thylakoid membranes are isolated as outside-out vesicles [13]. The same was found to be the case for isolated outer envelope membranes [14]. These findings are important to interpret the described data correctly. Outside-out envelope vesicles were also used to study their interaction with pSSU, a normal precursor [15]. As in the organellar system, pSSU requires ATP and protease sensitive receptors to bind to the envelope surface (Fig. 6). The interaction between pSSU and the isolated envelope membrane does not halt at the binding stage but pSSU is partly inserted into the translocation apparatus as characterized by protease protected translocation intermediates [15]. Isolated outer envelope membranes therefore contain at least part of the chloroplast import machinery in a functionally active manner. Early events in binding and translocation can thus be analyzed in this isolated and partially purified system in vitro.

Our results indicated that the precursor was not translocated into the inside of the vesicle but was stuck in the transport apparatus. Solubilization of precursor loaded outer envelope membrane vesicles followed by sucrose density centrifugation resulted in the isolation of a membrane fraction with precursor protein still bound to it (Fig. 7)

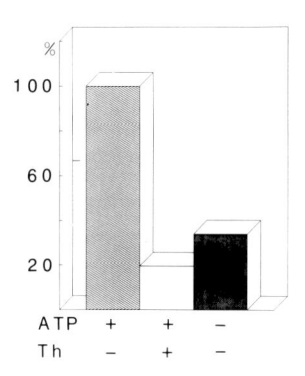

Fig. 6. Isolated outer envelope membrane vesicles represent a bonafide system to study pSSU binding. pSSU binding to isolated outer envelope was analyzed (column 1). Interaction is dependent on thermolysin (Th) sensitive receptors (column 2) and the presence of ATP (column 3). Results were quantified by laser densitometry of an exposed X-ray film. Methods as in Waegemann and Soll [15].

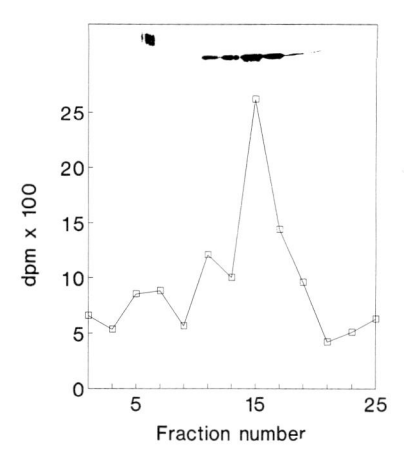

Fig. 7. A membrane complex loaded with pSSU can be isolated from outer envelope membranes. Purified outer envelopes were incubated with pSSU translation product, re-isolated, solubilized by digitonin and subjected to fractionation on a linear sucrose density gradient. pSSU distribution was determined by liquid scintillation counting (graph) or SDS-PAGE and fluorography (insert). Free pSSU stays on top of the gradient while complex bound pSSU migrates to higher density in the sucrose density gradient. Methods as described by Waegemann and Soll [15].

[15]. When the same experiment was carried out using OEP 7 translation product, no radiolabelled protein was detectable in fractions 14–18 of the sucrose gradient (compare Fig. 7) (not shown). This might indicate that OEP 7 does not enter the common import apparatus to be inserted into the outer envelope.

The membrane fraction recovered from the sucrose density gradient was shown to contain all the proteins necessary for a transit sequence and ATP dependent insertion of pSSU into the isolated complex [16]. The interaction of pSSU with the isolated complex also gave rise to the transport intermediates described for the chloroplast system. Isolation of an active import apparatus represents a major advantage to study the function of single components in the translocation event. So far we have identified an outer envelope localized hsc70 homologue and OEP 86 as constituents of the isolated import complex. The hsc70 homologue localized in the import apparatus could act in sequence with its cytosolic and stromal counterparts in an unidirectional import process [15,16].

The polypeptide composition of the isolated import apparatus together with results from crosslink studies imply the involvement, either direct or indirect, of a number of proteins in the translocation event. A schematic view of the different transport pathways into chloroplasts is depicted in Fig. 8.

The major envelope protein which was described as the master receptor for chloroplast protein import [17] and subsequently found to be identical to the phosphate-triose phosphate translocator of the inner envelope [18,19] is neither found in isolated outer envelope membranes which are active in pSSU recognition and insertion nor in the isolated import complex. Together with data presented in [20]

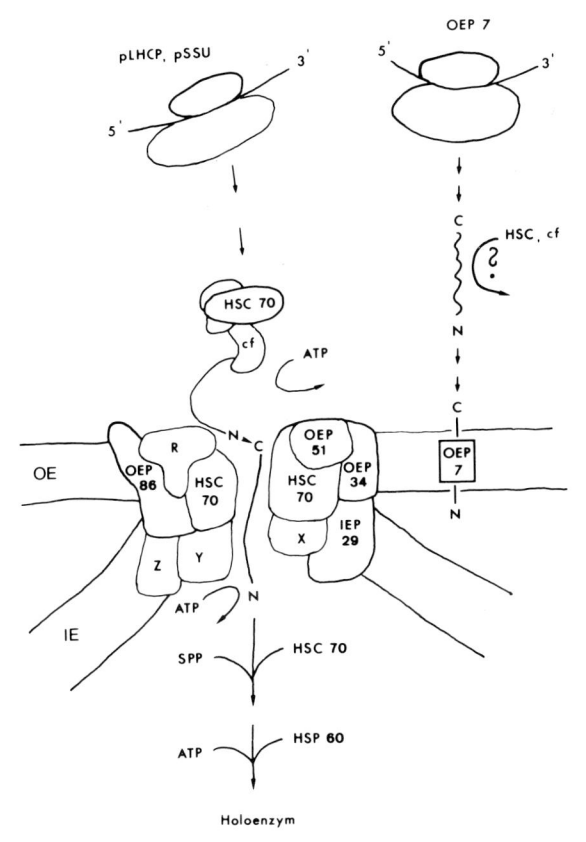

Fig. 8. Schematic view of two import pathways into chloroplasts. The scheme comprises proteins which have been implied to function in the import process in papers mentioned in the text. The number of proteins involved in the import and shown in the scheme is probably underestimated. We propose that the import apparatus forms a proteinaceous pore in the membranes which could be coated by hsc70 homologues to guide the passage of a precursor through the membrane. Other proteins of the complex are most likely also in close contact with the precursor protein on route to the inside of the organelle. This is not represented in the drawing. Much less is known about the insertion pathway of OEP 7. No envelope components have been identified so far which influence the insertion of OEP 7.

we conclude that a receptor for chloroplastic precursor proteins still remains to be identified.

Acknowledgements

We thank U.-I. Flügge, Würzburg, for part of the data presented in Fig. 2, K. Keegstra, Madison, for providing cDNA clones of pSSU and SSU. This work was supported by the Deutsche Forschungsgemeinschaft.

References

1. Douce, R., Block, M.A., Dorne, A.J. and Joyard, J. (1984) Subcell. Biochem. 10, 1–84.
2. Keegstra, K., Olsen, L.J. and Theg, S.M. (1989) Annu. Rev. Plant Physiol. 40, 471–501.
3. Salomon, M., Fischer, K., Flügge, U.-I. and Soll, J. (1990) Proc. Natl. Acad. Sci. USA 87, 5778–5782.
4. Waegemann, K., Paulsen, H. and Soll, J. (1990) FEBS Lett. 261, 89–92.
5. Olsen, L.J., Theg, S.M., Selman, B.R. and Keegstra, K. (1989) J. Biol. Chem. 264, 6724–6729.
6. Li, H.-M., Moore, T. and Keegstra, K. (1991) The Plant Cell 3, 709–717.
7. Joyard, J., Billecocq, A., Bartlett, S.G., Block, M.A., Chua, N.H. and Douce, R. (1983) J. Biol. Chem. 258, 10000–10006.
8. Cline, K., Werner-Washburne, M., Andrews, J. and Keegstra, K. (1984) Plant Physiol. 75, 675–678.
9. Hartl, F.-U., Pfanner, N., Nicholson, D.W. and Neupert, W. (1989) Biochim. Biophys. Acta 988, 1–45.
10. Attardi, G. and Schatz, G.A. (1988) Annu. Rev. Cell. Biol. 4, 289–333.
11. Deshaies, R.J., Koch, B.D., Werner-Washburne, M., Craig, E. and Schekmann, R. (1988) Nature 332, 800–805.
12. Chirico, W.J., Waters, M.G. and Blobel, G. (1988) Nature 332, 805–810.
13. Andersson, B. (1986) Methods Enzymol. 118, 325–338.
14. Waegemann, K., Eichacker, S. and Soll, J. (1992) Planta 187, 88–94.
15. Waegemann, K. and Soll, J. (1991) The Plant J. 1, 149–158.
16. Soll, J. and Waegemann, K. (1992) The Plant J. 2, 253–256.
17. Pain, D., Kanwar, Y.S. and Blobel, G. (1988) Nature 331, 232–237.
18. Schnell, D.J., Blobel, G. and Pain, D. (1990) J. Cell. Biol. 111, 1825–1838.
19. Flügge, U.-I., Fischer, K., Gross, A., Sebald, W., Lottspeich, F. and Eckerskorn, C. (1989) EMBO J. 8, 39–46.
20. Flügge, U.I., Weber, A., Fischer, K., Lottspeich, F., Eckerskorn, C., Waegemann, K. and Soll, J. (1991) Nature 353, 364–367.

Part G

Chaperones

CHAPTER 25

DnaJ homologs and protein transport

TAKAO KURIHARA and PAMELA A. SILVER

Department of Molecular Biology, Princeton University, Princeton, NJ 08544, USA

Abstract

A variety of intracellular protein transport processes require or are stimulated by members of the evolutionarily conserved 70 kDa heat shock protein (HSP70) family and other unidentified factors. The ability of HSP70s to control protein folding and assembly is thought to be the basis for their role in protein transport. In *Escherichia coli*, genetic and biochemical evidence suggests that the unique bacterial HSP70, DnaK, does not act alone but functions in concert with two other heat shock proteins, DnaJ and GrpE, to mediate such protein folding and assembly. Complete and partial DnaJ homologs have now been identified in other bacterial species as well as in the yeast, *Saccharomyces cerevisiae*. Genetic evidence suggests a role in protein transport for three of the four yeast homologs. Based on the functional complex of DnaJ and DnaK in *E. coli*, we propose that DnaJ homologs will interact with HSP70s via an evolutionarily conserved 'J-region' to mediate protein folding and assembly reactions important for protein transport processes. It will be of great interest to determine if the unidentified factors that are required along with HSP70s for maximum stimulation of some transport processes correspond to DnaJ homologs.

1. Introduction

1.1. Stimulation of protein transport by HSP70s and additional factors

The accurate and efficient transport of precursor proteins into cellular organelles is essential for proper cell function. For many organelles, transport accuracy is ensured by specific receptor proteins that recognize and bind targeting sequences encoded within precursor proteins (for reviews, see [1–3]). Transport efficiency is mediated, at least in part, by members of a major class of stress-induced proteins of ≈ 70 kDa molecular weight (for reviews, see [4–7]). Some members of this protein family are expressed constitutively (HSCs) in the cell, while others are only expressed transiently during stressful conditions (HSPs). In this review, these proteins are collectively termed HSP70s, since they are members of the same family. The amino acid sequences of HSP70s are highly conserved throughout evolution, indicating a fundamental function for these proteins. Distinct HSP70s are localized to several different cellular compartments and may have functions specific for those compartments. However, a common function that has emerged for all HSP70s is control of protein folding and complex assembly (for reviews, see [6,8]).

The ability of HSP70s to affect protein folding and complex assembly most likely explains their essential roles in several protein transport processes. In the yeast *Saccharomyces cerevisiae*, cytosolic and organelle-associated HSP70s are required for precursor protein import into both the ER lumen and the mitochondria [9–14]. In *E. coli*, the bacterial HSP70, DnaK, has been shown to facilitate export of lacZ hybrid proteins [15]. HSP70s may stimulate these transport processes at several levels. First, they may bind to precursor proteins and maintain them in translocation-competent conformations. Second, they may catalyze the assembly or disassembly of proteinaceous translocator channels embedded within organellar membranes. Finally, they may catalyze the refolding and assembly of proteins that have been translocated across membranes.

HSP70s are necessary but not always sufficient for efficient protein transport. In vitro transport experiments have demonstrated that precursor protein translocation into yeast microsomes and mitochondria requires HSP70s as well as another cytosolic component [9–11]. Similar assays have shown a requirement for HSP70s and an additional component in reticulocyte lysate for stimulation of protein import into canine pancreas microsomes [16]. These additional components may act with HSP70s to directly stimulate transport, may indirectly affect transport by modulating HSP70 function, and, like HSP70s, may also be involved in controlling protein folding and assembly. The identity of these eukaryotic factors that act with HSP70s is not yet known, but experiments with *E. coli* have revealed two prokaryotic proteins, DnaJ and GrpE, that fulfill such a role of functioning with an HSP70 (DnaK) to control protein folding and assembly (see below). Thus, the additional factors required along with HSP70s for maximum stimulation of protein transport in eukaryotes may have DnaJ- or GrpE-like function. The identification of such a family of DnaJ homologs in *S. cerevisiae* is the focus of this review.

1.2. E. coli DnaJ and GrpE function with and regulate bacterial HSP70 (DnaK)

There is much genetic and biochemical evidence for interactions between the three *E. coli* heat-shock proteins DnaK(HSP70), DnaJ, and GrpE (for reviews, see [17,18]). *dnaK*, *dnaJ*, and *grpE* were originally identified as *E. coli* host genes required for bacteriophage λ replication (for review, see [19]). Mutations in any of the three genes often result in similar pleiotropic defects in a variety of cellular processes, suggesting that the corresponding proteins perform similar functions in the cell. A variety of genetic and biochemical analyses have implicated DnaK, DnaJ, and GrpE in the kind of protein folding and disassembly processes thought to be important for protein transport (see below). While these experiments do not address protein transport per se, the conclusions that can be drawn about the role of DnaK, DnaJ, and GrpE in protein folding and assembly provide insight into how HSP70s and DnaJ- and GrpE-like proteins might function to stimulate protein transport.

1.2.1. Bacteriophage λ and P1 replication and protein complex disassembly
E. coli cells mutated in the *dnaK*, *dnaJ*, or *grpE* genes do not support bacteriophage λ replication [19]. Elegant biochemical reconstitution assays have demonstrated a requirement for the DnaK and DnaJ proteins at the initiation of λ replication [20–22]. In the absence of DnaK and DnaJ, host DnaB helicase activity is inhibited by phage λP protein in a pre-primed complex at the λ DNA replication origin. DnaK and DnaJ allow replication initiation to occur by catalyzing the partial disassembly of this pre-initiation complex, thus freeing DnaB helicase from λP. Based on DnaJ's affinity for λP and DnaB, Zylicz et al. [22] proposed that DnaJ might target DnaK to the complex acted upon by the two proteins.

Bacteriophage P1 exists as a low-copy-number plasmid in its prophage form (for review, see [23] and probably requires the DNA binding activity of phage-encoded RepA protein for P1 plasmid replication initiation [24–26]. E. coli cells with mutations in the *dnaK*, *dnaJ*, or *grpE* genes are deficient for P1 plasmid replication, suggesting that these three proteins affect RepA DNA-binding activity [27,28]. This genetic prediction is supported by biochemical assays in which DnaK and DnaJ activate RepA DNA-binding activity [29]. Wickner et al. [29] propose that DnaK and DnaJ may activate the DNA-binding activity of RepA by converting RepA from dimers to monomers. Since DnaJ and RepA form a complex [28], Wickner et al. [29] also propose that DnaJ could act to target DnaK to RepA substrate.

The λ and P1 replication systems suggest that DnaK and DnaJ act together in protein complex disassembly processes, as only the simultaneous presence of DnaK and DnaJ can catalyze such reactions. Interestingly, in both systems, DnaJ has been proposed to target DnaK to the substrate complexes by interacting with members of the complexes. GrpE was not required for activation of RepA DNA-binding activity, although GrpE decreased by 10-fold the concentration of DnaK required in the λ replication system [21,22], possibly due to a recycling effect (see below). Extrapolating from the role of DnaK and DnaJ in protein disassembly processes, the following models are possible for HSP70 and DnaJ-like protein function in protein transport. If proteinaceous complexes exist for translocation of proteins across membranes, then HSP70s and DnaJ-like proteins might catalyze their assembly or disassembly. HSP70s and DnaJ-like proteins might also indirectly stimulate protein transport by preventing aggregation of precursor or translocated proteins. The ER-localized HSP70 BiP has been proposed to bind to some translocated proteins in the ER lumen to prevent aggregation prior to their assembly into oligomeric complexes (for review, see [30].

1.2.2. Refolding of thermally inactivated λcI857 repressor

The temperature-sensitive mutant λcI857 repressor protein is inactivated at 42°C but rapidly renatures if returned to 32°C, as assayed both in vivo and in vitro by sequence-specific binding to λ DNA. In *E. coli* cells mutated for the *dnaK*, *dnaJ*, or *grpE* genes, thermally inactivated λcI857 repressor either renatures with delayed kinetics or not at all at 32°C, suggesting that those heat-shock proteins can stimulate protein refolding [31]. By analogy, it is possible that HSP70s and DnaJ-like proteins might function together to stimulate protein refolding during transport.

One pair of proteins already known to be involved in protein refolding during protein transport are the yeast mitochondrial HSP70 (SSC1) [32] and HSP60 proteins [33,34]. SSC1 function is required both for protein translocation into the mitochondria as well as for subsequent refolding of translocated proteins [13], while HSP60 is required both for refolding of proteins translocated into mitochondria as well as for assembly of those proteins into oligomeric complexes within the mitochondria [35,36]. Kang et al. [13] propose that the mitochondrial HSP70 may only be indirectly involved in protein refolding by mediating the transfer of unfolded, translocated proteins to the HSP60 'foldase'. If HSP70s and DnaJ-like proteins assist in protein refolding, they may do so directly by providing a structural scaffold [31] upon which protein refolding might occur. On the other hand, HSP70s and DnaJ-like proteins might only indirectly assist protein refolding by preventing denatured polypeptides from aggregating before they are acted upon by a 'foldase', like HSP60.

1.2.3. Proteolysis of puromycin-generated polypeptide fragments

Addition of puromycin aborts protein translation in *E. coli* cells and releases unstable polypeptide fragments from ribosomes. *E. coli* mutated in the heat-shock genes encoding DnaK, DnaJ, and GrpE are defective in proteolytic degradation of such unstable polypeptides, while overproduction of those three heat-shock genes increases the degradation rate [37]. Protease activity has not been demonstrated for DnaK, DnaJ, or GrpE, so it is likely that they are stimulating proteolysis of puromycyl fragments by rendering those polypeptides more susceptible to protease action.

Similar to the proposed DnaK-DnaJ scaffold in protein refolding, Straus et al. [37] propose that DnaK, DnaJ, and GrpE might have a structural role in proteolysis. One interpretation of this model is that DnaK and DnaJ are binding to unstable proteins and facilitating their proteolysis by maintaining them in a protease-sensitive conformation. A parallel idea in protein transport is that HSP70s and DnaJ-like proteins bind precursor proteins and facilitate their translocation by maintaining them in translocation-competent conformations. In yeast, the cytosolic HSP70s are proposed to fulfill such a function during protein translocation into the ER and mitochondria

[9–11]. The same studies indicated that an additional cytosolic factor acts synergistically with HSP70s to stimulate mitochondrial and ER translocation. Based on evidence from the *E. coli* systems, it is tempting to speculate that this factor might be a DnaJ-like protein.

1.2.4. Stimulation of DnaK ATPase activity by DnaJ and GrpE
All of the processes described above require DnaK, DnaJ, ATP, and possibly GrpE. The ATP-dependence is most likely due to DnaK's ATPase activity [38]. Purified DnaK, in addition to all other HSP70s tested so far, has a very low rate of ATP hydrolysis, measuring between 0.1 and 1.0 molecules ATP hydrolyzed per minute per HSP70 monomer (for review, see [8]). It has now been demonstrated that DnaJ and GrpE together can stimulate DnaK's ATPase activity by 20-fold, on average [39]. Since the V_{max}, but not the K_m, for DnaK's ATPase activity was affected, DnaJ and GrpE do not appear to increase DnaK's affinity for ATP. The roles of DnaJ and GrpE could be separated, as DnaJ alone stimulated hydrolysis of ATP, while GrpE stimulated release of bound ADP or ATP.

The control of HSP70 ATPase activity is crucial for its proposed role in protein folding and complexing. Based on experiments from several labs, it is likely that HSP70s bind substrate polypeptides in an ATP-independent manner [40] and that the energy of ATP hydrolysis elicits a conformational change both in the HSP70 and in the substrate polypeptide, which is simultaneously released [41,42]. It has been suggested that DnaJ and GrpE act to recycle DnaK function by stimulating its ATPase activity and promoting release of bound substrates [42]. A similar recycling scheme is proposed for the pair of heat-shock proteins GroEL and GroES [43]. Thus, DnaJ-like proteins might stimulate transport processes by similarly recycling and freeing HSP70s for substrate binding. In a more regulated scheme, DnaJ-like proteins might fine-tune the intrinsic HSP70 ATPase activity so that substrate proteins are released at a rate most efficient for transport.

2. Results

2.1. DnaJ homologs

Complete or partial homologs of *E. coli* DnaJ have been identified in three other bacterial species, the yeasts *S. cerevisiae* and *Schizosaccharomyces pombe*, and in *Drosophila melanogaster* (Table I). The regions of homology among these proteins are well conserved, suggesting that this family of DnaJ-like proteins has a fundamental role in cells, much like the highly conserved family of HSP70s. With reference to *E. coli* DnaJ [44,45], three distinctive regions of similarity can be identified in the other homologs (Fig. 1). All of the homologs contain a region that is very highly conserved to the 70 N-terminal amino acids of *E. coli* DnaJ. These highly conserved

TABLE I

The DnaJ family

Organism and protein	Functions and comments
E. coli DnaJ	Interacts and functions with DnaK in a variety of cellular processes involving protein folding and complexing (for reviews, see [17,18])
M. tuberculosis DnaJ	Both DnaJ homologs identified by sequencing downstream of DnaK
C. crescentus DnaJ	analogs [46,47]
R. fredii USDA 257 nolC	Transposon insertion into genomic nolC affects host range specificity of USDA257 strain [48]; genomic Southern blots with nolC gene probe identified nolC-related genes in other *R. fredii* strains and in *A. tumefaciens* strain
S. cerevisiae SCJ1	Overexpression alters localization of fusion protein between nucleus and mitochondria [49]; localization to mitochondria, possibly ER; SCJ1-disrupted haploid strains viable; genomic Southern blotting with SCJ1 gene probe identified several cross-hybridizing genes in *S. cerevisiae*
S. cerevisiae YDJ1/MAS5	Identified as component of matrix lamina pore complex [53] and in genetic selection scheme for mitochondrial import mutants [55]; cytoplasmic and associated with ER membrane; growth defect of YDJ1/MAS5-disrupted strains partially rescued by SIS1 overproduction
S. cerevisiae SIS1	High copy number suppressor of yeast mutants in SIT4 protein phosphatase [58]; cytoplasmic and concentrated around nucleus; interaction with 40 kDa protein dependent on gly/met-rich region; growth defect of SIS1-disrupted haploid strains not rescued by YDJ1/MAS5-overproduction
S. cerevisiae NPL1/SEC63	Identified in genetic selection schemes for nuclear localization [50] and ER translocation mutants [60,61]; similar to *E. coli* DnaJ only in 70 amino acid 'J-region'; localization to ER/nuclear membrane; genetic evidence for KAR2 interaction; assembles into multi-subunit membrane-bound complex along with other proteins required for ER translocation [68]
D. melanogaster csp29/32	Related proteins of unknown function expressed in retina and brain [65]; 'J-region' in csp32 identified by Tfasta comparison of NPL1/SEC63 'J-region' to Genbank sequences (T. Kurihara, unpublished data); contains no other region of similarity to *E. coli* DnaJ
S. pombe DnaJ	Uncharacterized open reading frame containing 'J-region'; identified by Tfasta comparison of NPL1/SEC63 'J-region' to translated Genbank sequences (T. Kurihara, unpublished data); ORF is located 3' of PMA1 gene [66] and extends through the end of the Genbank database sequence; partial amino acid sequence similar to *E. coli* DnaJ only in 'J-region'

regions, which we collectively term 'J-region's', are discussed in more detail later. The middle part of the *E. coli* DnaJ protein is a very glycine/phenylalanine-rich region that is conserved in five of the homologs. Finally, following the gly/phe-rich region in *E. coli* DnaJ is a repeated motif, Cys-X-X-Cys-X-Gly-X-Gly, which is conserved in four of the homologs.

```
                                 MVIRCSTDKTWWIGQKSVHWLAKRSRTMIPKLYIHLILSLLLLPL    SCJ1
                                                                          MAQRE    MTBJ
```

CONSERVED 'J-REGION'

```
MAKQDYYEILGVSKTAEEREIRKAYKRLAMKYHPDR------NQGDKEAEAKFKEIKEAYEVLTDSQKRAAYDQYGHA    ECOJ
ILA****A*EID*D*T*K**KS**RQ*SK*****K------*A*SE**HQ**I*VG***D**S*PE*KKI***F*AD    SCJ1
MVKETKF*D****PV**TDV**K***RKC*L*****K------*PSEEA**-****ASA***I*S*PE**DI***F*ED   YDJ1/MAS5
WVEK*F*QF****SD*SPE**KR**RK**RDL**A------*P*NPA*GER**AVS**HN**S*PA**KE**ETRRL     MTBJ
MVKETKL*DL****PS*N*Q*LK*G*RKA******K------PT**T*---*****S**F*IIN*P***EI*****LE    SIS1
M*R*L**T***ARN*D*K*LKS*FRK***Q*****------P**Q****KS****NQ****T*K*P*******R****    NOLC
MR*********TR*ID*AGLKSRVRK***EH****------*G*CEN*AGR****N****S**S**********RF**    CCRJ
TKLF*P*****I*TS*SD*D*KS**RK*SV*F****KLAKGLTPDEKSVM*ETYVQ*TK****S***ELV*QN*LK***P   NPL1/SEC63
MDI*P*SV***E*D*SDEL**R**RKK*LQH****---IHDEEKKV**RIE*DKVAI**G**S*KKR*KH**KT*QL      SSPJ
SGDSL*****LP**TGDD*K*T*RK**L*****K------*PDNVD*AD****VNR*HSI*S*QT**NI**N**SL       CSP29
```

GLYCINE/PHENYLALANINE-RICH REGION

```
AFEQGGMGGGGFGGGA-DFSDIFGDVFGDIF-------GGGRG----------------------------RQRAARGA   ECOJ
*VKN**G****PG*P**GG*H*P**-*I*ERM*QGGHGGP***F*--------------------------QRQRQ**P   SCJ1
GLSGA*GA**FP***FGFGD***SQF**A---------**AQR---------------------------PRGPQ**K   YDJ1/MAS5
FAGG**FG**RRFDS**FGGG**GGFGVGGD**AE*------NLNDLFDAASRTGGTTIGDLFGGLFGRGGSARPS*PR**N  MTBJ
*ARS**PSF*PG*P*GAGGAGG**PGGA*GFS-------**HHAFS-------------------------NED*FNIF   SIS1
********A*FGN*F*GGSAGGTSRH*RRHL-------RRDD---------------------------*RSS*PLL    NOLC
*GQR*RN*P*****QGF*A****N*****V*GEMF---***A--------------------------V**PQ**Q     CCRJ
```

```
                CxxCxGxG          CxxCxGxG                    CxxCxGxG
DLRYNMELTLEEAVRGVTKEIRIPTLEECDVCHGSGAKPGTQPQTCPTCHGSGQV--QMRQGFFA--VQQTCPHCQGRG    ECOJ
MIKVQEK*S*KQFYS**SSI*FTLNLND****SAD**KLAQ-**D*Q*R*VIIQVL*M*IMTQQI**M*GR*G*T*      SCJ1
*IKHEISAS***LYK*R*AKLALNKQIL**KE*E*R*G*K*-AVKK**TS*N*Q*IKFVTRQM*PMIQRF*TE*DV*H*T*  YDJ1/MAS5
*LETET**DFV**AK**AMPL*LTSPAP**TN******R***SPKV****N****VI--NRN**A*G--FSEP*TD*R*S*  MTBJ
SQFFGGSSPFGG*DDSGFSFSSY*SGGGAGMGGMP*GMG*MHGGMGGMPG*FRSA--SSSPTYPEET**VNLPVSLEDL   SIS1
GRSRTRCGPSLQHGDHPRGGLFRQDGADPRADV*HLRRLHGLGREAGHQPEDLRH--LPGLRPYP--RRPGLLLDRTHL   NOLC
****DL*I****Q*YA*AEV**T**RH*P*E***E**********LCLR**G*A*P**-RAT*****----VEAAR*G*S  CCRJ
```

```
    CxxCxGxG
TLI--KDPCNKCHGHGRVERSKTLSVKIPAGVDTGDRIRLAGEGEAGEHGAPA-GDLYVQVQVK--QHPIFEREGNN     ECOJ
QI*---*NE*KT***KKVTKKN*FFH*DV*P*APRNYMDTRV--***EKGPDFDA***VIEFKE*DTENMGYR*R*D*     SCJ1
DIIDP**R*KS*N*KKVENER*I*E*HVEP*MKD*Q**V---FK****DQAPDVIP**VVFI*SER--P**KS*K*D*DD   YDJ1/MAS5
SII--EH**EE*K*T*VTT*TR*IN*R***P**ED*Q******Q*****LR***S-*****T*H*R---PDK**G*D*DD    MTBJ
FVGKK**SFKIGRK*PHGASEKTQIDIQLKP*WKA*TK*TYKNQ*DYNPQTGRR--KT*QFVI*E*--S***N*K*D*DD    SIS1
PDL--RRSRSDD*RSLQQMPWPGPGHRGAHA*GQYSDRHRGRHAYPPLRRGRT-****IFLS*R--P*EFYQ*D*AD      NOLC
```

```
LY-CEVPINFAMAALGGE--IEVPTLDGR---VKLKVPGETQTGKLFRMRGKGVKSVRGGAQ-GDLLCRVVVETPVGL    ECOJ
**LSA*E*LY**W-QRTIEF**ENKP-***SR*AHVVVSNGEVEVV**FGMPK*SKGY***YIDY**VM*KTF          SCJ1
*V-Y*AE*DLLT*IA***FAL*HVSG*WLKVG-IVPGEVIAPGMRKVIE-***MPIPKY*GY-*N*IIKFTIKF*ENH     YDJ1/MAS5
*T-VT**VS*TEL***ST--LS******T---*GVR**KG*AD*RIL**V**RVCP*AV*V*ATYLSP                MTBJ
*I-YTL*LS*KESL**FS--KTIQ*I***TL--P*SRVQPV*PSQTSTYP*Q**MPTPKNPS*R*N*IVKYK*DY*IS*    SIS1
**-*S****SMTT*T***K--FD*T****TKSR*TVPEGTQAGKQFRLKGKMH*RALQPD*RPLYPDPD*DAA*AHQAP    NOLC
```

```
NERQKQLLQELQESFGGPTGEHNSPRSKSFFDGVKKFFDDLTR                                        ECOJ
KSG**NM*KD**                                                                      SCJ1
FTSEEN**KKLEEILPPRIVPAIPKKATVDECVLADFDPAKYNRTRASRGGANYDSEEEQGGEGVQCASQ             YDJ1/MAS5
*DA**RAID*NF                                                                      SIS1
ARIAAGVRAD**VQGEQSAIDGLLL**HERFLRYTERIATPGFRLCPSP                                  NOLC
```

Fig. 1. Alignment of DnaJ homologs. Abbreviations and references for protein sequences, ECOJ, *E. coli* DnaJ [44,45]; SCJ1, *S. cerevisiae* SCJ1 [49]; YDJ1/MAS5, *S. cerevisiae* YDJ1/MAS5 [53,55]; MTBJ, *M. tuberculosis* DnaJ [46]; SIS1, *S. cerevisiae* SIS1 [58]; NOLC, *R. fredii* USDA 257 nolC [48]; CCRJ, *C. crescentus* DnaJ [47]; NPL1/SEC63, *S. cerevisiae* NPL1/SEC63 [50]; SSPJ, uncharacterized ORF downstream of *S. pombe* PMA1 [66]; CSP29, *D. melanogaster* csp29 [65]. For NPL1/SEC63, SSPJ and CSP29, only 'J-region's are shown. Conserved 'J-region's, gly/phe-rich regions, and CxxCxGxG regions are denoted. *, identities with *E. coli* DnaJ.

2.1.1. Bacterial DnaJ homologs

In *E. coli*, the *dnaK* and *dnaJ* genes form an operon, with the *dnaJ* gene downstream of the *dnaK* gene [44,45]. In both *Mycobacterium tuberculosis* and *Caulobacter crescentus*, sequencing downstream of DnaK homologs has also identified DnaJ homologs [46,47]. The *M. tuberculosis* DnaJ homolog is similar to *E. coli* DnaJ in all three regions mentioned above (Fig. 1). The partial sequence of the *C. crescentus* DnaJ homolog is similar to *E. coli* DnaJ in the N-terminal and gly/phe-rich regions (Fig. 1). In *Rhizobium fredii*, a DnaJ homolog termed NolC was identified by a transposon insertion into the bacterial chromosome which alters the strain's host range [48]. The NolC amino acid sequence is homologous to *E. coli* DnaJ throughout the length of the protein ($\approx 30\%$ identity; Fig. 1). It is not known whether this DnaJ homolog is directly or indirectly involved in determining host-range specificity of *R. fredii* USDA257. Extrapolating from DnaJ's role in *E. coli*, NolC might be involved in protein folding or assembly of proteins that determine host-range specificity. Using the *nolC* gene as a probe, the authors have detected other potential NolC/DnaJ homologs in several other *R. fredii* strains, as well as in a *Agrobacterium tumefaciens* strain.

2.1.2. DnaJ homologs in the yeast Saccharomyces cerevisiae

In the yeast *S. cerevisiae*, four DnaJ homologs have been identified (NPL1/SEC63, SCJ1, YDJ1/MAS5, and SIS1; see Table I). Based on DnaK and DnaJ function in *E. coli*, we propose that the yeast DnaJ homologs interact and function with corresponding HSP70s that co-localize to the same cellular compartments. Since HSP70s are involved in protein transport, we predict one function of the DnaJ homologs will be to act with HSP70s in protein transport processes. As described below, there is currently evidence for a role in protein transport for three of the DnaJ homologs.

2.1.3. SCJ1

SCJ1 was identified [49] as a gene whose overexpression altered the intracellular localization of a fusion protein containing the nuclear localization sequence from SV40 large T-antigen fused to the precursor form of the mitochondrial inner membrane protein, cytochrome c_1 (SV40-CYT1) [50]. This fusion protein is normally nuclear-associated [50], presumably because the nuclear localization sequence acts as a stronger targeting signal than the mitochondrial cytochrome c_1 presequence in such a context. However, in yeast cells overexpressing the SCJ1 gene, the SV40-CYT1 fusion protein can assemble into the mitochondria. The protein sequence of SCJ1 is homologous to *E. coli* DnaJ throughout the whole protein, but also contains a 45-amino acid extension at the N-terminus and the sequence KDEL at the C-terminus. Amino acids 1–25 are rich in basic and hydroxylated residues and could potentially fold into an amphiphilic helix, similar to known mitochondrial presequences (reviewed in [1]). The mitochondrial localization suggested by the putative presequence was confirmed by cell fractionation. Amino acids 26–45 contain motifs common to ER signal sequences [51], while the C-terminal KDEL-sequence has been proposed to act as a retention signal for ER lumen proteins in animal cells [52]. Thus, if translation were to initiate at the methionine at amino acid 26, an ER form of SCJ1 could be produced. The N-terminal 200 amino acids of SCJ1 fused to invertase lacking its own signal sequence resulted in some secretion of invertase activity, suggesting that SCJ1 could contain an ER signal sequence. However, SCJ1 was not detected in microsomal fractions in cell fractionation experiments, so it is not known if any native SCJ1 might reside in the ER.

Two models have been proposed to explain how SCJ1 overproduction allows SV40-CYT1 assembly into the mitochondria. First, overproduction of mitochondrial SCJ1 might stimulate mitochondrial import. Second, overproduction of an ER form of SCJ1 might affect assembly of ER or nuclear membrane proteins, causing a defect in nuclear transport and indirectly favoring mitochondrial import. The latter model is based on the identification of nuclear transport defective yeast strains that can assemble SV40-CYT1 into mitochondria [50]. SCJ1 is not essential for normal cell growth, as haploid yeast cells disrupted for the *SCJ1* gene are viable. SCJ1 might not be essential because functionally redundant proteins exist in the yeast cell. Southern blotting of yeast genomic DNA with the *SCJ1* gene as probe has revealed four to five cross-hybridizing bands. These cross-hybridizing bands may correspond to other yeast DnaJ homologs (discussed below), some of which are known to be involved in nuclear, ER, or mitochondrial transport. Based on DnaK-DnaJ function in *E. coli*, we predict that the mitochondrial form of SCJ1 would interact with the yeast mitochondrial HSP70, SSC1, to stimulate mitochondrial import.

2.1.4. YDJ1/MAS5

The *YDJ1* gene was isolated by screening a yeast expression library with polyclonal antisera generated against a yeast nuclear subfraction termed the matrix lamina pore complex (MPLC) [53]. Mono-specific antibodies were purified from anti-MPLC crude serum using β-galactosidase fusion protein from *YDJ1*-containing phage. These mono-specific antibodies reacted with the YDJ1 protein, a major 46 kDa and minor 49 kDa band on Western blots of nuclear and MPLC extract. The YDJ1 protein sequence is homologous to *E. coli* DnaJ for the entire length of the protein, with a C-terminal extension that ends in C-A-S-Q. The C-A-S-Q sequence fits the consensus CaaX motif found at the C-terminus of farnesylated proteins (for review, see [54]).

Various antibodies prepared against the YDJ1 protein were used to immunolocalize YDJ1 to both the cytoplasm and nuclear envelope, with a concentration at the nuclear envelope. In biochemical fractionation experiments, up to 80% of YDJ1 in the cell was recovered in the soluble fraction, suggesting that it is predominantly a cytoplasmic protein. Some YDJ1 protein co-fractionated with nuclei and microsomal membranes, supporting the concentrated perinuclear YDJ1 staining observed by immunofluorescence. YDJ1 associated with microsomal membranes is sensitive to protease treatment in the absence of detergent, indicating that it does not reside in the ER lumen. It has not been determined if YDJ1 membrane association is due to lipid modification at the C-terminal C-A-S-Q sequence. Haploid cells disrupted for the *YDJ1* gene grow slowly on plates but cannot grow in liquid media, indicating that YDJ1 is required for normal cell growth. Overexpression of another yeast DnaJ homolog, *SIS1* (discussed below), partially suppresses the growth defect of *YDJ1*-disrupted haploid cells, suggesting a conservation of DnaJ-like function between those two proteins.

One clue into the nature of this conserved DnaJ-like function comes from studies with the yeast *mas5* mutant [55]. The *mas5* mutant was isolated by screening a collection of temperature-sensitive yeast strains [56] for mutants defective in mitochondrial protein import. *mas5* mutants accumulated precursor of the $F_{1\beta}$-subunit of the mitochondrial ATPase and mitochondrial citrate synthase precursor at the nonpermissive temperature. This defect was only evident in vivo, as mitochondria isolated from *mas5* cells imported $F_{1\beta}$-precursor and citrate synthase precursor as efficiently as mitochondria from wild-type cells in an in vitro import assay. The *MAS5* gene has been cloned [55], and its sequence is identical to *YDJ1*. How might YDJ1/MAS5 be involved in mitochondrial protein import? Above, we proposed that the mitochondrial-localized DnaJ homolog, SCJ1, interacts with the mitochondrial HSP70 (SSC1), to stimulate mitochondrial import. Similarly, we propose that YDJ1/MAS5 interacts with cytoplasmic HSP70s (SSA1,2), to stimulate mitochondrial import. In the cell fractionation experiments described above for YDJ1, very little YDJ1 protein co-fractionated with mitochondrial membranes. If the fractionation data reflect the localization of YDJ1/MAS5 in vivo, it would suggest that mitochondrial association of YDJ1/MAS5 is not required for its mitochondrial import function. Perhaps a predominantly cytoplasmic YDJ1/MAS5 could indirectly stimulate mitochondrial import by interacting with and recycling cytoplasmic HSP70 transport function.

2.1.5. SIS1

Yeast strains mutated in the *SIT4* gene, which encodes a protein homologous to the catalytic subunit of a protein phosphatase, grow very slowly [57]). The *SIS1* gene was identified as a high-copy number suppressor of this slow growth phenotype [58]. Unlike the SCJ1 and YDJ1/MAS5 proteins, the SIS1 protein is only homologous to *E. coli* DnaJ in the N-terminal and C-terminal thirds of the protein. The very glycine/methionine-rich middle third of SIS1 is not similar to the corresponding region in *E. coli* DnaJ, the Cys-x-x-Cys-x-x-Gly-x-Gly repeat region. Similar glycine/methionine-rich regions are also found at the C-terminus of the chaperonin proteins, *E. coli* groEL [59] and *S. cerevisiae* HSP60 [34], but the functional significance of these regions is unknown. Like YDJ1 discussed above, SIS1 was localized throughout the cytoplasm and in a concentrated manner at the nucleus. The *SIS1* gene is essential for normal cell growth, although cells were able to divide several times in the absence of SIS1. While SIS1 overexpression partially suppressed the growth defect of *YDJ1*-disrupted strains, YDJ1 overexpression did not suppress the growth defect of *SIS1*-disrupted strains.

SIS1 co-immunoprecipitated with a protein of apparent molecular mass 40 kDa (p40), and this interaction depended on the glycine/methionine-rich region unique to SIS1 among DnaJ homologs. The glycine-rich region common to several DnaJ homologs and immediately N-terminal to the glycine/methionine-rich region in SIS1 was also required for SIS1 interaction with p40. How might SIS1 function in yeast cells? Luke et al. [58] suggest that SIS1 might somehow be involved in mRNA transcription, processing, or transport, since $\approx 25\%$ of SIS1 was released from a nuclear fraction by RNAse treatment and since high copy *SIS1* suppresses mutant *sit4* strains, in which the amount or length of mRNA for several genes is altered compared to wild-type [57]. As with YDJ1/MAS5, the cytoplasmic and nuclear-associated localization of SIS1 leads to the prediction that it interacts and functions with cytoplasmic HSP70s. Based on proposals in *E. coli* that DnaJ-bound substrates are targets for DnaK action [22,29], it will be interesting to determine if the SIS1-associated p40 is a substrate for HSP70 action.

2.1.6. NPL1/SEC63

Like the *YDJ1/MAS5* gene, the *NPL1/SEC63* gene was identified by two different approaches. *NPL1* was identified by a genetic selection scheme for mutants defective in nuclear protein localization [50], while *SEC63* was identified by a genetic selection scheme for mutants defective in precursor protein translocation into the ER [60]. Another genetic selection scheme for mutants defective in ER translocation resulted in the identification of the *ptl1* strain [61], which is also mutated in the *NPL1/SEC63* gene. The *NPL1* selection scheme depended on competition between nuclear transport and mitochondrial assembly of a SV40-CYT1 fusion protein described earlier for the *SCJ1* gene isolation. The mutants isolated by this scheme are termed *npl1* mutants. The temperature-sensitive *npl1-1* mutant strain had a defect in nuclear localization not only for the SV40-CYT1 fusion protein but also for an endogenous nucleolar protein at the non-permissive temperature [50]. The *SEC63* selection scheme depended on a defect in ER translocation retaining a HIS4-fusion protein in the cytoplasm and allowing growth on histidinol. The mutants isolated in the ER translocation scheme are termed *sec63* mutants. The temperature-sensitive *sec63-1* strain accumulated precursor forms of several normally secreted proteins at the non-permissive temperature, confirming a defect in ER translocation [60].

Interestingly, the *sec63-1* strain also had a strong nuclear localization defect at the non-permissive temperature, while a very weak defect in ER translocation was also detected in the *npl1-1* strain at the non-permissive temperature [50]. Thus, NPL1/SEC63 might be a bifunctional protein involved in two transport processes, nuclear localization and ER translocation.

The primary amino acid sequence of the NPL1/SEC63 protein is 663 amino acids

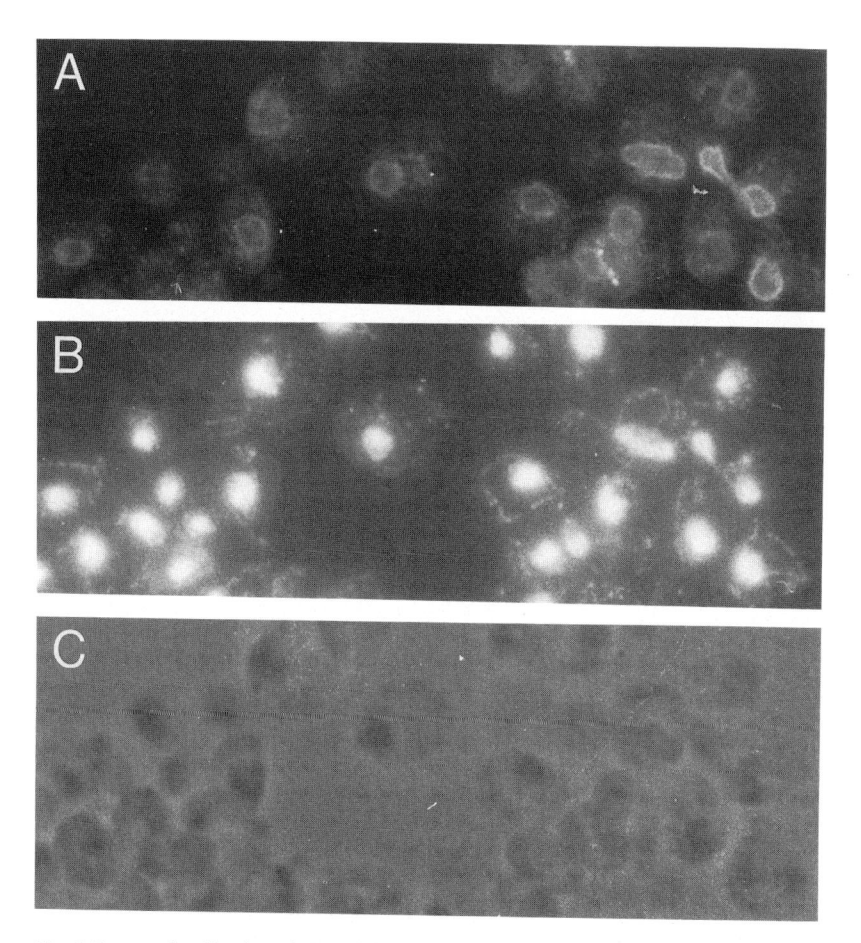

Fig. 2. Immunolocalization of NPL1/SEC63. Anti-NPL1/SEC63 polyclonal antiserum 7341 was obtained from New Zealand white rabbits injected with peptide AG10, containing the 20 C-terminal amino acids from the deduced NPL1/SEC63 protein sequence [50]. NPL1/SEC63-specific antibodies were affinity purified from serum by binding AG10 peptide coupled to Affigel 401 (BioRad) and eluted with glycine buffer (pH 2.8). The affinity purified antibodies were concentrated (Amicon) and used for immunofluorescence on permeabilized yeast cells (Panel A) according to Sadler et al. [50]. Panel B: cells stained with DAPI to visualize nuclear DNA. Panel C: cells viewed by phase contrast microscopy.

long, with three hydrophobic regions in the N-terminal half predicting a membrane localization. Unlike all of the other DnaJ homologs discussed so far, only a small portion of NPL1/SEC63 (73 amino acids out of 663) is similar to E. coli DnaJ. To determine the intracellular localization of NPL1/SEC63 by indirect immunofluorescence (Fig. 2), rabbit antiserum was prepared and affinity purified against a peptide corresponding to the 20 C-terminal amino acids (T. Kurihara and P. Silver, unpublished results). These affinity-purified antibodies decorated yeast cells in a perinuclear staining pattern similar to that seen for the ER membrane protein, SEC62 [62], as well as the ER lumenal protein, KAR2 [63]. Other antibodies directed against SEC63 also decorated yeast cells at the ER [64]. Since the ER membrane is continuous with the nuclear membrane in yeast cells, the ER localization of NPL1/SEC63 is consistent with its role in both ER translocation and nuclear localization.

2.2. The 'J-region'

The 70-amino acid region within NPL1/SEC63 that is similar to the N-terminus of *E. coli* DnaJ corresponds to the most highly conserved region among the homologs. For example, *R. fredii* nolC is ≈30% identical to *E. coli* DnaJ throughout the length of the protein but is 68% identical to the N-terminal region of DnaJ [48]. SCJ1 is 37% identical to *E. coli* DnaJ throughout its protein sequence but is 54% identical in the region corresponding to the N-terminus of DnaJ [49]. Caplan and Douglas [53] have also noted the high degree of conservation (66% identity) between the corresponding regions in YDJ1 and SIS1. A homology search comparing this region of NPL1/SEC63 to GenBank sequences (T. Kurihara, unpublished data) revealed that similar regions are also found in two related *D. melanogaster* proteins (csp29/32) [65] as well as in an uncharacterized open reading frame downstream of the *S. pombe PMA1* gene [66].

We propose that all these regions comprise members of an evolutionarily conserved structural motif which we term the 'J-region'. These 'J-regions' probably fold into distinct structural domains, as deduced from the following lines of evidence. The NPL1/SEC63 'J-region' is located between two hydrophobic, membrane-spanning regions. The SCJ1, YDJ1, and bacterial 'J-regions' are separated from other regions of the proteins by glycine/phenylalanine-rich regions. Like NPL1/SEC63, the only regions within *D. melanogaster* csp29/32 and the potential *S. pombe* DnaJ homolog that are similar to *E. coli* DnaJ are their respective 'J-regions'. What is the function of these 'J-regions'? Based on the evidence for DnaK–DnaJ interaction in *E. coli*, we predict that 'J-regions' are an evolutionarily conserved structure for mediating protein interactions with HSP70s.

2.2.1. The NPL1/SEC63 'J-region': localization to the ER lumen
Since the NPL1/SEC63 'J-region' lies between two hydrophobic regions, the membrane topology of NPL1/SEC63 was determined by Feldheim et al. [64]. Similar results are shown in Fig. 3. The glycosylation pattern of invertase during secretion was utilized as a marker for the intracellular localization of NPL1/SEC63 regions. Normally, cytoplasmic forms of invertase are not glycosylated, while ER lumenal forms of invertase are glycosylated. Thus, the glycosylation state of invertase would indicate a cytoplasmic or ER lumen localization of NPL1/SEC63 regions fused to invertase lacking its own signal sequence. Two such fusion proteins were expressed in yeast cells. In fusion protein TK10, invertase was fused to NPL1/SEC63 lacking its 14 C-terminal amino acids, while in fusion protein TK23, invertase was fused to

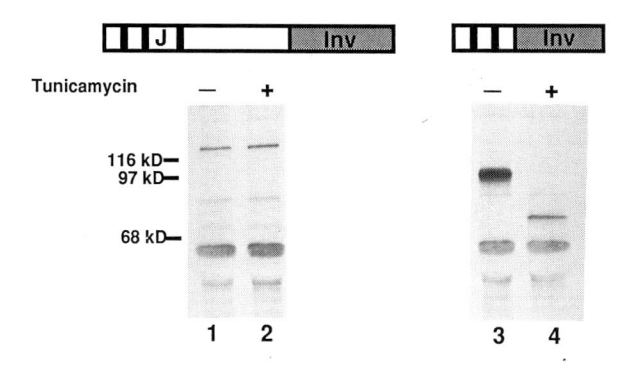

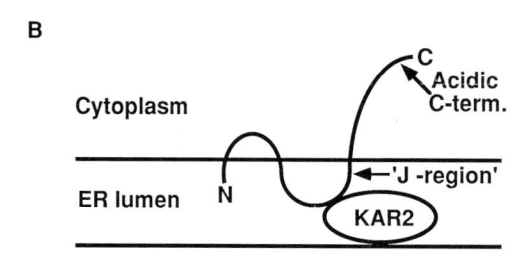

Fig. 3. (A) Immunoprecipitation of invertase fusion proteins. Plasmids TK10 and TK23 were constructed by subcloning fragments of NPL1/SEC63 into pMN10 [70] cut with HindIII and XhoI. To construct TK10, a StuI-EcoRV fragment with HindIII and XhoI linkers attached, respectively, at both ends was cloned into pMN10. To construct TK23, a StuI-SpeI fragment with HindIII and XhoI linkers attached, respectively, was cloned into pMN10. In protein schematics, hydrophobic regions are denoted by wide vertical lines, the 'J-region' by a J, and invertase by Inv. Cell labeling and immunoprecipitation were performed as in Nelson and Silver [70], except that 50 μl of anti-invertase serum was used. For tunicamycin treated samples, cells were incubated at 30°C for 15 min with 10 μg/ml tunicamycin before labeling. (B) Model of NPL1/SEC63 membrane topology. Predicted interaction of 'J-region' with KAR2 is shown (see text for details)

NPL1/SEC63 within the 'J-region'. Cells expressing the fusion proteins were labeled with ^{35}S in the absence or presence of the glycosylation inhibitor tunicamycin, and detergent-solubilized cell extracts immunoprecipitated with anti-invertase antibody. Increased mobility of fusion protein in extracts labeled in the presence of tunicamycin would indicate that the protein is normally glycosylated. While NPL1/SEC63 has three potential glycosylation sites, none of these sites is normally glycosylated (T. Kurihara and P. Silver, unpublished results), so only invertase glycosylation would be affected by tunicamycin treatment. Invertase fused to NPL1/SEC63 after the second hydrophobic region (TK23) but not the third hydrophobic region (TK10) was glycosylated, indicating an ER lumen localization for the 'J-region' and a cytoplasmic localization for the C-terminal end of the protein. Feldheim et al. [64] have extended this type of analysis to predict that the N-terminus is located in the ER lumen. The predicted membrane topology of NPL1/SEC63 is illustrated in Fig. 3B. Since 'J-region' domains are proposed to interact with HSP70s, the ER-localized 'J-region' of NPL1/SEC63 is predicted to interact with the ER-localized HSP70 homolog, KAR2 [63,67].

2.2.2. Genetic evidence for 'J-region' role in KAR2 interaction and ER translocation
The ER-lumenal location of the NPL1/SEC63 protein 'J-region' predicts an interaction with KAR2, the yeast ER-localized HSP70 homolog [63,67]. KAR2 is required for translocation of proteins into the ER, as evidenced by the accumulation of secretory protein precursors in *kar2-159*^{ts} mutant cells at a non-permissive temperature or in cells depleted of wild-type KAR2 protein [12]. Like the *kar2-159*^{ts} mutant, *sec63-1*^{ts} mutant cells also accumulate precursor proteins at the non-permissive temperature, indicating a requirement for NPL1/SEC63 function in protein translocation into the ER [60]. While *kar2-159* or *sec63-1* mutant cells can grow at 25°C, *kar2-159 sec63-1* double-mutant cells display synthetic lethality (M. Scidmore and M. Rose, pers. commun.). Synthetic lethal combinations such as *kar2-159 sec63-1* sometimes identify proteins that physically interact. Dominant *kar2* alleles have been generated that can suppress the growth defect of *sec63* mutants in an allele-specific manner (M. Scidmore, H. Okamura and M. Rose, pers. commun.), suggesting that KAR2 and NPL1/SEC63 physically interact, as predicted. Since the growth-defect of mutant *sec63-1* cells can be suppressed by mutant *kar2* alleles and the region of interaction between KAR2 and NPL1/SEC63 must be at the ER-lumenal 'J-region', it was predicted that a mutation is located in the 'J-region' of the NPL1/SEC63 gene in the *sec63-1* mutant cells. Genomic DNA sequencing has confirmed the prediction of a mutation in the 'J-region' of the NPL1/SEC63 gene in *sec63-1* mutant cells (T. Kurihara and P. Silver, unpublished data).

3. Discussion

3.1. Model for NPL1/SEC63 function

The NPL1/SEC63 protein has been demonstrated to be a component of a membrane-bound, multi-subunit complex [68] along with two proteins required for ER translocation, SEC61 [60] and SEC62 [69] and two other unidentified proteins. Deshaies et al. [68] suggest that this complex may comprise a part of a translocation machine in yeast for transferring precursor proteins across the ER membrane into the ER lumen. Based on the genetic evidence for NPL1/SEC63 interaction with KAR2 and the function of the *E. coli* DnaJ–DnaK complex, we proposed that KAR2 interacts with the 'J-region' extending into the ER lumen to mediate protein folding or assembly processes. How might NPL1/SEC63 function in ER translocation? If the multi-subunit complex containing NPL1/SEC63 were indeed a translocator through which precursor proteins passed into the ER lumen, then the 'J-region' might interact with KAR2 to help maintain precursor proteins in a translocation-competent conformation. Deshaies et al. [68] also propose that 'J-region' interaction with KAR2 might modify the composition of or catalyze the assembly/disassembly of the translocator complex. On the other hand, it is possible that the 'J-region' has a less direct role in ER translocation by stimulating substrate release from KAR2 thereby recycling KAR2 function in translocation.

While the assembly of NPL1/SEC63 into a complex with other proteins involved in ER translocation is a clue toward its function in secretion, its role in nuclear protein import is much less clear. It is possible that proteins important for nuclear import (e.g. nuclear pore proteins) are assembled into the nuclear envelope via the secretory pathway, in which case NPL1/SEC63 would play an indirect role in nuclear import. A more interesting possibility is that the extremely acidic C-terminal region of NPL1/SEC63 that extends into the cytoplasm (see Fig. 3B) might interact with basic nuclear localization sequences. This more direct role of NPL1/SEC63 in nuclear import does not involve the 'J-region' but rather would suggest structurally and functionally separate domains for ER translocation and nuclear transport. Preliminary evidence for a separation in nuclear transport and ER translocation domains comes from the mapping and sequencing of mutations in the C-terminal half of the *NPL1/SEC63* gene in *npl1*-type mutants, as opposed to the 'J-region' mutation in *sec63-1* cells (T. Kurihara and P. Silver, unpublished data).

3.2. DnaJ homologs, 'J-regions' and protein transport

In this review, we have described the identification of a family of proteins homologous to the *E. coli* heat shock protein DnaJ and proposed a role for yeast DnaJ homologs in protein transport. Both eukaryotic and prokaryotic homologs are well conserved in their similar regions, indicating a fundamental role for these regions which have been conserved throughout evolution. The 70-amino acid 'J-region' of NPL1/SEC63 is the only region of similarity shared among all members of the DnaJ family. This 'J-region' is also the most well conserved region among the homologs.

Insight into the possible role of DnaJ homologs comes from experiments in *E. coli*. DnaJ and another heat-shock protein, GrpE, have been shown to stimulate the in vitro ATPase activity of DnaK, the *E. coli* HSP70 [39]. Genetic and biochemical evidence indicates that DnaJ, DnaK, and GrpE function together to mediate protein folding or assembly processes in *E. coli* (for reviews, see [17,18]). Based on the identification of a family of DnaJ homologs both in eukaryotes and prokaryotes, we propose a conservation of *E. coli* DnaJ-like function throughout evolution. Specifically, we predict that all members of the DnaJ family will interact with HSP70s and that these interactions will be important for protein folding or assembly processes. Accordingly, all DnaJ homologs identified so far are localized to cellular compartments that also contain HSP70s. We also propose that the highly conserved 'J-region' is important for interaction with HSP70s. The localization of the 'J-region' in NPL1/SEC63 to the yeast ER lumen and the genetic evidence for physical interaction between NPL1/SEC63 and the ER lumenal HSP70, KAR2, support such a model (see Results section).

The evidence that DnaJ homolog–HSP70 interactions are important for protein folding and assembly processes comes from genetic and biochemical analyses of protein transport in the yeast *S. cerevisiae*. Many of the HSP70s and some DnaJ homologs that co-localize to the same cellular compartments are known or thought to be required for similar protein transport processes. Extrapolating from the *E. coli* model, it is likely that DnaJ–HSP70 interactions control protein folding and assembly processes required for efficient protein transport. The recent discovery of NPL1/SEC63 in a multi-subunit complex important for ER translocation [68] offers a promising start for the role of DnaJ homologs in protein folding and transport. Many exciting experiments lie ahead, including testing the role of DnaJ homologs in in vitro transport assays, testing if 'J-regions' can interact with and stimulate the ATPase activity of HSP70s, searching for mammalian DnaJ homologs, and determining the roles of the non-'J-regions' in the various homologs. It will also be interesting to see if eukaryotic homologs of GrpE exist that may interact and function with DnaJ homologs and HSP70s.

Acknowledgements

The authors thank Mark Osborne for comments on the manuscript and acknowledge D. Feldheim, R. Schekman, M. Scidmore and M. Rose for kindly communicating their results prior to publication. Parts of the work presented here were supported by grants from the National Institutes of Health, the American Tobacco Council and the National Institutes of Health Genetics Training Grant (TK)

References

1. Hartl, F.-U., Pfanner, N., Nicholson, D.W. and Neupert, W. (1989) Biochim. Biophys. Acta 988, 1–45.
2. Rapoport, T.A. (1990) in: J.A.F. Op den Kamp (Ed.), NATO ASI Series, Vol. H40, Dynamics and Biogenesis of Membranes, Springer-Verlag, Berlin, pp. 231–245.
3. Silver, P.A. (1991) Cell 64, 489–497.
4. Morimoto, R.I., Tissieres, A. and Georgopoulos, C. (1990) in: R.I. Morimoto, A. Tissieres and C. Georgopoulos (Eds.), Stress Proteins in Biology and Medicine, Cold Spring Harbor Laboratory, Cold Spring Harbor, NY, pp. 1–36.
5. Craig, E.A. (1990) in: R.I. Morimoto, A. Tissieres and C. Georgopoulos (Eds.), Stress Proteins in Biology and Medicine, Cold Spring Harbor Laboratory, Cold Spring Harbor, NY, pp. 279–286.
6. Pelham, H.R.B. (1990) in: R.I. Morimoto, A. Tissieres and C. Georgopoulos (Eds.), Stress Proteins in

Biology and Medicine, Cold Spring Harbor Laboratory, Cold Spring Harbor, NY, pp. 287–299.
7. Craig, E.A. and Gross, C.A. (1991) Trends Biochem. Sci. 16, 135–140.
8. Rothman, J.E. (1989) Cell 59, 591–601.
9. Deshaies, R.J., Koch, B.D., Werner-Washburne, M., Craig, E.A. and Schekman, R. (1988) Nature 332, 800–805.
10. Chirico, W.J., Waters, M.G. and Blobel, G. (1988) Nature 332, 805–810.
11. Murakami, H., Pain, D. and Blobel, G. (1988) J. Cell Biol. 107, 2051–2057.
12. Vogel, J.P., Misra, L.M. and Rose, M.D. (1990) J. Cell Biol. 110, 1885–1895.
13. Kang, P.-J., Ostermann, J., Shilling, J., Neupert, W., Craig, E. A. and Pfanner, N. (1990) Nature 348, 137–143.
14. Nguyen, T.H., Law, D.T.S. and Williams, D.B. (1991) Proc. Natl. Acad. Sci. 88, 1565–1569.
15. Phillips, G.J. and Silhavy, T.J. (1990) Nature 344, 882–884.
16. Zimmermann, R., Sagstetter, M., Lewis, M.J. and Pelham, H.R.B. (1988) EMBO J. 7, 2875–2880.
17. Gross, C.A., Straus, D.B., Erickson, J.W. and Yura, T. (1990) in: R.I. Morimoto, A. Tissieres and C. Georgopoulos (Eds.), Stress Proteins in Biology and Medicine, Cold Spring Harbor Laboratory, Cold Spring Harbor, NY, pp. 167–189.
18. Georgopoulos, C., Ang, D., Liberek, K. and Zylicz, M. (1990) in: R.I. Morimoto, A. Tissieres and C. Georgopoulos (Eds.), Stress Proteins in Biology and Medicine, Cold Spring Harbor Laboratory, Cold Spring Harbor, NY, pp. 191–221.
19. Friedman, D.E., Olson, E.R., Georgopoulos, C., Tilly, K., Herskowitz, I. and Banuett, F. (1984) Microbiol. Rev. 48, 299–325.
20. Liberek, K., Georgopoulos, C. and Zylicz, M. (1988) Proc. Natl. Acad. Sci. 85, 6632–6636.
21. Alfano, C. and McMacken, R. (1989) J. Biol. Chem. 264, 10709–10718.
22. Zylicz, M., Ang, D., Liberek, K. and Georgopoulos, C. (1989) EMBO J. 8, 1601–1608.
23. Yarmolinsky, M. and Sternberg, N. (1988) in: R. Calender (Ed.), The Bacteriophages, Vol. 1, Plenum Press, New York, pp. 291–438.
24. Abeles, A. (1986) J. Biol. Chem. 261, 3548–3555.
25. Abeles, A.L., Reaves, L.D. and Austin, S.J. (1989) J. Bacteriol. 171, 43–52.
26. Tilly, K., Sozhamannan, S. and Yarmolinsky, M. (1990) New Biol. 2, 812–817.
27. Tilly, K. and Yarmolinsky, M. (1989) J. Bacteriol. 171, 6025–6029.
28. Wickner, S.H. (1990) Proc. Natl. Acad. Sci. 87, 2690–2694.
29. Wickner, S., Hoskins, J. and McKenney, K. (1991) Nature 350, 165–167.
30. Pelham, H.R.B. (1989) Annu. Rev. Cell Biol. 5, 1–23.
31. Gaitanaris, G.A., Papavassiliou, A.G., Rubock, P., Silverstein, S.J. and Gottesman, M.E. (1990) Cell 61, 1013–1020.
32. Craig, E.A., Kramer, J., Shilling, J., Werner-Washburne, M., Holmes, S., Kosic-Smithers, J. and Nicolet, C.M. (1989) Mol. Cell. Biol. 9, 3000–3008.
33. McMullin, T.W. and Hallberg, R.L. (1988) Mol. Cell. Biol. 8, 371–380.
34. Reading, D.S., Hallberg, R.L. and Myers, A.M. (1989) Nature 337, 655–659.
35. Cheng, M.Y., Hartl, F.-U., Martin, J., Pollock, R.A., Kalousek, F., Neupert, W., Hallberg, E.M., Hallberg, R.L. and Horwich, A.L. (1989) Nature 337, 620–625.
36. Ostermann, J., Horwich, A.L., Neupert, W. and Hartl, F.-U. (1989) Nature 341, 125–130.
37. Straus, D.B., Walter, W.A. and Gross, C.A. (1988) Genes Dev. 2, 1851–1858.
38. Zylicz, M., LeBowitz, J.H., McMacken, R. and Georgopoulos, C. (1983) Proc. Natl. Acad. Sci. 80, 6431–6435.
39. Liberek, K., Marszalek, J., Ang, D., Georgopoulos, C. and Zylicz, M. (1991) Proc. Natl. Acad. Sci. 88, 2874–2878.
40. Flynn, G.C., Chappell, T.G. and Rothman, J.E. (1989) Science 245, 385–390.
41. Flaherty, K.M., DeLuca-Flaherty, C. and McKay, D.B. (1990) Nature 346, 623–628.
42. Liberek, K., Skowyra, D., Zylicz, M., Johnson, C. and Georgopoulos, C. (1991) J. Biol. Chem. 266,

14491–14496.
43. Goloubinoff, P., Christeller, J.T., Gatensby, A.A. and Lorimer, G.H. (1989) Nature 342, 884–889.
44. Ohki, M., Tamura, F., Nishimura, S. and Uchida, H. (1986) J. Biol. Chem. 261, 1778–1781.
45. Bardwell, J.C.A., Tilly, K., Craig, E., King, J., Zylicz, M. and Georgopoulos, C. (1986) J. Biol. Chem. 261, 1782–1785.
46. Lathigra, R.B., Young, D.B., Sweetser, D. and Young, R.A. (1988) Nucl. Acids Res. 16, 1636.
47. Gomes, S.L., Gober, J.W. and Shapiro, L. (1990) J. Bacteriol. 172, 3051–3059.
48. Krishnan, H.B. and Pueppke, S.G. (1991) Mol. Microbiol. 5, 737–745.
49. Blumberg, H. and Silver, P. (1991) Nature 349, 627–630.
50. Sadler, I., Chiang, A., Kurihara, T., Rothblatt, J., Way, J. and Silver, P. (1989) J. Cell Biol. 109, 2665–2775.
51. von Heijne, G. (1985) Curr. Top. Membr. Transp. 24, 151–179.
52. Munro, S. and Pelham, H.R.B. (1986) Cell 46, 291–300.
53. Caplan, A.J. and Douglas, M.G. (1991) J. Cell Biol. 114, 609–621.
54. Glomset, J.A., Gelb, M.H. and Farnsworth, C.C. (1990) Trends Biochem. Sci. 15, 139–142.
55. Atencio, D.P. and Yaffe, M.P. (1992) Mol. Cell Biol. 12, 283–291.
56. Yaffe, M.P. and Schatz, G. (1984) Proc. Natl. Acad. Sci. 81, 4819–4823.
57. Arndt, K.T., Styles, C.A. and Fink, G.R. (1989) Cell 56, 527–537.
58. Luke, M.M., Sutton, A. and Arndt, K.T. (1991) J. Cell Biol. 114, 623–638.
59. Hemmingsen, S.M., Woolford, C., van der Vies, S.M., Tilly, K., Dennis, D.T., Georgopoulos, C.P., Hendrix, R.W. and Ellis, R. J. (1988) Nature 333, 330–334.
60. Rothblatt, J.A., Deshaies, R.J., Sanders, S.L., Daum, G. and Schekman, R. (1989) J. Cell Biol. 109, 2641–2652.
61. Toyn, J., Hibbs, A.R., Sanz, P., Crowe, J. and Meyer, D.I. (1988) EMBO J. 7, 4347–4353.
62. Deshaies, R.J. and Schekman, R. (1990) Mol. Cell. Biol. 10, 6024–6035.
63. Rose, M.D., Misra, L.M. and Vogel, J.P. (1989) Cell 57, 1211–1221.
64. Feldheim, D., Rothblatt, J. and Schekman, R. (1992) Mol. Cell Biol. 12, 3288–3296.
65. Zinsmaier, K.E., Hofbauer, A., Heimbeck, G., Pflugfelder, G.O., Buchner, S. and Buchner, E. (1990) J. Neurogenet. 7, 15–29.
66. Serrano, R., Kielland-Brandt, M.C. and Fink, G.R. (1986) Nature 319, 689–693.
67. Normington, K., Kohno, K., Kozutsumi, Y., Gething, M.-J. and Sambrook, J. (1989) Cell 57, 1223–1236.
68. Deshaies, R.J., Sanders, S.L., Feldheim, D.A. and Schekman, R. (1991) Nature 349, 806–808.
69. Deshaies, R.J. and Schekman, R. (1989) J. Cell Biol. 109, 2653–2664.
70. Nelson, M. and Silver, P. (1989) Mol. Cell. Biol. 9, 384–389.

CHAPTER 26

Chaperonin-mediated protein folding

ARTHUR L. HORWICH[1], SHARI CAPLAN[1], JOSEPH S. WALL[2] and F.-ULRICH HARTL[3]

[1]*Howard Hughes Medical Institute and Department of Genetics, Yale School of Medicine, 333 Cedar Street, New Haven, CT 06510, USA,* [2]*Brookhaven National Laboratory, Biology Department, Upton, NY 11973, USA and* [3]*Program in Cellular Biochemistry and Biophysics, Rockefeller Research Laboratory, Sloan-Kettering Institute, 1275 York Avenue, New York, NY 10021, USA*

1. Introduction

The central dogma of molecular biology defines the major avenue for the transfer of genetic information encoded in linear genomic DNA to three-dimensional effectors of function and structure, the proteins. Until recently it has been hypothesized that this transfer of information required only two types of machinery, one to transcribe a DNA template into an RNA message, and a second to translate RNA into protein. It has been postulated that newly translated polypeptide chains contain sufficient information in their primary amino acid sequences to direct spontaneous folding into active tertiary structures. Recent experiments, however, indicate that in the intact cell both newly synthesized proteins and proteins translocated through several biological membranes do not in general fold spontaneously but rather utilize a third type of machinery, a folding machinery, to reach biologically active conformations [1,2]. The best studied class of components catalyzing such folding are the so-called chaperonins, groEL, hsp60, and RUBISCO binding protein, found, respectively, in the eubacterial cytoplasm and inside the evolutionarily related organelles, mitochondria and chloroplasts. At the level of primary structure, these three components share nearly 60% of their amino acids [3,4]. At the level of quaternary structure, each is a homo-oligomeric complex composed of two stacked rings each containing seven radially arranged members [5–7]. The functions of groEL and hsp60 have been shown to be essential for cell viability [8,9].

2. In vivo analysis of chaperonin function

In the 1970s there were intimations for a role of these components in the acquisition of protein structure. The first came from an observation that mutations in the *Escherichia coli* groE operon affected bacteriophage head assembly in lambda phage-infected cells [10,11]. Several years later, an abundant protein in the chloroplast stroma was found to be associated with newly synthesized large subunits of the CO_2-fixing enzyme RUBISCO [12]. The so-called RUBISCO binding protein was not associated with the mature hetero-oligomeric RUBISCO enzyme, suggesting a role in RUBISCO assembly. More recently, the functions of groEL and RUBISCO binding protein in protein assembly have been related to each other by the observation that overexpression in *E. coli* of the groE operon could enhance the assembly of a co-expressed cyanobacterial RUBISCO [13].

A more general role for chaperonins in mediating not only assembly but polypeptide chain folding of most or possibly all of the proteins within a cellular compartment, was indicated by the characterization of a mutant of yeast that affected the mitochondrial chaperonin, hsp60 [9]. The mutant was isolated from a genetic screen aimed at identifying components involved with import of proteins from the cytosol to the mitochondrial matrix. In the mitochondrial import function defective mutant, mif4, each of four different precursor proteins was imported into the mitochondrial matrix and correctly processed to a mature form, but in every case the protein failed to achieve its biologically active form. The matrix protein OTC failed to reach its trimeric form; the $F_{1\beta}$-ATPase subunit failed to assemble into the F_1-ATPase stalk; citrate synthase failed to reach its active (dimeric) form (S. Caplan, unpublished results); and newly imported hsp60 subunits themselves failed to assemble to form a new 14mer complex [14]. This indicated a defect of either or both polypeptide chain folding and subunit assembly.

A role for hsp60 in chain folding was indicated by the effect of the mif4 mutation on the group of imported proteins that are conservatively sorted. These proteins normally undergo two steps of biogenesis: first they are imported from the cytosol to the mitochondrial matrix; then they are re-exported from the matrix to either the inner mitochondrial membrane or the intermembrane space [15]. This two-step targeting pathway can be considered to join an evolutionarily new step, targeting to the matrix, with a relatively ancient step, re-export, that resembles bacterial protein export. Notably, conservatively sorted proteins proceed through the pathway of import and re-export in an unassembled state, as protein monomers. Two such proteins were initially examined in mif4 cells, the Rieske Fe/S protein and cytochrome b_2 [9]. Both proteins were found inside the matrix compartment in intermediate-sized forms produced in this compartment during normal biogenesis. They had apparently been unable to achieve conformations permitting further biogenesis and re-export. Rather, these proteins, along with the matrix-localized proteins that were examined, had apparently become misfolded; they were found in insoluble aggregates.

A direct demonstration that hsp60 could fold polypeptide chains to the native state was provided by examining import into intact isolated mitochondria of a fusion protein joining a mitochondrial signal peptide with a monomeric cytosolic enzyme, dihydrofolate reductase (DHFR) [16]. When the fusion protein was imported into mitochondria in the absence of ATP, the signal peptide was proteolytically cleaved, and the DHFR moiety became associated with the hsp60 complex. The DHFR was exquisitely protease-susceptible, suggesting that it was bound in an unfolded conformation. After subsequent addition of ATP to the import mixture, the DHFR dissociated from hsp60 and exhibited the same degree of protease resistance as native DHFR, indicating that it had been folded to the native form. If the DHFR–hsp60 complex was first crudely purified from the organelles and then ATP was added, the protein remained associated with the hsp60 complex, and only partial protease resistance was acquired. This indicated that folding occurs in association with the hsp60 complex and that another component is apparently involved. In *E. coli*, such a component has been identified: groEL shares an operon with the gene for a 10 kDa protein, groES, that is found as a homo-oligomeric seven-member single ring [17]. More recently, a component that is functionally homologous to groES has been identified in mammalian mitochondria [18].

3. Role of hsp60 in biogenesis of mitochondrial-encoded proteins

While it appears that most or possibly all imported proteins require the action of hsp60 for folding and assembly into biologically active forms, an important question asks whether hsp60 also plays a role in the folding and assembly of the class of proteins that is encoded by the mitochondrial genome and translated on mitochondrial ribosomes. A recent study suggests that this might be the case, observing an association in maize mitochondria between newly synthesized mitochondrial-encoded ATPase $F_{1\alpha}$ subunit and hsp60 [19]. The mif4 mutant provides a further system in which the interaction of hsp60 with mitochondrial-encoded products can be addressed, by studying the physiology of mitochondrial protein biogenesis after shift to nonpermissive conditions. We sought to address, in particular, whether hsp60 function is required for the process of mitochondrial protein translation, and whether newly translated mitochondrial gene products require the function of hsp60 to reach their active conformations.

First, the relative levels of mitochondrial translation in wild-type and mif4 cells were measured at various times after shift to 37°C, by pulse-radiolabeling in the presence of the inhibitor of cytosolic translation, cycloheximide. At 2 h after shift, mitochondrial translation in mif4 cells proceeds at a rate nearly that in wild-type cells (Fig.1). Previous studies have demonstrated that, by this time after shift, the phenotype of defective folding and assembly inside mif4 mitochondria is fully evident [9]. Hsp60 itself becomes insoluble. Presuming that the functional translation machinery

MITOCHONDRIAL TRANSLATION

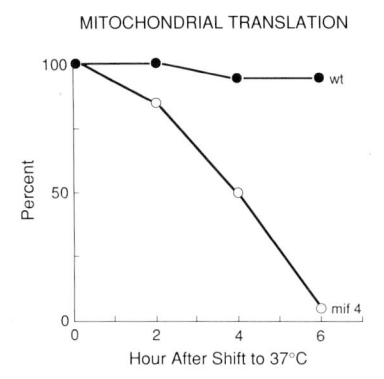

Fig. 1. Mitochondrial translation in wild-type and mif4 cells measured by pulse-radiolabeling in the presence of cycloheximide. Wild-type and mif4 strains were grown at 23°C in synthetic minimal medium containing 2% galactose. Cells were shifted to 37°C for the times indicated. For each time, cells (10 × 8) were radiolabeled with [^{35}S]methionine (100 μCi/ml) in the presence of cycloheximide (100 μg/ml) that had been added to the culture 10 min earlier. After 20 min, the cells were harvested and lysed at 4°C by addition of sodium hydroxide to a final concentration of 0.2 N. Trichloroacetic acid was then added to a final concentration of 5% and the sample spun at 14000 × g for 15 min. The precipitates were washed twice with acetone, resuspended in 10 mM Tris (pH 6.8), and protein concentration determined (Biorad). Equal amounts of protein were added to Laemmli buffer and analyzed in an 11% SDS-polyacrylamide gel [29]. The gel was fluorographed and the radiolabeled mitochondrial translation products quantitated by densitometric scanning. The amount of translation at the designated times is expressed as a percentage of the amount of translation observed prior to shift to 37°C (equivalent for wild-type and mif4 cells).

remains soluble, it seems that hsp60 would be physically unavailable to interact with it. At later times after shift, mitochondrial translation in mif4 cells became substantially impaired (Fig. 1). This probably reflects lack of ability, in the absence of hsp60 function, to supply functional versions of imported components involved with mitochondrial transcription and translation that are needed to replace losses from normal turnover. Assuming that this is an indirect impairment of translation, we conclude, based on the observation of near-normal translation at earlier times, that the hsp60 complex is not required for mitochondrial translation. Apparently it does not, for example, interact co-translationally with nascent chains of mitochondrial proteins. This seems analogous to the apparent lack of requirement for hsp60 in the process of translocation of proteins through the mitochondrial membranes [9,14,16].

To determine whether hsp60 could act at a post-translational level in folding or assembly of mitochondrial gene products, we selected a specific translation product for study, the var1 protein. This component of the mitochondrial ribosomes is the only one translated inside the mitochondria of *Saccharomyces cerevisiae* [20]; the other ribosomal proteins are imported from the cytosol. To assess the fate of var1, wild-type or mif4 cells were once again shifted to 37°C, then after 2 h were pulse-radiolabeled in the presence of cycloheximide. Mitochondria were prepared and extracted with Triton X-100. Equal portions of soluble and insoluble fractions of the extract were analyzed. We observed, as expected, that the total amount of var1 synthesized in the two cell types during the radiolabeling period was equal, as would be expected from the foregoing translation study. However, the distributions into soluble and insoluble fractions were strikingly different. In wild-type mitochondrial extract, as reported previously, var1 behaves as a soluble protein [20]. In contrast, in the mif4 extract, var1 was virtually completely insoluble. This indicates that following translation, it failed to associate with the mitochondrial ribosomes, which were in the soluble fraction. This behavior of var1 most likely reflects the same misfolding and aggregation observed with imported proteins in the absence of hsp60 function. However, we cannot exclude that the fate of var1 in mif4 cells is an indirect result of hsp60 deficiency, since var1 must normally assemble with a host of imported ribosomal protein components, whose folding and assembly, like that of other imported proteins examined to date, is likely to be impaired in mif4. Failure of such imported components to become folded and assembled into a nascent ribosomal structure would leave var1 without a target for the completion of its own biogenesis and liable to misfolding.

4. Chaperonin-mediated folding reconstituted in vitro

In order to examine the mechanism of chaperonin-mediated protein folding, in vitro reconstitution of a folding reaction has been particularly desirable. Initial experiments demonstrated the reactivation by purified groEL and groES of a dimeric bacterial RUBISCO enzyme diluted from 6 M guanidine HCl [21]. It was shown that RUBISCO subunits diluted from guanidine HCl were prevented from spontaneous aggregation by association with groEL 14mer. Subsequent addition of Mg-ATP and groES resulted in assembly of active dimeric RUBISCO. More recently, the folding of two monomeric enzymes diluted from guanidine HCl [22] has been reconstituted. Here, binding by groEL was shown to occur with a stoichiometry of approximately one polypeptide per groEL 14mer. In the case of monomeric DHFR, binding prevented spontaneous refolding; in the case of the two domain mitochondrial sulfur transferase, rhodanese, binding prevented spontaneous aggregation.

When groES and Mg-ATP were added to groEL-bound DHFR or rhodanese, the proteins were folded to their active forms with a $t_{1/2}$ of 3 and 10 min, respectively [22,23]. The polypeptides remained at the surface of groEL during the folding reaction and were released in conformations nearer to or at the native state. In contrast, if only Mg-ATP was added to groEL-bound DHFR, it caused DHFR to undergo cycles of release and rebinding by groEL, ultimately producing the native conformation. Similarly, when Mg-ATP was added to groEL-bound rhodanese, cycles of release and rebinding were observed, but instead of reaching the native state, a portion of the released protein became aggregated. Thus, groES plays a critical role in mediating folding, coupling the reaction to groEL [22].

Folding mediated by groEL was shown to be associated with a burst of ATP hydrolysis [22]. Alone, the groEL 14mer exhibits ATPase activity measured at 25°C as 7 molecules ATP hydrolyzed/min per 14mer [22,24]. This activity is nearly completely inhibited when groES is added [17,22,25]. In contrast, activity of the groEL ATPase is approximately tripled by the binding of unfolded rhodanese. A burst of activity of the groEL ATPase is observed upon addition of unfolded rhodanese to a mixture of groES, groEL, and Mg-ATP. Activity is increased initially by 40-fold, then declines to a negligible level in a temporal association with the refolding of rhodanese. Approximately 100 ATP molecules were hydrolyzed per monomer folded. Thus surface-mediated protein folding involves a controlled expenditure of energy, presumably translated into conformational changes of groEL that mediate folding of the bound polypeptide.

Because both groEL and groES are devoid of tryptophan, the conformations of the protein substrates could be monitored by measuring tryptophan fluorescence [22]. In their groEL-bound state, both DHFR and rhodanese exhibited fluorescence maxima at wavelengths between those of unfolded and folded conformations, suggesting that their tryptophans reside in an environment that is not completely polar, as in the unfolded state, nor as hydrophobic as the core of the folded protein. The complex of groEL and bound protein was also shown to specifically bind the fluorescent dye, 1-anilino-naphthalene-8-sulfonate, a fluorescent probe that has been shown to bind selectively to the loosely packed hydrophobic core of early folding intermediates known as molten globule intermediates, containing native-like secondary structures and unorganized tertiary structure. The set of observations, together with the protease sensitivity of the bound proteins, led to the conclusion that proteins bound to groEL are stabilized in molten globule-like conformations.

Upon initiation of the folding reaction by addition of Mg-ATP and groES, the molten globule conformation was rapidly lost [22]: ANS fluorescence decreased within seconds and tryptophan fluorescence was simultaneously quenched, reflecting a compacting of the hydrophobic core. This was followed by a slower (minutes) shift of tryptophan fluorescence to the wavelength maximum of the native state. The latter changes were associated with acquisition of increasing protease resistance and finally with release of the enzyme.

5. Models for physical interactions of components in chaperonin-mediated folding

Electron microscopic inspection of chaperonin molecules typically reveals a complex of two stacked 7-fold symmetric rings, measuring approximately 130 Å in diameter and 120 Å in height [5–7]. In top views, individual component members cannot be distinguished except by star-like points projecting outward (Fig. 2a). A central hole measures approximately 40 Å in diameter. In lateral views of chaperonin complexes,

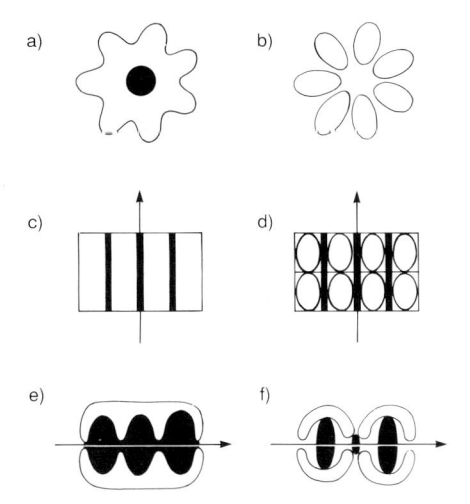

Fig. 2. Interpretation of electron microscopic views of chaperonin complexes. (a), (c), (e) Illustration of electron microscopic images. Blackened areas correspond to electron-dense regions; (b), (d), (f) interpretations of the respective images, with oval shapes representing chaperonin monomers. Arrows designate axes of 7-fold symmetry.

four vertical columns are visible, separated by three dark vertical bars (Fig. 2c). Hendrix originally suggested that the columns represent component groEL monomers from opposite rings, in register with each other, while the dark bars represent grooves between the component monomers [5] (Fig. 2d). This would place the dark bars parallel to the axis of 7-fold symmetry. However, more recent studies suggest that a continuous midline dark column parallels the axis of 7-fold symmetry, comprising the hole observed in top views, and that the three dark bars lie perpendicular to the axis of symmetry [26] (see Fig. 2e). In this interpretation, the dark bars would correspond to two outside cavities within the apposing rings and to a space between the two rings (Fig. 2f). Interpreted in this manner, our own recent images in scanning transmission electron microscopy suggest that the rings are more like flowers with petals reaching up and down from the equatorial zone (Wall, unpublished results) (Fig. 2f).

How might such a putative structure interact with unfolded proteins and with the cooperating component groES? Biochemical studies measuring the relative amount of groES required either to completely inhibit groEL ATPase activity [17,22,25] or to promote complete refolding of a groEL-bound polypeptide [21,22] suggest that the groES ring binds to the groEL complex with a stoichiometry of 1 groES 7mer per groEL 14mer. Given a similar 7-fold axis of symmetry of the groES complex, it seems likely that its ring would appose its planar surface with the outside surface of one of the groEL rings. In support of such an interaction is a recent electron microscopic image of groES-bound groEL in which two of the dark bars usually observable in

a) Surface

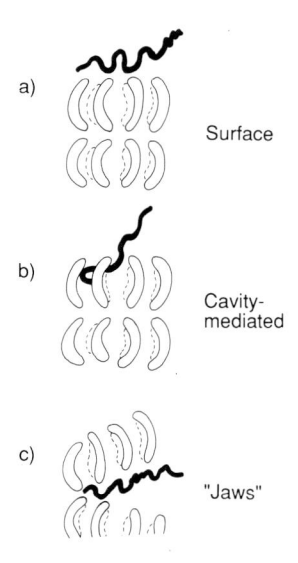

b) Cavity-mediated

c) "Jaws"

Fig. 3. Models for interaction of unfolded polypeptides with chaperonin.

lateral views of groEL are obliterated [27]; this could be the result of altered conformation of the petals either in the ring bound by groES or perhaps in the opposite ring. Considering the likely geometry of the groES interaction, three major models for molecular interaction can be distinguished, illustrated in Fig. 3a–c.

One model involves surface folding, wherein the polypeptide chain binds at the outer surface of one of the groEL rings (Fig. 3a), at the outer aspect of the petals formed by the monomers. GroES could bind either at the same aspect or at the outside aspect of the opposite ring. The latter type of interaction would require communication from the outer surface of one ring through the opposite ring to its outer surface, an allosteric type of interaction. A second model has recently been proposed [28], involving what is termed here cavity-mediated folding but which has also recently been termed by Ellis as caged folding. Here, the bound polypeptide chain enters the hole in the groEL complex, and interacts with surrounding monomeric members (Fig. 3b). While the diameter of the hole is only approximately 40 Å, this represents a narrowest passage in top views. A considerable additional amount of volume could be available between the monomeric petals, particularly if they bend outward (e.g. Fig. 2e,f). As suggested, conceivably a polypeptide could pass from the cavity of one ring into that of the apposing one, and ultimately find its way all the way through the hole [28].

Finally, a third model, dubbed jaws, places the polypeptide chain in the equator between the rings (Fig. 3c). GroES binding to the surface of one of the rings would alter conformation of that ring and in turn control conformational changes in the equatorial-localized polypeptide. A jaws model might involve opening and closing of the space between the rings during the reaction in a hinge-like fashion. The three foregoing models may now be resolvable as further experiments are carried out using electron microscopic techniques, biochemical manipulations, and structural analyses including X-ray crystallography.

References

1. Rothman, J.E. (1989) Cell 59, 591–601.
2. Ellis, R.J. and van der Vies, S.M. (1991) Annu. Rev. Biochem. 60, 327–347.
3. Hemmingsen, S.M., Woolford, C., van der Vies, S.M., Tilly, K., Dennis, D.T., Georgopoulos, C.P., Hendrix, R.W. and Ellis, J. (1988) Nature 333, 330–334.
4. Reading, D.S., Hallberg, R.L. and Myers, A.M. (1989) Nature 337, 655–659.
5. Hendrix, R.W. (1979) J. Mol. Biol. 129, 359–373.
6. McMullen, T.W. and Hallberg, R.L. (1988) Mol. Cell. Biol. 8, 371–380.
7. Pushkin, A.v., Tsupryn, V.L., Solovjeva, N.A., Shubin, V.V., Eustigneeva, Z.G. and Kretovich, W.L. (1982) Biochim. Biophys. Acta 704, 379–384.
8. Fayet, O., Ziegelhoffer, T. and Georgopoulos, C. (1989) J. Bacteriol. 171, 1379–1385.
9. Cheng, M.Y., Hartl, F.-U., Martin, J., Pollock, R.A., Kalousek, F., Neupert, W., Hallberg, E.M., Hallberg, R.L. and Horwich, A.L. (1989) Nature 337, 620–625.
10. Sternberg, N. (1973) J. Mol. Biol. 76, 1–23; 76, 25–44.
11. Georgopoulos, C.P., Hendrix, R.W., Casjens, S.R. and Kaiser, A.D. (1973) J. Mol. Biol. 76, 45–60.
12. Barraclough, R. and Ellis, R.J. (1980) Biochim. Biophys. Acta 608, 19–31.
13. Goloubinoff, P., Gatenby, A.A. and Lorimer, G. (1989) Nature 337, 44–47.
14. Cheng, M.Y., Hartl, F.-U. and Horwich, A.L. (1990) Nature 348, 455–458.
15. Hartl, F.-U. and Neupert, W. (1990) Science 247, 930–938.
16. Ostermann, J., Horwich, A.L., Neupert, W. and Hartl, F.-U. (1989) Nature 341, 125–130.
17. Chandrasekhar, G.N., Tilly, K., Woolford, C., Hendrix, R. and Georgopoulos, C. (1986) J. Biol. Chem. 261, 12414–12419.
18. Lubben, T.H., Gatenby, A.A., Donaldson, G.K., Lorimer, G.H. and Viitanen, P.V. (1990) Proc. Natl. Acad. Sci. USA 87, 7683–7687.
19. Prasad, T.K., Hack, E. and Hallberg, R.L. (1990) Mol. Cell. Biol. 10, 3979–3986.
20. Terpstra, P., Zanders, E. and Butow, R.A. (1979) J. Biol. Chem. 254, 12653–12661.
21. Goloubinoff, P., Christeller, J.T., Gatenby, A.A. and Lorimer, G.H. (1989) Nature 342, 884–889.
22. Martin, J., Langer, T., Boteva, R., Schramel, A., Horwich, A.L. and Hartl, F.-U. (1991) Nature 352, 36–42.
23. Mendoza, J.A., Rogers, E., Lorimer, G.H. and Horowitz, P.M. (1991) J. Biol. Chem. 266, 13044–13049.
24. Hendrix, R.W. (1979) J. Mol. Biol. 129, 375–392.
25. Viitanen, P.V., Lubben, T.H., Reed, J., Goloubinoff, P., O'Keefe, D.P. and Lorimer, G.H. (1990) Biochemistry 29, 5665–5671.
26. Hutchinson, E.G., Tichelaar, W., Hofhaus, G., Weiss, H. and Leonard, K.R. (1989) EMBO J. 8, 1485–1490.
27. Saibil, H., Dong, Z., Wood, S., Auf der Mauer, A. (1991) Nature (Sci. Corresp.) 353, 25–26.
28. Creighton, T.E. (1991) Nature (News and Views) 352, 17–18.
29. McKee, E.E. and Poyton, R.O. (1984) J. Biol. Chem. 259, 9320–9331.

INDEX

AAC, 254, 255, 259–261, 265–269
acidic phospholipid, 4, 5, 81, 87
ADP/ATP carrier, 244, 249, 254, 255, 265, 267, 268
alcohol oxidase, 157, 209–214, 219
alkaline phosphatase fusions, 49, 59
amber mutation, 42, 58
amber suppressor, 58
amphipathic α-helices, 91, 249, 265, 287
amphipathic β-strand, 76
1-anilino-naphthalene-8-sulfonate, 334
ANS fluorescence, 334
anti-idiotypic antibody, 244, 283
anti-SKL antibody, 224
antimycin A, 272
apocytochrome *c*, 93–97
apyrase, 271, 273, 300
ATP-binding proteins, 197
ATP hydrolysis, 5, 10, 18, 51, 89, 95, 150, 267, 271, 272, 313, 334
autophagosome, 152
8-azido ATP, 142
azido-ATP sensitive protein, 137

bacteriophage M13, 33, 36
band 1, 4
β-barrels, 75
basic amino acids, 9, 12, 13, 51, 53, 91, 171
basicity, 12, 13
BiP, 141, 311
bifunctional reagents, 120, 121
bipartite signal sequence, 249, 290

C element, 10–12, 35, 54
carbonate extraction, 195, 196
carboxypeptidase Y, 166
cardiolipin, 5, 86, 91–93, 95, 97
catabolite inactivation, 156–158, 160
catalase, 157, 185, 189, 195, 202, 210, 211, 216, 217, 222, 223, 231, 232, 234–236
CCCP, 272
cell cycle, 7, 150, 161, 186, 243
chaperone protein, 4, 6, 18, 27, 36, 242, 247, 265, 272, 275, 299, 302
chaperonins, 46, 95, 234, 246, 329, 330
charge density, 55

chloroplast envelope, 97, 98, 291, 302, 303
chloroplasts, 3, 75, 82, 86, 99, 200, 277, 279–287, 289–295, 299–305, 329
cholate, 23, 123
circular dichroism, 90, 96
class I protein, 77
class II protein, 77–80
class III protein, 77, 79, 80
class IV protein, 78
clathrin heavy chain, 177, 178
co-translational translocation, 24
coated pits, 172, 179
complementation group, 187, 201–203, 223, 225, 226, 231–233, 235, 236
concanavalin A, 123
conformational change, 18, 25, 40, 58, 88, 89, 133, 313, 334, 336
consensus sequence, 106, 131, 193, 197, 241
conservative sorting model, 249
conservative sorting pathway, 254, 255, 258, 267, 271, 272, 286
contact site, 247, 248, 255, 258–262, 267, 268, 272, 274, 277, 282–284
counterflux of protons, 29
cytochrome *c*, 91–97, 195, 249, 255, 268, 269, 271, 272, 316
cytochrome *c* heme lyase (CCHL), 269, 271–274
cytochrome *c*1 heme lyase (CC$_1$HL), 271
cytochrome *c* oxidase subunit Va, 268
cytoplasmic domains, 14, 17, 49, 50, 53–56
cytoplasmic membrane, 9, 14, 26, 34, 49, 51, 52, 56, 57, 59, 60, 254, 259

default pathway, 166
DHFR, 94, 95, 256, 331, 333, 334
digalactosyl diglyceride, 97
dihydrofolate reductase, 94, 226, 256, 260, 261, 331
dipeptidyl aminopeptidase A, 166, 167
dipeptidyl aminopeptidase B, 166
distal stop-transfer sequence, 253, 257, 259, 262
disulfide bridge, 72, 122
DnaB helicase, 311

DnaJ, 275, 309–321, 323–325
DnaJ homolog-HSP70 interaction, 325
DnaK, 4, 309–314, 316, 317, 319, 321, 323, 324
docking protein, 111, 112, 115
dynamin, 171, 172, 179

electrochemical potential, 5, 6, 40, 241, 248, 249, 292
electrophoretic mechanism, 40
electrostatic binding, 37
encapsidation, 41
endoplasmic reticulum, 3, 58, 91, 105, 106, 119, 129, 130, 166, 223
endosome, 105, 130
envelope membranes, 277, 279, 281–285, 287, 289, 291, 299–301, 303, 304
N-ethylmaleimide, 142
N-ethylmaleimide-sensitive component, 142

F_1-ATPase β-subunit, 244
FBPase degradation, 158–162
filamentous bacteriophage fd, 43
fox mutant, 187, 200, 201
fructose-1,6-bisphosphatase (FBPase), 149, 150, 156–162
functional complementation, 191, 192, 202, 203

β-galactosidase fusions, 51, 52
gene fusion analysis, 50
general insertion pore (GIP), 267
glycosome, 210, 221–224, 226
glycosyltransferase, 126
glyoxysome, 205, 210, 221, 222, 224, 226
Golgi apparatus, 119, 130, 156, 160, 165–167, 170, 173, 178, 179
gp36, 123, 125, 126
GroEL, 4, 46, 246, 313, 318, 329–331, 333–336
GroEL ATPase, 334, 335
GroES, 275, 313, 331, 333–336
GrpE, 309–313, 324, 325
GTP-binding domain, 165
GTP hydrolysis, 115, 129–132, 134, 135, 165, 167, 171, 173, 174, 177, 179
GTPase domain, 130–132, 135
guanine nucleotide release factors, 129

H element, 9, 11, 12
α-helical, 39, 50, 90–92, 95, 96, 259
helical bundles, 75
α-helical conformation, 39, 95, 96
heme attachment, 96, 269

homotypic interaction, 134
hsc70, 302, 304, 305
hsp60, 242, 243, 246, 247, 266–268, 270, 329–333
hsp70, 155, 242, 243, 245–247, 265–268, 270–272, 284, 309–313, 316–319, 321, 323–325
hydrophobic membrane spanning segments, 49, 50
hydrophobic region, 34, 35, 38–40, 46, 63, 65, 76, 105, 194, 249, 320–322
hydrophobicity, 38, 39, 42, 67, 76, 77, 194
hydroxylamine mutagenesis, 175
hydroxylated amino acids, 91, 97

immunoelectron microscopy, 161, 224, 232
immunofluorescence, 149, 159, 167–170, 202, 215, 224, 231, 232, 234–236, 317, 320
import receptor, 93, 203, 242–244, 247, 248, 250, 258, 262, 268, 269, 296
import signal, 269, 273, 274
independent assembly of membrane components, 58
initiation of λ replication, 311
inner boundary membrane, 266, 267
inner envelope membrane, 277, 286, 287
inner membrane (IM), 13, 78, 82, 97, 241, 245, 246, 255, 259, 266–268, 272–274, 286, 330
inner membrane protease, 242, 267, 295
insertion domain, 254, 259
intermembrane space, 92–94, 97, 241, 246, 249, 255, 266–269, 271–275, 277, 282, 284, 286, 287, 295, 301, 330
inverted micelle structure, 119
ionophore, 272, 292
IPTG, 86–88
ISP42, 242, 245, 248, 249, 267, 270

J-region, 309, 314, 315, 321–325

KAR2, 314, 321–325
KDEL-sequence, 317
Kex2p, 170, 177, 178

LacY protein, 56
leader peptidase, 6, 34, 38, 41, 42, 44, 45, 52, 78, 79, 86, 259
leader sequence, 4, 34–36, 38, 44
light-driven import, 292
lipid transfer protein, 87, 88
loop model, 69, 108–110
luciferase, 223–225

lysosome, 105, 119, 130, 149, 151–155, 161, 166
lysyl-tRNA, 120

M13 procoat, 6, 33, 34, 36–41, 44, 46
macroautophagy, 152
malE, 58
MalF, 6, 49, 51–60, 82
MalG, 49, 51, 57–60
MalK, 49, 51, 57–59
maltose binding protein, 51, 58
maltose transport complex, 49, 57, 59, 60
MAS1, 242, 246
MAS2, 242, 246
MAS70, 242, 244, 255, 256, 265, 268
mas5 mutant, 318
Mas70p, 242–244, 248, 249
matrix-targeting signal, 249, 253, 254, 256–262
membrane anchor sequence, 38
membrane ghost, 231, 232, 235
membrane potential, 5, 41, 80, 81, 91, 92, 246, 248, 267, 268, 272, 274, 283, 284
membrane proteins, type I, 108
membrane proteins, type II, 108
mhsp70, 242, 245, 246
microautophagy, 151, 152
microbodies, 185, 210, 221–224, 226, 227
microinjection, 149, 153, 154, 225
microtubule, 171–173, 179, 180
MIR1, 242, 244
mitochondrial outer membrane, 93, 98, 241–243, 247, 248, 257, 261, 265, 266
mitochondrial translation, 331, 332
molecular chaperone, 18, 141, 299, 302
molten globule, 334
MOM19, 243, 244, 248, 265–270, 272, 274
MOM38, 245, 253, 257, 258, 262, 266, 267, 270, 273
MOM72, 243, 244, 247, 255, 257, 265–268, 270, 272
monogalactosyl diglyceride, 97
morphological contact site, 267
multi-spanning protein, 83, 109, 110
mutagenesis, 9, 41, 55, 107, 175, 180, 196, 197, 224, 247

N element, 10, 11, 171
N-end rule, 161, 162
N-glycosylation, 106
N(m) element, 9, 10, 12, 13
nitrogen starvation, 156
nocodazole, 172, 173

non-conservative import pathway, 268, 271–273
non-hydrolyzable ATP analog, 142
non-permissive temperature, 14, 16, 44, 150, 246, 319, 323
NPL1/SEC63, 319

octylglucoside, 23
oleic acid, 186–189, 191, 195–197, 201–205, 213, 215
oligomycin, 272
OM vesicle, 273
OMM70, 255–258
OMM signal-anchor sequence, 253, 256–259, 261
outer envelope membrane, 282, 284–286, 301–304
outer envelope protein 7, 299–305
outer envelope protein 86, 304
outer membrane, 3, 4, 6, 7, 75, 76, 86, 93–99, 241–243, 245, 247–250, 253–259, 261, 262, 265, 266, 277, 280, 283–286, 301
outer membrane proteins, 4, 6, 86, 242, 248, 249, 255, 257, 258, 262, 283–286
overexpression of *PAS4*, 189, 196
overproduction of Sec proteins, 22
β-oxidation pathway, 185, 186, 188
oxygen-evolving complex (OEC), 289

PEP4, 149, 156, 158
periplasm, 6, 10, 34, 36, 38, 46, 51–53, 56
periplasmic domain, 16, 45, 52–57, 79
peroxisomal ghost, 190, 219, 233
peroxisomal mutant, 186, 187, 191, 192, 197, 200–202, 205
peroxisome, 3, 157, 185–191, 195–205, 209–219, 221–227, 229, 231–236
peroxisome assembly (*pas*) mutant, 187–191, 197, 201–204
– type I, 189, 202–204
– type II, 191
– type III, 188, 191
peroxisome ghost, 223, 226, 227, 235
peroxisome import (PIM) mutations, 231, 234
peroxisome proliferation, 191, 197, 204, 205
Pf3 coat protein, 34–36, 46
pgsA gene, 86
phage-encoded RepA protein, 311
phosphate translocator, 244, 283, 304
phosphatidylcholine, 37, 38, 88, 92, 94, 98, 284, 302
phosphatidylglycerol, 5, 36, 38, 86

phospholipid, 5, 25, 26, 81, 87–90, 93, 96, 119, 121, 126, 200, 203, 222, 232, 234
phosphorylation, 68, 159–161, 283
photoactivatable cross-linking, 120, 283
photocrosslinking, 111–113, 115, 120, 121, 123, 124
plasma membrane, 3, 4, 6, 58, 78, 82, 105, 119, 130, 153, 157, 159, 165, 166, 173, 177–179, 267
plastocyanin, 282, 289, 290, 292–294, 296
PMP31, 209, 211, 214, 215, 217, 218
PMP32, 209, 211, 214, 215, 217
PMP47, 209, 211, 213–219
polarity, 12, 13
polymyxin B, 38
porin, 215, 218, 253, 254, 257, 258, 268, 272
positive charge, 12, 13, 37, 38, 41, 43, 54, 57, 63, 80, 91, 257
positive inside-rule, 75, 78, 81, 82
post-translational translocation, 24
prePhoE, 46, 86–90
preprocecropin A, 138
prepromelittin, 138
prepropeptide GLa, 138
presequence–lipid interaction, 92
primary structure, 91, 122, 234, 329
prlA, 9, 14, 16
prlG, 9, 16
prolipoprotein (pLpp), 14
protease inhibitor, 23, 152
protease-sensitive surface receptor, 265, 268–270
protein-conducting channel, 116, 120
protein degradation, 149–153, 155–158, 161, 162
protein folding, 49, 59, 275, 309–311, 313, 314, 316, 323–325, 329, 333, 334
protein topologies, 106
protein–lipid interaction, 45, 46, 85, 86, 91, 99
protein–protein interactions, 43, 45, 116
proteinase A, 156
proteinase B, 156
proteoliposomes, 27, 123
proteosome, 150
proton gradient, 289, 292
proton motive force, 15, 18, 29, 86, 93, 287, 292, 296
pSSU, 299, 300, 303–305
PTS-1, 221, 226, 227
PTS-2, 221, 226, 227
pulse-labelled cell, 38, 45

purification of Sec proteins, 23
puromycin, 120, 312

quaternary structure, 51, 59, 329

receptor/GIP complex, 265–267, 272
reconstituted translocation, 21
reconstitution of membrane vesicles, 114
reversal of topology, 79
reverse signal peptide, 35, 77, 78
rhodanese, 333, 334
ribonuclease A (RNase A), 153
ribonucleoparticle-dependent pathway, 139
ribonucleoparticle-independent transport, 139
ribophorin I, 125
ribophorin II, 125
rough endoplasmic reticulum, 105, 130
RUBISCO binding protein, 329, 330

S-peptide, 153–155
S-protein, 153
Sarcosyl, 23
scanning transmission electron microscopy, 335
SCJ1, 314–319, 321
SEC61, 114, 323
SEC62, 114, 160, 320, 323
SEC63, 114, 314–316, 319–325
Sec-dependence or independence, 44, 45, 78, 79, 81, 82, 259
sec gene products, 14, 41, 50, 53
SEC protein, 114
SecA, 4–6, 9, 10, 15, 17, 18, 33, 36, 43–45, 52, 81, 87–89
SecA dimer, 28
SecA fragment, 25
SecB, 4, 5, 9, 10, 17, 18, 33, 36, 45, 46, 88
SecD, 6, 9, 10, 14–18, 52
SecF, 6, 9, 10, 14–18
secFcs, 16, 17
SecI, 17, 18
secondary structure, 50, 96, 334
Sec4p, 173
Sec61p, 124–126
Sec62p, 124–126
Sec18p/NSF, 192
secretory pathway, 76, 105, 129, 149, 150, 159, 160, 165, 166, 173, 177, 178, 180, 192, 324
SecY/E, 4–6, 33
SecY/PrlA, 9, 10, 14, 15
signal-anchor sequence, 106, 107, 122, 253, 254, 256–259, 261

signal peptidase I, 69
signal peptidase II, 69
signal peptide, 9–12, 14–16, 18, 35, 46, 63, 76–79, 81, 89, 90, 132, 210, 244, 265, 331
signal peptide-independent mechanism, 138
signal peptide mutant, 64
signal recognition particle (SRP), 18, 22, 33, 46, 50, 66, 83, 111, 112, 115, 120, 121, 125, 129, 130, 132–134
signal sequence cleavage, 105, 107, 115, 126, 168
signal sequence receptor α (SSRα), 113, 114, 121–123, 125, 126, 131–134
SIS1, 318
site directed mutagenesis, 107, 197
SKL signal, 203, 204, 210, 217, 218, 223–225, 234
small secretory protein, 115
sodium azide, 52
specific lipid classes, 85, 99
spheroplast, 52, 55–58, 186, 195, 198, 212, 219
Spo15p, 172, 180
SRP receptor, 111, 129, 130, 132–134
SRP/SRP receptor complex, 133
SSC1, 242, 246
SSR-complex, 121, 122
stop-transfer, 45, 46, 49, 50, 71, 76–78, 106, 107, 109, 110, 126, 249, 253–259, 261, 262, 286
stop-transfer model, 249, 259
stop-transfer sequence, 49, 50, 76–78, 106, 107, 109, 110, 126, 253, 254, 257–259, 262
β-strand, 75, 254
stroma, 82, 277, 280–284, 286, 287, 290–293, 299, 330
stromal factor, 289, 293, 296
stromal processing peptidase (SPP), 290
subunit VI of yeast QH_2 cytochrome *c* reductase (Sub VI), 268
sulfolipid sulfoquinovosyl diglyceride, 97
suppression of defective signal peptides, 15, 16
suppressors of signal sequence, 14
surface pressure, 88–90, 96–99
synthetic lethality, 323

temperature-sensitive strain, 44, 45
tertiary structure, 59, 130, 329, 334
thermolysin-sensitive receptor, 277, 279, 285
thiolase, 157, 187, 189, 191, 196, 198, 203, 204, 210, 219, 221, 223–226, 232, 234, 269

thylakoid, 75, 78, 82, 280–282, 286, 287, 289–296, 299, 302, 303
thylakoid lumen, 280–282, 287, 289–293, 295, 299
thylakoid membrane, 75, 78, 82, 280–282, 286, 287, 289–295, 299, 303
thylakoid transfer signal, 290, 292, 295
thylakoidal processing peptidase, 290
topogenic signal, 50, 51, 53–55, 234, 254, 256, 258, 261, 287
topological titration, 79
topology, 49–51, 53–56, 58, 59, 75–84, 105, 107, 109, 116, 122, 169, 196, 199, 215, 254, 256, 321–323, 327
transit peptide, 277, 280–283, 285–287
transit sequence, 97, 98, 290, 300, 304
translocase, 4–6, 34, 267
translocation ATPase, 5
translocation intermediate, 113, 114, 116, 120, 125, 247, 248, 259, 274, 282, 303, 304
translocon, 130, 132–134
transmembrane orientation, 67
transmembrane segment, 78, 79, 82, 132, 169, 254, 256, 258–261
transthylakoid proton gradient, 289
Triton X-114, 211, 214
tryptophan fluorescence, 92, 334
type II non-bilayer lipid structures, 91

ubc mutants, 161, 162
ubiquitin, 150, 151, 161, 163, 194, 197, 203
ubiquitin conjugation, 150, 161, 194, 197, 203
uncoupling protein (UCP), 254, 255, 257, 259–262

vacuolar H^+-ATPase, 167
vacuole, 3, 149–152, 156, 158–161, 165–170, 172, 173, 177–180, 189
var1 protein, 332
VPS1, 165, 167, 170, 172, 173, 175, 176, 180
vps mutants, 167
Vps1p, 165, 167, 170–177, 179–181

water–micelle interface, 96

YDJ1, 317–319, 321
YDJ1/MAS5, 319
Ypt1p, 173

Zellweger fibroblast, 190, 202, 234
Zellweger's syndrome, 210, 219, 221, 223, 225, 231, 234
zinc-finger motif, 194